Recent Titles in This Series

166 **O. A. Ladyzhenskaya, Editor,** Proceedings of the St. Petersburg Mathematical Society, Volume III
165 **Yu. Ilyashenko and S. Yakovenko, Editors,** Concerning the Hilbert 16th Problem
164 **N. N. Uraltseva, Editor,** Nonlinear Evolution Equations
163 **L. A. Bokut', M. Hazewinkel, and Yu. G. Reshetnyak,** Third Siberian School "Algebra and Analysis"
162 **S. G. Gindikin, Editor,** Applied Problems of Radon Transform
161 **Katsumi Nomizu, Editor,** Selected Papers on Analysis, Probability, and Statistics
160 **K. Nomizu, Editor,** Selected Papers on Number Theory, Algebraic Geometry, and Differential Geometry
159 **O. A. Ladyzhenskaya, Editor,** Proceedings of the St. Petersburg Mathematical Society, Volume II
158 **A. K. Kelmans, Editor,** Selected Topics in Discrete Mathematics: Proceedings of the Moscow Discrete Mathematics Seminar, 1972–1990
157 **M. Sh. Birman, Editor,** Wave Propagation. Scattering Theory
156 **V. N. Gerasimov, N. G. Nesterenko, and A. I. Valitskas,** Three Papers on Algebras and Their Representations
155 **O. A. Ladyzhenskaya and A. M. Vershik, Editors,** Proceedings of the St. Petersburg Mathematical Society, Volume I
154 **V. A. Artamonov et al.,** Selected Papers in K-Theory
153 **S. G. Gindikin, Editor,** Singularity Theory and Some Problems of Functional Analysis
152 **H. Draškovičová et al.,** Ordered Sets and Lattices II
151 **I. A. Aleksandrov, L. A. Bokut', and Yu. G. Reshetnyak, Editors,** Second Siberian Winter School "Algebra and Analysis"
150 **S. G. Gindikin, Editor,** Spectral Theory of Operators
149 **V. S. Afraĭmovich et al.,** Thirteen Papers in Algebra, Functional Analysis, Topology, and Probability, Translated from the Russian
148 **A. D. Aleksandrov, O. V. Belegradek, L. A. Bokut', and Yu. L. Ershov, Editors,** First Siberian Winter School "Algebra and Analysis"
147 **I. G. Bashmakova et al.,** Nine Papers from the International Congress of Mathematicians, 1986
146 **L. A. Aĭzenberg et al.,** Fifteen Papers in Complex Analysis
145 **S. G. Dalalyan et al.,** Eight Papers Translated from the Russian
144 **S. D. Berman et al.,** Thirteen Papers Translated from the Russian
143 **V. A. Belonogov et al.,** Eight Papers Translated from the Russian
142 **M. B. Abalovich et al.,** Ten Papers Translated from the Russian
141 **H. Draškovičová et al.,** Ordered Sets and Lattices
140 **V. I. Bernik et al.,** Eleven Papers Translated from the Russian
139 **A. Ya. Aĭzenshtat et al.,** Nineteen Papers on Algebraic Semigroups
138 **I. V. Kovalishina and V. P. Potapov,** Seven Papers Translated from the Russian
137 **V. I. Arnol'd et al.,** Fourteen Papers Translated from the Russian
136 **L. A. Aksent'ev et al.,** Fourteen Papers Translated from the Russian
135 **S. N. Artemov et al.,** Six Papers in Logic
134 **A. Ya. Aĭzenshtat et al.,** Fourteen Papers Translated from the Russian
133 **R. R. Suncheleev et al.,** Thirteen Papers in Analysis
132 **I. G. Dmitriev et al.,** Thirteen Papers in Algebra
131 **V. A. Zmorovich et al.,** Ten Papers in Analysis
130 **M. M. Lavrent'ev, K. G. Reznitskaya, and V. G. Yakhno,** One-dimensional Inverse Problems of Mathematical Physics
129 **S. Ya. Khavinson,** Two Papers on Extremal Problems in Complex Analysis
128 **I. K. Zhuk et al.,** Thirteen Papers in Algebra and Number Theory

(*Continued in the back of this publication*)

Proceedings of the
St. Petersburg
Mathematical Society
Volume III

American Mathematical Society

TRANSLATIONS

Series 2 • Volume 166

Proceedings of the
St. Petersburg
Mathematical Society
Volume III

O. A. Ladyzhenskaya
Editor

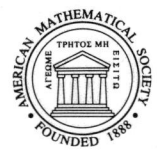

American Mathematical Society
Providence, Rhode Island

Translated from an original Russian manuscript
Translation edited by A. B. SOSINSKY

1991 *Mathematics Subject Classification.* Primary 11Sxx.

ISBN 0-8218-0387-5
ISSN 0065-9290

Copying and reprinting. Material in this book may be reproduced by any means for educational and scientific purposes without fee or permission with the exception of reproduction by services that collect fees for delivery of documents and provided that the customary acknowledgment of the source is given. This consent does not extend to other kinds of copying for general distribution, for advertising or promotional purposes, or for resale. Requests for permission for commercial use of material should be addressed to the Assistant Director of Production, American Mathematical Society, P. O. Box 6248, Providence, Rhode Island 02940-6248. Requests can also be made by e-mail to reprint-permission@math.ams.org.

Excluded from these provisions is material in articles for which the author holds copyright. In such cases, requests for permission to use or reprint should be addressed directly to the author(s). (Copyright ownership is indicated in the notice in the lower right-hand corner of the first page of each article.)

© Copyright 1995 by the American Mathematical Society. All rights reserved.
The American Mathematical Society retains all rights
except those granted to the United States Government.
Printed in the United States of America.

∞ The paper used in this book is acid-free and falls within the guidelines
established to ensure permanence and durability.
♻ Printed on recycled paper.
This volume was typeset using $\mathcal{A}_{\mathcal{M}}\mathcal{S}$-TEX,
the American Mathematical Society's TEX macro system.

10 9 8 7 6 5 4 3 2 1   00 99 98 97 96 95

The volume consists of papers by students of D. K. Faddeev and by students of his students. The book is dedicated to the memory of D. K. Faddeev.

Contents

Multidimensional Complete Fields: Topology and Other Basic Constructions
A. I. MADUNTS and I. B. ZHUKOV ... 1

A Ramification Filtration of the Galois Group of a Local Field
V. A. ABRASHKIN ... 35

Class Field Theory for Multidimensional Complete Fields with Quasifinite Residue Fields
B. M. BEKKER ... 101

On p-adic Representations Arising from Formal Groups
D. G. BENOIS ... 107

The Pairing on K-Groups in Fields of Valuation of Rank n
S. V. VOSTOKOV ... 111

Artin-Hasse Exponentials and Bernoulli Numbers
S. V. VOSTOKOV ... 149

Some Approaches to the Construction of Abelian Extensions for $\mathfrak{p}$-adic Fields
S. V. VOSTOKOV and I. B. ZHUKOV ... 157

Structure Theorems for Complete Fields
I. B. ZHUKOV ... 175

Unit Fractions
O. IZHBOLDIN and L. KURLIANDCHIK ... 193

Collections of Multiple Sums
D. V. FOMIN and O. T. IZHBOLDIN ... 201

On Convergence of Series over Local Fields
A. I. MADUNTS ... 215

On Köck's Conjecture about Shuffle Products
A. NENASHEV ... 235

Multidimensional Complete Fields: Topology and Other Basic Constructions

A. I. MADUNTS AND I. B. ZHUKOV

The present collection is essentially devoted to various points of the theory of multidimensional complete fields. To make it easier for the reader to orient himself in this subject matter, we present the key notation in this introductory paper and give an exposition of more or less known (at least in the folklore) initial facts about such fields.

§0. Main definitions

It is said that the structure of an *n-dimensional complete field* is defined on the field K if K is complete with respect to the given discrete valuation, and on the residue field there is a structure of an $(n-1)$-dimensional complete field; a 0-dimensional field is simply a perfect field. In other words, the structure of an n-dimensional complete field on K is a chain of fields $K^{(n)} = K$, $K^{(n-1)}, \ldots,$ $K^{(1)}$, $K^{(0)}$, where $K^{(i)}$ is a complete discretely valued field with residue field $K^{(i-1)}$, $i = 1, \ldots, n-1$; $K^{(0)}$ is a perfect field.

The field $K^{(n-1)}$, denoted also by $\overline{K}$, is referred to as *the first residue field* of K, and $K^{(0)}$ as the *last residue field*. By $\overline{a}$ we shall always denote the image in $K^{(n-1)}$ of the integral element a in $K^{(n)}$.

K is said to be an *equal characteristic field* if $\operatorname{char} K = \operatorname{char} \overline{K}$, and a *mixed characteristic field* if this is not so, i.e., $\operatorname{char} K = 0$, $\operatorname{char} \overline{K} = p > 0$.

Let t_n be a prime element with respect to the discrete valuation of $K = K^{(n)}$; t_{n-1} a unit in $K^{(n)}$ whose residue class is a prime element of $K^{(n-1)}$; ...; t_1 a unit in $K^{(n)}$, $K^{(n-1)}, \ldots, K^{(2)}$ that becomes a prime element when one passes to the next-to-last residue field $K^{(1)}$. The n-tuple $(t_1, \ldots, t_n)$ is called the *system of local parameters* in K. It determines a valuation of rank n on K

$$\overline{v}_K = \overline{v} = (v^{(1)}, \ldots, v^{(n)}) \colon K \to \mathbb{Z}^n \cup \{\infty\},$$

1991 *Mathematics Subject Classification.* Primary 12J10; Secondary 19F05, 11S70, 12F05.

Key words and phrases. Multidimensional local field, multidimensional complete field, topology, K-theory, convergence, higher class field theory.

These studies were supported by Russian Fund of Fundamental Studies (project No. 94-01-00753-a).

© 1995, American Mathematical Society

where $\bar{v}(0) = \infty$, and for $a \neq 0$

$$v^{(i)}(a) = v_{K^{(i)}}\left(\overline{at_n^{-v^{(n)}(a)} \ldots t_{i+1}^{-v^{(i+1)}(a)}}\right) \quad \text{for } 1 \leqslant i \leqslant n-1;$$

$$v^{(n)}(a) = v_{K^{(n)}}(a).$$

(Here overline means taking the image in $K^{(i)}$.)

The group $\mathbb{Z}^n$ is assumed to be lexicographically ordered in the following nonstandard sense: $\bar{r}_1 = (r_1^{(1)}, \ldots, r_1^{(n)}) < \bar{r}_2 = (r_2^{(1)}, \ldots, r_2^{(n)})$ means $r_1^{(m)} < r_2^{(m)}$, $r_1^{(m+1)} = r_2^{(m+1)}, \ldots, r_1^{(n)} = r_2^{(n)}$ for some $m \leqslant n$.

The valuation $\bar{v}'$ of rank n corresponding to another system of local parameters $t_1', \ldots, t_n'$, is related to $\bar{v}$ by the formula

(1) $$\bar{v}'(a) = \bar{v}(a)T$$

for all $a \in K$, where $T = (v'^{(j)}(t_i))_{i,j}$ is a lower triangular $n \times n$ matrix with units on the main diagonal.

Now, on a certain field K, let a valuation $\bar{v}$ with value group $\mathbb{Z}^n$ be determined. Then the last r components of $\bar{v}$ ($1 \leqslant r \leqslant n$) determine a valuation of rank r, whose residue field we denote by $K^{(n-r)}$. We obtain fields $K^{(n)} = K$, $K^{(n-1)}$, $\ldots$, $K^{(1)}$ under discrete valuations induced by the corresponding components of $\bar{v}$. If all these fields are complete, and the residue field of $\bar{v}$ is perfect, we get the structure of an n-dimensional complete field on K. The chain of fields under discrete valuations does not change if one replaces the valuation $\bar{v}$ by an equivalent one, i.e., by a valuation which is related to $\bar{v}$ via (1).

Now we construct examples of n-dimensional complete fields. If F is an $(n-1)$-dimensional complete field, then the field of Laurent series $F((X))$ is an n-dimensional field. Recall that an arbitrary equal characteristic complete discretely valued field with residue field F is isomorphic to $F((X))$.

Further, let F be an arbitrary complete field under the discrete valuation $w: F \to \mathbb{Z} \cup \{\infty\}$. Put

$$F\{\{t\}\} = \left\{ \sum_{i=-\infty}^{\infty} c_i t^i : c_i \in F,\ w(c_i) \geqslant c > -\infty,\ w(c_i) \xrightarrow[i \to -\infty]{} \infty \right\}.$$

One can add and multiply formal power series in $F\{\{t\}\}$, and $F\{\{t\}\}$ is a field. Let

$$v\left(\sum_{i=-\infty}^{\infty} c_i t^i\right) = \min_i w(c_i).$$

Then v is a discrete valuation on $F\{\{t\}\}$, and $F\{\{t\}\}$ is a complete field with residue field $\bar{F}((\bar{t}))$. Thus, if F is an $(n-1)$-dimensional field, then $F\{\{t\}\}$ is an n-dimensional complete field.

Notice that for an equal characteristic field F, i.e., for $F = F_0((X))$, the field $F\{\{t\}\}$ is canonically isomorphic to $F_0((t))((X))$. Thus, $F\{\{t\}\}$ is of interest only for a mixed characteristic field F.

In the mixed characteristic case, the fields $F\{\{t\}\}$ play an important role in view of the structure theorem (see below).

Elements $a \in K$ such that $\overline{v}(a) \geq 0$ form a ring $\mathcal{O}_K = \mathcal{O}$ that does not depend on the choice of local parameters. The only maximal ideal of this ring is

$$\mathfrak{M}_K = \mathfrak{M} = \{a \in \mathcal{O}_K : \overline{v}(a) > 0\}.$$

Let $1 \leq l \leq n$; $r^{(l)}, \ldots, r^{(n)} \in \mathbb{Q}$. By $\mathfrak{p}_K(r^{(l)}, \ldots, r^{(n)}) = \mathfrak{p}(r^{(l)}, \ldots, r^{(n)})$ we denote the set of all $a \in K$ with $(v^{(l)}(a), \ldots, v^{(n)}(a)) \geq (r^{(l)}, \ldots, r^{(n)})$. If $r^{(l)}, \ldots, r^{(n)} \in \mathbb{Z}$ and $(r^{(l)}, \ldots, r^{(n)}) > 0$ in $\mathbb{Z}^{n-l+1}$, then this set is an ideal of $\mathcal{O}_K$ and all nonzero ideals of $\mathcal{O}_K$ are of this kind. We shall also write $\mathfrak{p}(\overline{r}+)$ for $\mathfrak{p}(r^{(1)}+1, r^{(2)}, \ldots, r^{(n)})$.

By $\mathcal{R}_K$ (or $\mathcal{R}$) we denote the subgroup in K^* consisting of Teichmüller representatives of all nonzero elements of the last residue field. The Teichmüller representative of θ is denoted by $[\theta]$. Put $U_K = \mathcal{O}_K^*$ (the group of units); $V_K = 1 + \mathfrak{M}_K$ (the group of principal units); $U_K(\overline{r}) = 1 + \mathfrak{p}_K(\overline{r})$ for $\overline{r} \in \mathbb{Z}^l$, $1 \leq l \leq n$, $\overline{r} > 0$. We have

(2) $$K^* = \mathbb{Z}t_n \oplus \cdots \oplus \mathbb{Z}t_1 \oplus U_K; \qquad U_K = \mathcal{R}_K \oplus V_K.$$

Now consider extensions of n-dimensional complete fields. Let L/K be a finite extension. The structure of an n-dimensional complete field on K uniquely determines such a structure on L which conforms to K. If $t_{1,K}, \ldots, t_{n,K}$; $t_{1,L}, \ldots, t_{n,L}$ are the corresponding systems of local parameters that determine the valuations $\overline{v}_K$ and $\overline{v}_L$, then for any $a \in K$:

$$\overline{v}_L(a) = \overline{v}_K(a) T(L/K),$$

where $T(L/K) = T(L/K; t_{1,K}, \ldots, t_{n,K}; t_{1,L}, \ldots, t_{n,L}) = (v_L^{(j)}(t_{i,K}))_{i,j}$ is a lower triangular $n \times n$ matrix. In particular, in (1)

$$T = T(K/K; t_1, \ldots, t_n; t_1', \ldots, t_n').$$

The diagonal elements of $T(L/K)$ do not depend on the choice of local parameters and will be denoted by $e_1(L/K), \ldots, e_n(L/K)$.

Let $K = K^{(n)}, K^{(n-1)}, \ldots, K^{(0)}$; $L = L^{(n)}, L^{(n-1)}, \ldots, L^{(0)}$ be the chains of residue fields for K and L. Let $f(L/K) = [L^{(0)}:K^{(0)}]$. Then it is easy to see that $[L^{(i)}:K^{(i)}] = f(L/K)e_1(L/K) \ldots e_i(L/K)$, $e(L^{(i)}/K^{(i)}) = e_i(L/K)$, $i = 1, \ldots, n$. In the case $[L:K] = f(L/K)$, the extension L/K is said to be *purely unramified*. L/K is *totally ramified* if $K^{(n-1)} = L^{(n-1)}$.

We now state the structure theorem for n-dimensional complete fields which extends the Parshin theorem ([P1]).

THEOREM. *Let $K = K^{(n)}$ be an n-dimensional complete field, $K^{(n-1)}, \ldots, K^{(0)} = F$ be the residue fields. In the case $\operatorname{char} K = 0$, $\operatorname{char} F = p$, denote by $k_0 \hookrightarrow K$ the quotient field of $W(F)$.*

1. *If $\operatorname{char} K = \operatorname{char} F$, then $K \approx F((t_1)) \ldots ((t_n))$.*

2. *If* char $K^{(m)} = p$, char $K^{(m+1)} = 0$, $1 \leqslant m \leqslant n-1$, *then* K *is a finite totally ramified extension of the "standard" field of the form*

$$k\{\{t_1\}\}\ldots\{\{t_m\}\}((t_{m+2}))\ldots((t_n)),$$

where k *is a finite extension of* k_0. *Besides, there exists a finite extension* K'/K *such that* K' *is standard, and* $K' = K(a)$, *where* a *is algebraic over* k_0. *Moreover, one can take* $K' = K_1 K_2$, *where* K_1/K *is a cyclotomic extension, and* K_2/K *is a half-ramified extension, i.e.,* $[K_2:K] = [K_2^{(m)}:K^{(m)}]_{\text{sep}}$.

3. *If* char $K^{(1)} = 0$, char $K^{(0)} = p$, *then* $K \approx k((t_2))\ldots((t_n))$, *where* $k = K^{(1)}$ *is a finite extension of* k_0.

It is known that a complete discretely valued field K with residue field k, where char $K = $ char k, is isomorphic to $k((X))$ (cf. [Bou, Chapter VI, §9, Ex. 4]; [FV, Chapter II, §5]). In view of this fact, one needs to prove the structure theorem only for the mixed characteristic case (char $K = 0$, char $K^{(n-1)} = p$). A proof is contained in [Z].

§1. Topology

A topology on a multidimensional local field that takes into account the topology of the residue fields was introduced in the work of A. N. Parshin, S. V. Vostokov, and I. B. Fesenko. In similar way, such a topology can be introduced on a multidimensional complete field. The main properties of this topology will also be established.

We start with a "sketch" of the definition. We will use the recurrence on the dimension of the field, and introduce the usual discrete valuation topology on 1-dimensional fields.

Further, let char $K = $ char $\overline{K}$, i.e., $K = \overline{K}((X))$, and let the topology of $\overline{K}$ be already defined. For a sequence of neighborhoods of zero $\{U_i\}_{i \in \mathbb{Z}}$ in $\overline{K}$, where $U_i = \overline{K}$ for all sufficiently large i, put

$$\mathcal{U}_{\{U_i\}} = \left\{ \sum_{i \gg -\infty} a_i X^i \;\middle|\; a_i \in U_i \right\}.$$

The sets of the form $\mathcal{U}_{\{U_i\}}$ constitute a basis of neighborhoods of zero in the additive topological group K.

Now let char $K = 0$, char $\overline{K} = p$, and π be a prime in K. Once again, one takes all the sets $\mathcal{U}_{\{U_i\}}$ as a basis of neighborhoods of zero in K, but now $\mathcal{U}_{\{U_i\}}$ consists of all sums of the form $\sum_{i \gg -\infty} h(a_i) \pi^i$, where $h: \overline{K} \to K$ is a lifting (i.e., a map with the property $\overline{h(a)} = a$). The lifting h must be good enough, and here lies the main difficulty in constructing the topology on K. However, in the case of a standard field $K = k\{\{t_1\}\}\ldots\{\{t_{n-1}\}\}$, where k is 1-dimensional, we have $\overline{K} = \overline{k}((\overline{t_1}))\ldots((\overline{t_{n-1}}))$, and one may assume

$$h\left(\sum_{\overline{r}} \theta_{\overline{r}} \overline{t_1}^{r^{(1)}} \ldots \overline{t_{n-1}}^{r^{(n-1)}} \right) = \sum_{\overline{r}} [\theta_{\overline{r}}] t_1^{r^{(1)}} \ldots t_{n-1}^{r^{(n-1)}}.$$

Notice that even in the case of a standard field the "proper" lifting depends at least on the first $n-1$ local parameters. To define the "proper" lifting in the general case, we introduce generalized local parameters, which are, in fact, the first $n-1$ local parameters of some subfield.

In the first subsection, for any system of generalized local parameters $t_1, \ldots, t_{n-1}$ of K (char $\overline{K} > 0$), we convert K into a topological space $K_{t_1,\ldots,t_{n-1}}$ and verify that the topology cannot depend on anything but $t_1, \ldots, t_{n-1}$. In the second subsection the main properties of $K_{t_1,\ldots,t_{n-1}}$ are established. In the third subsection we make sure that this topology, in fact, does not depend on $t_1, \ldots, t_{n-1}$. Moreover, in the finite extension L/K, the topologies of L and K are compatible. We consider the mixed characteristic case and the easy case char $K =$ char $\overline{K} = p$ in parallel.

Finally, in the fourth subsection we consider the case char $K =$ char $\overline{K} = 0$. In this case the topology of K turns out to be noncanonical and to depend on the embedding $\overline{K} \hookrightarrow K$.

1°. Let K be an n-dimensional complete field. In the first three subsections we assume that char $\overline{K} = p > 0$. (Recall that $\overline{K}$ is the first residue field of K.) The topology of the ordinary discrete valuation on K will be referred to as the 1-topology. Put

$$\mathfrak{p}_i = \mathfrak{p}(1, \underbrace{0, \ldots, 0}_{n-i}), \qquad 1 \leq i \leq n.$$

We say that $t_1, \ldots, t_{n-1}$ are *generalized local parameters* of K if $t_i \in \mathfrak{p}_i$, $t_i \notin \mathfrak{p}_{i+1}$, $i = 1, \ldots, n-1$. In this subsection we associate with $t_1, \ldots, t_{n-1}$ a certain topological space with underlying set K. This space will be denoted by $K_{t_1,\ldots,t_{n-1}}$.

Let F_0 be the last residue field of K. Fix generalized local parameters $t_1, \ldots, t_{n-1} \in K$.

LEMMA 1.1. *The field $\overline{K}$ is a finite extension of its subfield $\widetilde{F}$ with local parameters $\overline{t_1}, \ldots, \overline{t_{n-1}} \in K$ and last residue field F_0. The subfield $\widetilde{F}$ is determined uniquely. There is a canonical isomorphism of $(n-1)$-dimensional complete fields*

$$i: F = F_0((X_1))\ldots((X_{n-1})) \xrightarrow{\sim} \widetilde{F}$$

such that $i(X_j) = \overline{t_j}$, $j = 1, \ldots, n-1$.

A map $h: F \to K$ will be called a *lifting* if $h(F) \subset \mathfrak{p}(0)$ and $\overline{h(a)} = i(a)$ for any $a \in F$.

LEMMA 1.2. *There exists a unique lifting* $H_{t_1,\ldots,t_{n-1}}$ *such that*

$$(1) \quad H_{t_1,\ldots,t_{n-1}}\left(\sum_{i_1=0}^{p-1}\cdots\sum_{i_{n-1}=0}^{p-1} X_1^{i_1}\ldots X_{n-1}^{i_{n-1}} a_{i_1,\ldots,i_{n-1}}^p\right)$$
$$= \sum_{i_1=0}^{p-1}\cdots\sum_{i_{n-1}=0}^{p-1} t_1^{i_1}\ldots t_{n-1}^{i_{n-1}}(H_{t_1,\ldots,t_{n-1}}(a_{i_1,\ldots,i_{n-1}}))^p$$

Furthermore,

$$H_\alpha(a)H_\alpha(b) = \sum_{j_1=0}^{p-1}\sum_{j_2=0}^{p-1} t_{(\alpha)}^{j_1+j_2} H_\alpha(a_{j_1})^p H_\alpha(b_{j_2})^p$$

$$\equiv \sum_{j_1=0}^{p-1}\sum_{j_2=0}^{p-1} t_{(\alpha)}^{j_1+j_2} \left(H_\alpha(p_0(a_{j_1}, b_{j_2})) + \cdots + p^{k-1} H_\alpha(p_{k-1}(a_{j_1}, b_{j_2})) \right)^p$$

$$\equiv \sum_{j_1=0}^{p-1}\sum_{j_2=0}^{p-1} t_{(\alpha)}^{j_1+j_2} H_\alpha(a_{j_1} b_{j_2})^p + p \sum_{l=0}^{p-1} p^l H_\alpha(f_l(a, b)) \mod \mathfrak{p}(ke+1),$$

where the f_l do not depend on α.

Then,

$$\sum_{j_1=0}^{p-1}\sum_{j_2=0}^{p-1} t_{(\alpha)}^{j_1+j_2} H_\alpha(a_{j_1} b_{j_2})^p$$

$$= \sum_{j=0}^{p-1} t_{(\alpha)}^j \left(\sum_{j_1=0}^{j} H_\alpha(a_j, b_{j-j_1})^p + \sum_{j_1=j+1}^{p-1} H_\alpha(Xa_j b_{j-j_1+p})^p \right).$$

Arguing in the same way as in the first part of the proof, we get the congruence

$$H_\alpha(a) \cdot H_\alpha(b) \equiv \sum_{j=0}^{p-1} t_{(\alpha)}^j H_\alpha\left(\sum_{j_1=0}^{j} a_{j_1} b_{j-j_1} + \sum_{j_1=j+1}^{p-1} Xa_{j_1} b_{j-j_1+p} \right)^p$$

$$+ p \sum_{l=0}^{k-1} p^l H_\alpha(\widetilde{f}_l(a, b)) \mod \mathfrak{p}(ke+1),$$

where the $\widetilde{f}_l$ do not depend on α.

The first summand, in view of (1), is equal to $H_\alpha(ab)$, and we get $P_{(k-1)e+1}^{(\alpha)} = \cdots = P_{ke-1}^{(\alpha)} = 0$, $P_{ke}^{(\alpha)} = \widetilde{f}_{k-1}$, i.e., $P_{ke}^{(1)} = P_{ke}^{(2)}$. □

LEMMA 1.4. *Let* $\operatorname{char} K = p$. *Then* $H_{t_1,\ldots,t_{n-1}}: F \to K$ *is an embedding of fields.*

PROOF. One easily concludes by induction on i that

$$H_{t_1,\ldots,t_{n-1}}(a+b) \equiv H_{t_1,\ldots,t_{n-1}}(a) + H_{t_1,\ldots,t_{n-1}}(b) \mod \mathfrak{p}(i),$$
$$H_{t_1,\ldots,t_{n-1}}(ab) \equiv H_{t_1,\ldots,t_{n-1}}(a) H_{t_1,\ldots,t_{n-1}}(b) \mod \mathfrak{p}(i). \quad \square$$

Let $\operatorname{char} K = 0$. Denote by k_0 the quotient field of $W(F_0)$. Then we see that $K_0 = k_0\{\{T_1\}\}\ldots\{\{T_{n-1}\}\}$ is an n-dimensional complete field with residue field $F_0((\overline{T_1}))\ldots((\overline{T_{n-1}})) \approx F$, and there is a lifting $H = H_{T_1,\ldots,T_{n-1}}: F \to K_0$. On the other hand, the lifting $h: F \to K_0$,

$$(6) \qquad h\left(\sum_{\overline{r} \in \mathbb{Z}^{n-1}} \theta_{\overline{r}} X_1^{r^{(1)}} \ldots X_{n-1}^{r^{(n-1)}} \right) = \sum_{\overline{r} \in \mathbb{Z}^{n-1}} [\theta_{\overline{r}}] T_1^{r^{(1)}} \ldots T_{n-1}^{r^{(n-1)}},$$

is also defined. It is clear that $H(F)$ is a system of representatives of $\overline{K}_0$ in K_0. Put

(7) $$h(a) = H(a) + pH(\lambda_1(a)) + p^2 H(\lambda_2(a)) + \ldots$$

for any $a \in F$. These equalities define mappings $\lambda_i : F \to F$.

Now define a new mapping $h_{t_1, \ldots, t_{n-1}} : F \to K$ by the formula

(8) $$\begin{aligned} h_{t_1, \ldots, t_{n-1}}(a) &= H_{t_1, \ldots, t_{n-1}}(a) + pH_{t_1, \ldots, t_{n-1}}(\lambda_1(a)) + \\ &\quad + p^2 H_{t_1, \ldots, t_{n-1}}(\lambda_2(a)) + \ldots. \end{aligned}$$

Notice that $h_{t_1, \ldots, t_{n-1}}(0) = 0$.

In the case char $K = p$, put $h_{t_1, \ldots, t_{n-1}} = H_{t_1, \ldots, t_{n-1}}$.

LEMMA 1.5. *Let $j : K \to L$ be an embedding of n-dimensional complete fields; $t_1, \ldots, t_{n-1}$ be generalized local parameters in K. Let $F_{0,K}$ and $F_{0,L}$ be the last residue fields of K and L respectively;*

$$F_K = F_{0,K}((X_1)) \ldots ((X_{n-1})), \qquad F_L = F_{0,L}((X_1)) \ldots ((X_{n-1}));$$

$\overline{j} : F_K \to F_L$ the natural embedding induced by the embedding of $F_{0,K}$ into $F_{0,L}$. Then the following diagram is commutative:

$$\begin{array}{ccc} K & \xrightarrow{j} & L \\ \uparrow h_{t_1, \ldots, t_{n-1}} & & \uparrow h_{j(t_1), \ldots, j(t_{n-1})} \\ F_K & \xrightarrow{\overline{j}} & F_L \end{array}$$

PROOF. Introduce the notations $h_K = h_{t_1, \ldots, t_{n-1}}$; $h_L = h_{j(t_1), \ldots, j(t_{n-1})}$; $H_K = H_{t_1, \ldots, t_{n-1}}$; $H_L = H_{j(t_1), \ldots, j(t_{n-1})}$. First let char $K = 0$. Then for $a \in K$

$$\begin{aligned} (j \circ h_K)(a) &= j(H_K(a) + pH_K(\lambda_1(a)) + p^2 H_K(\lambda_2(a)) + \ldots) \\ &= (j \circ H_K)(a) + p(j \circ H_K)(\lambda_1(a)) + p^2 (j \circ H_K)(\lambda_2(a)) + \ldots. \end{aligned}$$

Further,

$$(j \circ H_K)\left(\sum_{i_1=0}^{p-1} \ldots \sum_{i_{n-1}=0}^{p-1} X_1^{i_1} \ldots X_{n-1}^{i_{n-1}} a_{i_1, \ldots, i_{n-1}}^p \right)$$

$$= j\left(\sum_{i_1=0}^{p-1} \ldots \sum_{i_{n-1}=0}^{p-1} t_1^{i_1} \ldots t_{n-1}^{i_{n-1}} H_K(a_{i_1, \ldots, i_{n-1}})^p \right)$$

$$= \sum_{i_1=0}^{p-1} \ldots \sum_{i_{n-1}=0}^{p-1} j(t_1)^{i_1} \ldots j(t_{n-1})^{i_{n-1}} ((j \circ H_K)(a_{i_1, \ldots, i_{n-1}}))^p.$$

We have $F_L = F_K \otimes_{F_{0,K}} F_{0,L}$; put $\widetilde{H}(a \otimes \theta) = [\theta](j \circ H_K)(a)$. Then $\widetilde{H} : F_L \to L$ is a lifting possessing property (1), whence $\widetilde{H} = H_L$ and $j \circ H_K = H_L \circ \overline{j}$. It follows $j \circ h_K = h_L \circ \overline{j}$ since λ_i are obviously functorial.

When char $K = p$, we have

$$j(h_K(a)) = j(H_K(a)) = H_L(\overline{j}(a)) = h_L(\overline{j}(a)). \quad \square$$

PROPOSITION 1.1. *Let $t_1, \ldots, t_{n-1}$ be generalized local parameters in K, where* char $K = 0$. *For $a = \sum_{i=-\infty}^{\infty} p^i h(a_i) \in K_0$, let*

$$f_{t_1,\ldots,t_{n-1}}(a) = \sum_{i=-\infty}^{\infty} p^i h_{t_1,\ldots,t_{n-1}}(a_i).$$

Then $f_{t_1,\ldots,t_{n-1}}: K_0 \to K$ is an embedding of n-dimensional complete fields such that

(9) $\qquad f_{t_1,\ldots,t_{n-1}}(T_\nu) = t_\nu, \qquad \nu = 1, \ldots, n-1.$

It is natural to denote the image of this embedding by $k_0\{\{t_1\}\}\ldots\{\{t_{n-1}\}\} \subset K$.

PROOF. Define the mapping $f_1: K_0 \to K$ similarly to $f_{t_1,\ldots,t_{n-1}}$, but using H and $H_{t_1,\ldots,t_{n-1}}$ instead of h and $h_{t_1,\ldots,t_{n-1}}$; we have $f_1 \circ H = H_{t_1,\ldots,t_{n-1}}$.

It easily follows from Lemma 1.3 that f_1 is an embedding of fields, continuous with respect to the 1-topology. Then, (7) and (8) imply $f_1 \circ h = h_{t_1,\ldots,t_{n-1}}$, whence $f_1 = f_{t_1,\ldots,t_{n-1}}$. Furthermore, for $a \in \mathfrak{p}_{K_0}(0)$ we have $f_{t_1,\ldots,t_{n-1}}(a) \in \mathfrak{p}_K(0)$ and $\bar{a}$ is identified with $\overline{f_{t_1,\ldots,t_{n-1}}(a)}$ via

$$\overline{K_0} = F_0((\overline{T_1}))\ldots((\overline{T_{n-1}})) \approx F_0((X_1))\ldots((X_{n-1})) = F \xhookrightarrow{i} \overline{K}.$$

Hence $f_{t_1,\ldots,t_{n-1}}$ preserves the structure of n-dimensional complete field. The relations (9) are evident. □

To construct the topological space $K_{t_1,\ldots,t_{n-1}}$, temporarily fix $\pi_i \in K$ ($i \in \mathbb{Z}$), $v(\pi_i) = i$, and $e_j \in K$, $j = 1, \ldots, d$, so that $\{\overline{e_j}\}_{j=1}^{d}$ is a basis of the F-linear space $\overline{K}$ (via the isomorphism of Lemma 1.1).

Let $\{U_i\}_{i \in \mathbb{Z}}$ be a sequence of neighborhoods of zeros in F, $U_i = F$ for all sufficiently large i. Put

(10) $\qquad \mathcal{U}_{\{U_i\}} = \left\{ \sum_{i \gg -\infty} \pi_i \cdot \sum_{j=1}^{d} e_j h_{t_1,\ldots,t_{n-1}}(a_{ij}) \,\middle|\, a_{ij} \in U_i \right\}.$

The collection of all such sets $\mathcal{U}_{\{U_i\}}$ is denoted by B_U. B_U will play the role of the basis of neighborhoods of 0 in the topological group $K_{t_1,\ldots,t_{n-1}}$ that we are constructing. To prove the assertions below, we use induction on the dimension of the field and assume that the topology for $(n-1)$-dimensional fields has already been constructed and possesses the properties under consideration. For 1-dimensional fields we take the ordinary topology of discrete valuation.

Consider at first a particular case: a standard mixed characteristic field. Namely, let $K = k\{\{t_1\}\}\ldots\{\{t_{n-1}\}\}$, where k is a 1-dimensional complete field. Denote by K' the field $k\{\{t_1\}\}\ldots\{\{t_{n-2}\}\}$ and by F' the field $F_0((X_1))\ldots((X_{n-2})) \subset F$. For K we assume $\pi_i = \pi^i$, π is a prime of k, $e_1 = 1$.

Let $\{V_i\}_{i \in \mathbb{Z}}$ be a sequence of neighborhoods of zero in K' such that for some N we have $\mathfrak{p}_{K'}(N) \subset V_i$ for all $a \in \mathbb{Z}$, and for any $l \in \mathbb{Z}$ we have $\mathfrak{p}_{K'}(l) \subset V_i$ for all sufficiently large i. Put

$$\mathcal{V}_{\{V_i\}} = \left\{ \sum t_{n-1}^i \cdot b_i \,\middle|\, b_i \in V_i \right\}.$$

We denote the collection of all such sets $\mathcal{V}_{\{V_i\}}$ by B_V.

LEMMA 1.6. (i). *Any element of B_V contains (as a subset) an element of B_V which is a subgroup in K. Thus, B_V is a basis of neighborhoods of zero in some topology of the additive topological group on K.*

(ii) *Any element of B_V contains an element of B_U, and any element of B_U contains an element of B_V. Thus, B_U is a basis of neighborhoods of zero of the same topology on K.*

PROOF. (i). Consider $\mathcal{V}_{\{V_i\}} \in B_V = B_{V,K}$; $\mathfrak{p}_{K'}(N) \subset V_i$ for all i, and for any l we have $\mathfrak{p}_{K'}(l) \subset V_i$, $i \geq i_l$. Applying induction on n, we may assume that $V_i \supset V_i'$, $V_i' \in B_{V,K'}$, and, moreover, that V_i' is a subgroup in K'. Let $m_i = \min\{m : \mathfrak{p}_{K'}(m) \subset V_i\}$, and $V_i'' = V_i' + \mathfrak{p}_{K'}(m_i)$. Then $V_i'' \in B_{V,K'}$, V_i'' is a subgroup in K', $\mathfrak{p}_{K'}(N) \subset V_i''$ for all i, and for any l we have $\mathfrak{p}_{K'}(l) \subset V_i''$, $i \geq i_l$. Hence $\mathcal{V}_{\{V_i''\}}$ is a subgroup in K which belongs to $B_{V,K}$.

(ii). Consider $\mathcal{V}_{\{V_i\}} \in B_{V,K}$, $\mathfrak{p}_{K'}(N) \subset V_i$ for all i; $\mathfrak{p}_{K'}(l) \subset V_i$, $i \geq i_l$. We have $V_i \supset V_i' \in B_{U,K'}$. Having constructed V_i'' as in (i), we get $V_i'' \in B_{U,K'}$, $\mathcal{V}_{\{V_i''\}} \in B_{V,K}$. Let $V_i'' = \mathcal{U}_{\{U_{ij}\}}$. Then

$$\mathcal{V}_{\{V_i''\}} = \left\{\sum_i t_{n-1}^i \cdot b_i \,\Big|\, b_i \in V_i''\right\}$$
$$= \left\{\sum_i t_{n-1}^i \sum_j \pi^j h_{t_1,\ldots,t_{n-2}}(a_{ij}) \,\Big|\, a_{ij} \in U_{ij}\right\}$$
$$= \left\{\sum_j \pi^j \sum_i t_{n-1}^i X(a_{ij}) \,\Big|\, a_{ij} \in U_{ij}\right\}$$
$$= \left\{\sum_j \pi^j h_{t_1,\ldots,t_{n-1}}\left(\sum_i X_{n-1}^i a_{ij}\right) \,\Big|\, a_{ij} \in U_{ij}\right\} = \mathcal{U}_{\{U_j\}},$$

where $U_j = \mathcal{U}_{\{U_{ij}\}}$. Notice that $U_{ij} = \overline{F'}$ for $i \geq i_j$; thus, $\mathcal{U}_{\{U_{ij}\}}$ is well defined. On the other hand, $U_{ij} = \overline{F'}$ for $j \geq N$ and all i, whence $u_j = \overline{F}$ for $j \geq 1$, and $\mathcal{U}_{\{U_j\}}$ is well defined.

In a similar way one can prove that any element of B_U contains an element of B_V. □

LEMMA 1.7. *Let k be a mixed characteristic 1-dimensional complete field and $\{V_i\}_{i \in I}$ be a family of neighborhoods of zero in $k\{\{t_1\}\}\ldots\{\{t_{n-1}\}\}$. Suppose that for any $m \in \mathbb{Z}$ the set $\mathfrak{p}(m)$ is contained in all the V_i, except for a finite number of them. Then $\bigcap_{i \in I} V_i$ is a neighborhood of zero.*

PROOF. Construct V_i'' in the same way as in the proof of part (i) of Lemma 1.6. Then $V_i'' \in B_V$, and the family $\{V_i''\}_{i \in I}$ satisfies the conditions of Lemma 1.7. The fact that $\bigcap_{i \in I} V_i''$ is a neighborhood of zero is easily proved by induction on n. □

Now we return to the general case.

THEOREM 1.1. (i) *Any element of $B_U = B_{t_1,\ldots,t_{n-1},\{\pi_i\},\{e_j\}}$ contains an element of B_U which is a subgroup. Thus, B_U is a basis of neighborhoods of zero for the additive group of K. K acquires the structure of a topological group, which will be referred to as $K_{t_1,\ldots,t_{n-1},\{\pi_i\},\{e_j\}}$.*

(ii) *The multiplication by $c \in K^*$ is a homeomorphism of this topological group onto itself.*

(iii) *The topology of $K_{t_1,\ldots,t_{n-1},\{\pi_i\},\{e_j\}} = K_{t_1,\ldots,t_{n-1}}$ does not depend on the choice of $\{\pi_i\}$ and $\{e_j\}$.*

The theorem will be proved in several steps.

STEP 1. *Assertions* (i) *and* (ii) *of Theorem* 1.1 *are valid for the case* $K = F((X))$, $\pi_i = X^i$, $i \in \mathbb{Z}$, $e_1 = 1$.

PROOF. We get (i) immediately by induction on the dimension of the field. To prove (ii), it suffices to see that for any $\mathcal{U}_{\{U_i\}} \in B_U$ there exists a $\mathcal{U}_{\{U_i'\}} \in B_U$ such that
$$c\mathcal{U}_{\{U_i'\}} \subset \mathcal{U}_{\{U_i\}}.$$

Let $c = \sum_{j=s}^{\infty} c_j X^j$, $c_j \in F$; $u_i = F$ for $i \geqslant N$. Then one can put $U_i' = F$ for $i \geqslant N - s$, and
$$U_i' = \bigcap_{\substack{s \leqslant j \leqslant N-i-1 \\ c_j \neq 0}} c_j^{-1} U_{i+j}$$
for $i < N - s$. □

STEP 2. *Assertions* (i), (ii) *of Theorem* 1.1 *are valid for the case* $K = k_0\{\{t_1\}\}\ldots\{\{t_{n-1}\}\}$, $\pi_i = p^i$, $i \in \mathbb{Z}$, $e_1 = 1$. (*Here k_0 denotes the quotient field of $W(F_0)$, as above.*)

PROOF. Assertion (i) was proved in Lemma 1.6. To prove (ii), put
$$c = \sum c_i t_{n-1}^i, \quad c_i \in k_0\{\{t_1\}\}\ldots\{\{t_{n-2}\}\} = K'.$$

Consider a neighborhood W of zero. One may assume $W = \mathcal{V}_{\{V_i\}} \in B_V$. Put $V_i' = \bigcap_{c_j \neq 0} c_j^{-1} V_{i+j}$. It is easy to see that for any $m \in \mathbb{Z}$ we have $\mathfrak{p}_{K'}(m) \subset c_j^{-1} V_{i+j}$ for all sufficiently small or sufficiently large j. Therefore, in view of Lemma 1.7, V_i' is a neighborhood of zero in K'. Furthermore, if $\mathfrak{p}_{K'}(N) \subset V_i$ for all i, then $\mathfrak{p}_{K'}(N - v^{(n)}(c)) \subset V_i'$ for all i. Finally, $\mathfrak{p}_{K'}(l) \subset V_i'$ for all $i \geqslant i_0$, where i_0 is determined by the condition:
$$\mathfrak{p}_{K'}(l + v^{(n)}(c)) \subset V_{s+i}, \quad i \geqslant i_0,$$
and s, in its turn, is determined by the condition:
$$v^{(n)}(c_j) \geqslant N - l, \quad j < s.$$

Thus, $\mathcal{V}_{\{V_i'\}} \in B_V$ is well defined. Directly from the definition of multiplication in $K = K'\{\{t_{n-1}\}\}$ we get $c\mathcal{V}_{\{V_i'\}} \subset W$, whence multiplication by c is continuous and, moreover, is a homeomorphism for $c \neq 0$. □

The following lemma is practically evident.

LEMMA 1.8. *In the case* $\operatorname{char} K = p$ *assume* $\pi_i = X^i$, $i \in \mathbb{Z}$, *where X is an arbitrary uniformizing element. In the case* $\operatorname{char} K = 0$ *assume* $\pi_{i+e} = p\pi_i$, $i \in \mathbb{Z}$, $e = v^{(n)}(p)$. *Put*

$$K_0 = F_0((t_1))\ldots((t_{n-1})) \subset K \quad \text{for } \operatorname{char} K = p,$$
$$K_0 = k_0\{\{t_1\}\}\ldots\{\{t_{n-1}\}\} \subset K \quad \text{for } \operatorname{char} K = 0.$$

Then the map

$$\prod_{i=0}^{e-1}\prod_{j=1}^{d} K_0 \to K, \quad (a_{i,j}) \longmapsto \sum_{i=0}^{e-1}\sum_{j=1}^{d} \pi_i e_j a_{i,j}$$

is a homeomorphism (if $\operatorname{char} K = p$, *we put* $e = 1$).

STEP 3. *For the choice of π_i as in Lemma* 1.8 *assertions* (i) *and* (ii) *of Theorem* 1.1 *are valid.*

PROOF. Assertion (i) is evident in view of Step 1, Step 2, and Lemma 1.8. To prove (ii), notice that multiplication by $c \in K$ is a K_0-linear operator in the space $K \approx K_0^{de}$ and is continuous because of the continuity of addition and multiplication by a constant in K_0. □

One immediately proves:

COROLLARY. *Let elements* $\{\pi_i\}$ *be chosen as in Lemma* 1.8. *Then the structure of additive topological group is induced on* $\mathfrak{p}(l)$, $l \in \mathbb{Z}$. *For any sequence of neighborhoods of zero* $\{U_i\}_{i \geqslant l}$ *in* F, *consider the set*

$$\mathcal{U}_{\{U_i\}} = \left\{ \sum_{i=l}^{\infty} \pi_i \cdot \sum_{j=1}^{d} e_j h_{t_1,\ldots,t_{n-1}}(a_{ij}) \,\Big|\, a_{ij} \in U_i \right\}.$$

All such sets form a basis of neighborhoods of zero $B_U^{(l)}$ *in* $\mathfrak{p}(l)$. *Any element of* $B_U^{(l)}$ *contains an element of* $B_U^{(l)}$ *which is an open subgroup. Multiplication by* $c \in K^*$ *determines a homeomorphism of* $\mathfrak{p}(l)$ *onto* $\mathfrak{p}(l + v^{(n)}(c))$.

STEP 4. *Suppose* $v(\pi_i) = v(\pi_i') = i$, $i \in \mathbb{Z}$ *in* K; $\{e_j\}_{j=1}^{d}$ *and* $\{e_j'\}_{j=1}^{d}$ *be lifts of two bases of the F-space* $\overline{K}$. *Then any element of* $B_{t_1,\ldots,t_{n-1},\{\pi_i\},\{e_j\}}$ *contains an element of* $B_{t_1,\ldots,t_{n-1},\{\pi_i'\},\{e_j'\}}$ *and conversely.*

PROOF. In view of the "transitivity" of the property being proved, one may assume that elements π_i are chosen as in Lemma 1.8 and that $e_1 = 1$. Then the topology of $K_{t_1,\ldots,t_{n-1},\{\pi_i\},\{e_j\}}$ has already been defined in view of Step 3, and $h = h_{t_1,\ldots t_{n-1}}$ is continuous.

Let us verify that any neighborhood $U \in B_{t_1,\ldots,t_{n-1},\{\pi_i\},\{e_j\}}$ contains an element of $B_{t_1,\ldots,t_{n-1},\{\pi_i'\},\{e_j'\}}$. Assume $\mathfrak{p}(N) \subset U$, $N \in \mathbb{Z}$.

Taking into account the continuity of h and of the multiplication by a constant in $K_{t_1,\ldots,t_{n-1},\{\pi_i\},\{e_j\}}$, we see that for any $i < N$ one can construct a neighborhood of zero U'_i in F such that $\pi'_i e'_j h(U'_i) \subset U$ for all j. Put $U'_i = F$ for $i \geq N$. Then

$$\left\{ \sum_{i \gg -\infty} \pi'_i \sum_{j=1}^{d} e'_j h(a_{ij}) \,\middle|\, a_{ij} \in U_i \right\}$$

is the required element of $B_{t_1,\ldots,t_{n-1},\{\pi'_i\},\{e'_j\}}$.

Conversely, let $U' \in B_{t_1,\ldots,t_{n-1},\{\pi'_i\},\{e'_j\}}$. We shall construct neighborhoods of zero $U_i \subset F$ such that $\mathcal{U}_{\{U_i\}} \in B_{t_1,\ldots,t_{n-1},\{\pi_i\},\{e_j\}}$, $\mathcal{U}_{\{U_i\}} \subset U'$. Let $\mathfrak{p}(N) \subset U'$, $N \in \mathbb{Z}$. Put $U_i = F$, $i \geq N$, and start to construct U_i, $i < N$, by induction on $N - i$. Suppose U_i, $i > i_0$, have already been constructed. Decreasing U_i, we may assume that $\widetilde{U} = \mathcal{U}_{\{U_i\}}$ is a subgroup in $\mathfrak{p}(i_0 + 1)$, where $\mathcal{U}_{\{U_i\}}$ should be understood in the sense of the above Corollary.

It suffices to construct a neighborhood of zero U_{i_0} in F such that

$$(11) \qquad \pi_{i_0} \cdot \sum_{j=1}^{d} e_j h(U_{i_0}) \subset \pi'_{i_0} \cdot \sum_{j=1}^{d} e'_j h(U'_{i_0}) + \widetilde{U}.$$

Let $\pi_{i_0} e_j = \sum_{i=i_0}^{\infty} \pi'_i \sum_{l=1}^{d} \alpha_{jl}^{(i)} e'_l$, $\alpha_{jl}^{(i)} \in h(F)$. Then in order to satisfy (11), it suffices to satisfy the following conditions

$$(12) \qquad \pi'_{i_0} \alpha_{jl}^{(i_0)} e'_l h(U_{i_0}) \subset \pi'_{i_0} e'_l h(U'_{i_0}) \widetilde{U},$$

for $j, l = 1, \ldots, d$, and

$$(13) \qquad \pi'_i \alpha_{jl}^{(i)} e'_l h(U_{i_0}) \subset \widetilde{U},$$

for $i > i_0$, $j, l = 1, \ldots, d$.

Further, (12) is equivalent to

$$\alpha_{jl}^{(i_0)} h(U_{i_0}) \subset h(U'_{i_0}) + (\pi'_{i_0} e'_l)^{-1} \widetilde{U}$$

and to

$$(14) \qquad \alpha_{jl}^{(i_0)} h(U_{i_0}) \subset W = \sum_{k=1}^{d} e_k h(U'_{i_0}) + (\pi'_{i_0} e'_l)^{-1} \widetilde{U},$$

since $\alpha_{jl}^{(i_0)} h(U_{i_0}) \subset h(F) h(F) \subset h(F) + \wp(1) = e_1 h(F) + \wp(1)$. It is clear that W is a neighborhood of zero in $\wp(0)$. Now the continuity of h and of the multiplication by a constant implies both (13) and (14). □

This concludes the proof of Theorem 1.1.

REMARK. It is easy to see from the construction of B_U that in $K_{t_1,\ldots,t_{n-1}}$ one can choose a basis of neighborhoods of zero from open subgroups closed with respect to multiplication by elements of $\mathcal{R}_K$ (Teichmüller representatives).

2°. Now we establish the basic properties of $K_{t_1,\ldots,t_{n-1}}$.

LEMMA 1.9. *Let $a_i \to a$ in $K_{t_1,\ldots,t_{n-1}}$. Then $v^{(n)}(a_i)$ are bounded from below.*

PROOF. In view of Lemma 1.8, one may assume $K = K_0$, i.e., in the definition of the topology $d = 1$, $e_1 = 1$. One may also assume $a = 0$. Suppose that $v^{(n)}(a_i)$ are not bounded from below. Then one finds $m_1 > m_2 > \ldots$ such that $v^{(n)}(a_{i_\alpha}) = m_\alpha$ for some i_α, $\alpha = 1, 2, \ldots$ Choose neighborhoods of zero U_{m_α} in F such that $\overline{\pi_{m_\alpha}^{-1} a_{i_\alpha}} \notin i(U_{m_\alpha})$, where $i \colon F \xrightarrow{\sim} \overline{K}$ is the isomorphism of Lemma 1.1. Put $U_m = F$, if m is distinct from all m_α. Then $a_i \notin \mathcal{U}_{\{U_j\}}$ for $i = i_1, i_2, \ldots$ □

PROPOSITION 1.2. *The group $K_{t_1,\ldots,t_{n-1}}$ is complete and Hausdorff.*

PROOF. Let $(a_i)_{i=1}^\infty$ be a Cauchy sequence in $K_{t_1,\ldots,t_{n-1}}$. In the case $\operatorname{char} K = p$, one can assume $K = F((X))$,

$$a_i = \sum_{j \gg -\infty} a_{ij} X^j.$$

Then $(a_{ij})_{i=1}^\infty$ is a Cauchy sequence, and the convergence of $(a_i)_{i=1}^\infty$ can be verified by induction on n.

The case $\operatorname{char} K = 0$ can be reduced, in view of Lemma 1.8, to the subcase $K = K'\{\{t_{n-1}\}\} = k\{\{t_1\}\}\ldots\{\{t_{n-1}\}\}$. Let $a_i = \sum_{j=-\infty}^\infty b_{ij} t_{n-1}^j$. By Lemma 1.6, $(b_{ij})_{i=1}^\infty$ is a Cauchy sequence, and, by the assumption of induction, $b_{ij} \to b_j$. It is not difficult to see that $a = \sum_{j=-\infty}^\infty b_j t_{n-1}^j$ is well defined, and $a_i \to a$.

Similarly one verifies that $K_{t_1,\ldots,t_{n-1}}$ is Hausdorff. □

COROLLARY. *Suppose I is a countable set, and for any $i \in I$ there is an a_i belonging to $K_{t_1,\ldots,t_{n-1}}$. Let the set $\{i \in I \mid a_i \notin U\}$ be finite for any neighborhood of zero U. Then $\sum_{i \in I} a_i$ is well defined.*

A set of multi-indices $I \subset \mathbb{Z}^n$ will be called *admissible*, if for any $i^{(n)}, \ldots, i^{(l+1)}$ ($1 \leqslant l \leqslant n$) there is an integer i such that conditions $\overline{r} \in I$, $r^{(n)} = i^{(n)}, \ldots, r^{(l+1)} = i^{(l+1)}$ imply $r^{(l)} \geqslant i$. (In particular, an application of this definition for $l = n$ means that I is bounded from below.)

Notice the following obvious properties. (Here $\mathbb{Z}_+^n = \{\overline{r} \in \mathbb{Z}^n \mid \overline{r} > 0\}$.)

1. If $I \subset \mathbb{Z}_+^n$ is an admissible set, then

$$\hat{I} = \left\{ \sum_{j=1}^m \overline{r}_j \,\Big|\, \overline{r}_j \in I, \, m \geqslant 1 \right\}$$

is also an admissible set.

2. If $I \subset \mathbb{Z}_+^n$ is an admissible set, then any element of $\hat{I}$ can be expanded into a sum of elements of I in only finite number of ways.

LEMMA 1.10. *Let $I \subset \mathbb{Z}^n$ be an admissible set, U a neighborhood of zero, $v^{(n)}(\pi_i) = i$, $i \in \mathbb{Z}$. Then for almost all $\overline{r} \in I$ and all $\theta \in F_0$ we have*

$$[\theta] t_1^{r^{(1)}} \ldots t_{n-1}^{r^{(n-1)}} \pi_{r^{(n)}} \in U.$$

PROOF. The assertion is easily verified by induction on n since

$$[\theta] t_1^{r^{(1)}} \ldots t_{n-1}^{r^{(n-1)}} = h_{t_1,\ldots,t_{n-1}}(\theta \overline{t}_1^{r^{(1)}} \ldots \overline{t}_{n-1}^{r^{(n-1)}}). \quad \square$$

COROLLARY. *Let $I \subset \mathbb{Z}^n$ be an admissible set, $v(\pi_i) = i$, $i \in \mathbb{Z}$; $\theta_{\bar{r}} \in F_0$ for $\bar{r} \in I$. Then $\sum_{\bar{r} \in I} [\theta_{\bar{r}}] t_1^{r^{(1)}} \ldots t_{n-1}^{r^{(n-1)}} \pi_{r^{(n)}}$ is well defined.*

REMARK. It also follows from Lemma 1.10 that the sums

$$\sum_{\bar{r} \in I} \sum_{i=1}^{m_{\bar{r}}} \alpha_{\bar{r},i} t_1^{r^{(1)}} \ldots t_{n-1}^{r^{(n-1)}} \pi_{r^{(n)}}, \quad \alpha_{\bar{r},i} \in k_0;$$

$$\sum_{i=1}^{m} a_i \sum_{\bar{r} \in I} \alpha_{i,\bar{r}} t_1^{r^{(1)}} \ldots t_{n-1}^{r^{(n-1)}} \pi_{r^{(n)}}, \quad a_i \in K, \quad \alpha_{i,\bar{r}} \in k_0,$$

are well defined, and in such sums an arbitrary grouping of summands is possible. Here k_0 denotes the quotient field of $W(F_0)$ as a subfield in K in the case $\operatorname{char} K = 0$, and $k_0 = F_0$ in the case $\operatorname{char} K = p$.

THEOREM 1.2. *Let K be an n-dimensional complete field, $t_1, \ldots, t_{n-1}$ be generalized local parameters, $\pi_i \in K$, $v^{(n)}(\pi_i) = i$, $i \in \mathbb{Z}$, $\{e_j\}_{j=1}^d$ be elements of $\mathfrak{p}_K(0)$ such that $\{\overline{e_j}\}_{j=1}^d$ is a basis of $\overline{K}$ over $\widetilde{F} = F_0((\overline{t_1})) \ldots ((\overline{t_{n-1}})) \subset \overline{K}$, where F_0 is the last residue field of K. Then any $a \in K$ can be uniquely expanded into a sum*

$$a = \sum_{j=1}^d e_j \sum_{\bar{r} \in I_j} [\theta_{j,\bar{r}}] t_1^{r^{(1)}} \ldots t_{n-1}^{r^{(n-1)}} \pi_{r^{(n)}},$$

where I_j is an admissible set, $\theta_{j,\bar{r}} \in F_0$, $\theta_{j,\bar{r}} \neq 0$.

PROOF. Use induction on n and take into account the continuity of the residue map $\mathfrak{p}_K(0) \to \overline{K}$. □

PROPOSITION 1.3. *If $a_i \to a$, $b_i \to b$, then $a_i b_i \to ab$ in $K_{t_1, \ldots, t_{n-1}}$.*

PROOF. By Lemma 1.9 $v(a_i) \geq c$, $v(b_i) \geq c$, $i = 1, 2, \ldots$, $c \in \mathbb{Z}$. Applying Theorem 1.1, (ii), one may assume $a = b = 0$. Then, in view of Lemma 1.8, one may assume $K = K_0$. Fix an open subgroup U in K, $\mathfrak{p}_K(N) \subset U$.

Let $\operatorname{char} K = p$, i.e., $K = F((X))$. Then one can assume $U = \mathcal{U}_{\{U_j\}}$, U_j are open subgroups in F;

$$a_i b_i \stackrel{\mathfrak{p}_K(N)}{\equiv} \sum_{k=c}^{N-c} a_i^{(k)} X^k \cdot \sum_{l=c}^{N-c} b_i^{(l)} X^l.$$

We have $a_i^{(k)} \xrightarrow[i \to \infty]{} 0$, $b_i^{(l)} \xrightarrow[i \to \infty]{} 0$, whence $a_i^{(k)} b_i^{(l)} \to 0$ (induction on n), and $a_i^{(k)} b_i^{(l)} \in U_{k+l}$ for large i. Therefore, $a_i b_i \in \mathcal{U}_{\{U_j\}}$ for large j.

For $\operatorname{char} K = 0$, we have $K = K'\{\{t_{n-1}\}\}$. One can assume that $U = \mathcal{V}_{\{V_i\}}$, V_i are open subgroups in K', $\mathfrak{p}_{K'}(2c) \subset V_m$ for $m \geq m_0$. Let

$$a_i = \sum_{k=-\infty}^{\infty} a_i^{(k)} t_{n-1}^k, \quad b_i = \sum_{l=-\infty}^{\infty} b_i^{(l)} t_{n-1}^l, \quad a_i, b_i \in K'.$$

Notice that one can find a k_0 such that $a_i^{(k)} \in \mathfrak{p}_{K'}(N-c)$ for $k \leqslant k_0$ and all i. Indeed, if this were not so, one could choose a sequence $k_1 > k_2 > \ldots$ such that $a_{i_\alpha}^{(k_\alpha)} \notin \mathfrak{p}_{K'}(N-c)$ for some i_α. It is then easy to construct a neighborhood of zero V_{k_α} in K' such that $\mathfrak{p}_{K'}(N-c) \subset V_{k_\alpha}$, $a_{i_\alpha}^{(k_\alpha)} \notin V_{k_\alpha}$. Now put $V_k = K'$ when k does not coincide with any k_α, and $V = \mathcal{V}_{\{V_k\}}$. Then $a_{i_\alpha} \notin V$, $\alpha = 1, 2, \ldots$. The set $\{i_\alpha\}$ is infinite, since for any i we have $a_i^{(k)} \in \mathfrak{p}_{K'}(N-c)$ for sufficiently small k. We get a contradiction to the condition $a_i \to 0$.

Similarly, $b_i^{(l)} \in \mathfrak{p}_{K'}(N-c)$ for $l \leqslant l_0$ and all i. Therefore,

$$a_i b_i \equiv \sum_{k=k_0}^{m_0-l_0} a_i^{(k)} t_{n-1}^k \cdot \sum_{l=l_0}^{m_0-k_0} b_i^{(l)} t_{n-1}^l \mod U,$$

and the condition $a_i^{(k)} b_i^{(l)} \to 0$ for all k and l immediately implies $a_i b_i \to 0$. □

3°. Finally, we prove that the topology of $K_{t_1,\ldots,t_{n-1}}$ is independent of the the choice of generalized local parameters $t_1, \ldots, t_{n-1}$.

LEMMA 1.11. *If L/K is an extension of n-dimensional complete fields, then one can choose such a K-basis $\{f_1, \ldots, f_m\}$ in L that $O_L = O_K f_1 \oplus \cdots \oplus O_K f_m$.*

In the definition of $K_{t_1,\ldots,t_{n-1}}$ we may take $\pi_i = \pi^i$, where π is a prime in K. Further, choose $e_1, \ldots, e_d$ in such a way that $(\bar{e}_1, \ldots, \bar{e}_\alpha)$ satisfies the conditions of Lemma 1.11 for $\overline{K}/F$ (more accurately, for $\overline{K}/i(F)$). Let for $\alpha, \beta = 1, \ldots, d$:

$$e_\alpha e_\beta = \sum_{\gamma=1}^d c_{\alpha\beta}^{(\gamma)} e_i,$$

(15) $$c_{\alpha\beta}^{(\gamma)} = \sum_{\bar{r}\geqslant 0} [\theta_{\bar{r},\alpha,\beta,\gamma}] t_1^{r^{(1)}} \ldots t_{n-1}^{r^{(n-1)}} \pi^{r^{(n)}}, \quad \theta_{\bar{r},\alpha,\beta,\gamma} \in F_0.$$

Let I be an admissible set. Denote by $M(I)$ the closure of the subgroup generated by all $[\theta_{\bar{r},i}] e_i t_1^{r^{(1)}} \ldots t_{n-1}^{r^{(n-1)}} \pi^{r^{(n)}}$, $\theta_{\bar{r},i} \in F_0$, $\bar{r} \in I$, $i = 1, \ldots, d$.

LEMMA 1.12. *Let $1 \leqslant j \leqslant n-1$. For any admissible set $I \subset \mathbb{Z}^n$ and any open subgroup U in $K_{t_1,\ldots,t_{n-1}}$ there exists $A \in \mathbb{Z}$ such that*

(16) $$t_j^k t_1^{k_1} \ldots t_{j-1}^{k_{j-1}} M(I) \subset U$$

for all $k \geqslant A$, $k_1, \ldots, k_{j-1} \in \mathbb{Z}$.

PROOF. It is evident that (13) is equivalent to the condition

(17) $$[\theta] e_i t_1^{r^{(1)}+k_1} \ldots t_{j-1}^{r^{(j-1)}+k_{j-1}} t_j^{r^{(j)}+k} t_{j+1}^{r^{(j+1)}} \ldots t_{n-1}^{r^{(n-1)}} \pi^{r^{(n)}} \in U$$

for all $\theta \in F_0$, $i = 1, \ldots, d$, $\bar{r} \in I$.

One can assume $U = \mathcal{U}_{\{U_i\}}$. Use induction on $n-j$. Let $j = n-1$. We have $U_i = F$ for $i > R \in \mathbb{Z}$. On the other hand, $\bar{r} \in I$ implies $r^{(n)} \geqslant r \in \mathbb{Z}$. Thus, it suffices to establish the validity of (16) only for $r \leqslant r^{(n)} \leqslant R$.

Notice that for any i there exists m'_i such that $\mathfrak{p}_F(m'_i) \subset U_i$. Notice also that for any i there exists m_i such that the assumptions $\bar{r} \in I$, $r^{(n)} = i$ imply $r^{(n-1)} \geqslant m_i$. Then $A = \max_{r \leqslant i \leqslant R}(m'_i - m_i)$ is the required number.

The induction step is essentially similar. One has only to notice that in fact the assumption of induction means the following.

For all $(n-j)$-tuples $(r^{(j+1)}, \ldots, r^{(n)})$ (except for a finite number) one of the following two assertions is true.

(i) $\bar{r} \notin I$ for any $r^{(1)}, \ldots, r^{(j)}$.

(ii) (17) is valid automatically, for all integral $r^{(1)}, \ldots, r^{(j)}$, $k_1, \ldots, k_j$, all $\theta \in F_0$, and all $i = 1, \ldots, \alpha$.

Let $\{(r_i^{(j+1)}, \ldots, r_i^{(n)})\}_{i=1}^{l}$ be all the $(n-j)$-tuples for which both (i) and (ii) are not true. By induction on n we easily show that

$$[\theta]e_i t_1^{k_1} \ldots t_{j-1}^{k_{j-1}} t_j^{k_j} t_{j+1}^{r^{(j+1)}_i} \ldots t_{n-1}^{r^{(n-1)}_i} \pi^{r^{(n)}_i} \in U$$

for all $\theta \in F_0$, $i = 1 \ldots, d$, $k_1, \ldots, k_j \in \mathbb{Z}$, $k_j \geqslant m'_i$, where m'_i is a certain integer. On the other hand, the assumptions

$$\bar{r} \in I, \qquad r^{(j+1)} = r_i^{(j+1)}, \ldots, r^{(n)} = r_i^{(n)}$$

imply $r^{(j)} \geqslant m_i$. Again one may take $A = \max_{1 \leqslant i \leqslant l}(m'_i - m_i)$. $\square$

LEMMA 1.13. *For any admissible set $I \subset \mathbb{Z}_+^n$ there exists an admissible set $J \subset \mathbb{Z}_+^n$, $I \subset J$, such that*

$$M(J) \cdot M(J) \subset M(J).$$

PROOF. Take an admissible set $J_0 \subset \mathbb{Z}_+^n$ such that $I \subset J_0$ and $\theta_{\bar{r}, \alpha, \beta, \gamma} = 0$ in (15) for $\bar{r} \notin J_0$, $\alpha, \beta, \gamma = 1, \ldots, d$. Now it suffices to put $J = \hat{J}_0$.

PROPOSITION 1.4. *Let $1 \leqslant j \leqslant n$, $b_0 \in \mathfrak{p}_j$, $b_1, \ldots, b_m \in K$, $v^{(j)}(b_i) = v^{(j+1)}(b_i) = \cdots = v^{(n)}(b_i) = 0$, $i = 1, \ldots, m$. Let U be a neighborhood of zero in $K_{t_1, \ldots, t_{n-1}}$. Then there exists $N \in \mathbb{Z}$ such that $b_0^k b_1^{k_1} \ldots b_m^{k_m} \in U$ for any $k \geqslant N$, $k_1, \ldots, k_m \in \mathbb{Z}$.*

PROOF. Proposition 1.4 is evident for $j = n$. Let $j \leqslant n-1$. To begin with, assume $v^{(j)}(b_0) \geqslant v^{(j)}(t_j)$. Introduce $a_0, a_1, \ldots, a_{2m} \in \mathcal{O}_K$ such that

$$b_0 = t_{j-1}^{\alpha_0} t_j^{\beta} a_0, \beta > 0; \qquad b_i = t_{j-1}^{\alpha_i} a_i, \qquad b_i^{-1} = t_{j-1}^{\alpha_{m+i}} a_{m+i}, \quad i = 1, \ldots, m$$

(we formally put $t_0 = 1$). Then any element $b = b_0^k b_1^{k_1} \ldots b_m^{k_m}$, $k > 0$, can be written in the form

$$b = t_j^{k\beta} t_{j-1}^{\alpha} a_0^k a_1^{k'_1} \ldots a_{2m}^{k'_{2m}},$$

where $\alpha \in \mathbb{Z}$, $k'_1, \ldots, k'_{2m} \geqslant 0$.

Since all a_i belong to K, one can find an admissible set $I \subset \mathbb{Z}_+^n$ such that $a_i \in M(I)$, $i = 0, 1, \ldots, 2m$. By Lemma 1.13, there exists an admissible set $J \subset \mathbb{Z}_+^n$ such that $a_0^k a_1^{k'_1} \ldots a_{2m}^{k'_{2m}} \in M(J)$ for any $k > 0$, $k'_1, \ldots, k'_{2m} \geqslant 0$. For the admissible set J, determine A from Lemma 1.12 and choose N such that $k\beta \geq A$ for $k \geq N$.

Now consider an arbitrary $b_0 \in \wp_j$. Then $sv^{(j)}(b_0) \geq v^{(j)}(t_j)$ for some s, and we apply the preceding reasoning to $b_0' = b_0^s$ and $U' = U \cap b_0 U \cap \cdots \cap b_0^{s-1} U$. □

Now let $t_1', \ldots, t_{n-1}'$ be a new system of generalized local parameters in K. We continue to treat K as a topological group $K_{t_1, \ldots, t_{n-1}}$.

LEMMA 1.14. *Let* $1 \leq j \leq n-1$; $F_j = F_0((X_1)) \ldots ((X_j)) \subset F$. *Let* U *be an open subgroup in* $K_{t_1, \ldots, t_{n-1}}$. *Then one can find a neighborhood of zero* V *in* F_j *such that assumption* $\theta X_1^{r^{(1)}} \ldots X_j^{r^{(j)}} \in V$, *where* $\theta \in F_0$, $r^{(1)}, \ldots, r^{(j)} \in \mathbb{Z}$, *implies* $[\theta] t_1'^{r^{(1)}} \ldots t_j'^{r^{(j)}} \in U$.

PROOF. Use induction on j. By Proposition 1.4, we have $[\theta] t_1'^{r^{(1)}} \ldots t_j'^{r^{(j)}} \in U$ for $r^{(j)} \geq N$. Put $V_i = F_{j-1}$ when $i \geq N$. For $i < N$ by the assumption of induction $[\theta] t_1'^{r^{(1)}} \ldots t_{j-1}'^{r^{(j-1)}} \in t_j'^{-i} U$ for $\theta X_1^{r^{(1)}} \ldots X_{j-1}^{r^{(j-1)}} \in V_i$. Then $V = \mathcal{U}_{\{V_i\}}$ is as required. □

Let $I \subset \mathbb{Z}^{n-1}$ be an admissible set. Then for $a = \sum_{\bar{r} \in I} \theta_{\bar{r}} X_1^{r^{(1)}} \ldots X_{n-1}^{r^{(n-1)}} \in F'$, where $\theta_{\bar{r}} \in F_0$,

$$\widetilde{h}(a) = \sum_{\bar{r} \in I} [\theta_{\bar{r}}] t_1'^{r^{(1)}} \ldots t_{n-1}'^{r^{(n-1)}}$$

is well defined by Lemma 1.14.

LEMMA 1.15. $\widetilde{h}: F \to K$ *is a continuous lifting.*

PROOF. $\widetilde{h}$ is a lifting because of the continuity of the residue homomorphism. The continuity of $\widetilde{h}$ can be deduced from Lemma 1.14 for $j = n - 1$. □

Let $\operatorname{char} K = 0$. Put $K_0 = k_0\{\{T_1\}\} \ldots \{\{T_{n-1}\}\}$ where k_0 is the quotient field of $W(F_0)$;

$$\widetilde{f}\left(\sum_{\bar{r} \in I} [\theta_{\bar{r}}] T_1^{r^{(1)}} \ldots T_{n-1}^{r^{(n-1)}} p^{r^{(n)}}\right) = \sum_{\bar{r} \in I} [\theta_{\bar{r}}] t_1'^{r^{(1)}} \ldots t_{n-1}'^{r^{(n-1)}} p^{r^{(n)}}.$$

When $\operatorname{char} K = p$, we put $K_0 = F((X))$, and

$$\widetilde{f}\left(\sum_i a_i X^i\right) = \sum_i \widetilde{h}(a_i) \pi^i, \quad a_i \in F.$$

LEMMA 1.16. *The map* $\widetilde{f}: K_0 \to K$ *defined above is a continuous embedding of fields preserving valuation.*

PROOF. Continuity immediately follows from Lemma 1.15. Let us verify that $\widetilde{f}(x+y) = \widetilde{f}(x) + \widetilde{f}(y)$, $\widetilde{f}(xy) = \widetilde{f}(x)\widetilde{f}(y)$. For $\operatorname{char} K = p$ this is evident. Let $\operatorname{char} K = 0$. Using the sequential continuity of addition and multiplication, we may assume

$$x = [\theta] T_1^{r^{(1)}} \ldots T_{n-1}^{r^{(n-1)}} p^{r^{(n)}}, \quad y = [\theta'] T_1^{s^{(1)}} \ldots T_{n-1}^{s^{(n-1)}} p^{s^{(n)}},$$

The only thing we need to notice is that the relation

$$[\theta] + [\theta'] = [\theta_0] + p[\theta_1] + p^2[\theta_2] + \ldots$$

in k_0 implies the same relation in K. □

LEMMA 1.17. $\widetilde{h} = h_{t'_1, \ldots, t'_{n-1}}$.

PROOF. For $\operatorname{char} K = 0$, we have the canonical lifting $h \colon F \to K_0$ defined by (6). For $\operatorname{char} K = p$, we have the evident embedding $h \colon F \to K_0 = F((X))$. Obviously $\widetilde{h} = \widetilde{f} \circ h$. Now the result immediately follows from Lemma 1.5. □

THEOREM 1.3. *The topology of* $K_{t_1, \ldots, t_{n-1}}$ *coincides with that of* $K_{t'_1, \ldots, t'_{n-1}}$.

PROOF. Let U be an open subgroup in $K_{t_1, \ldots, t_{n-1}}$, $\mathfrak{p}_K(N) \subset U$. Put

$$U_i = \bigcap_{j=1}^{d'} h_{t'_1, \ldots, t'_{n-1}}^{-1} ((e'_j)^{-1} \pi_i^{-1} U),$$

where d', $e'_1, \ldots, e'_{d'}$, π_i are used to construct $K_{t'_1, \ldots, t'_{n-1}}$. By Lemmas 1.15, and 1.17 and Theorem 1.1 (ii) U_i is a neighborhood of zero in F and it is clear that $U_i = F$ for $i \geqslant N$.

Now

$$U' = \left\{ \sum_{i \gg -\infty} \pi_i \sum_{j=1}^{d'} e'_j h_{t'_1, \ldots, t'_{n-1}}(a_{ij}) \;\middle|\; a_{ij} \in U_i \right\},$$

π being prime in K, is a neighborhood of zero in $K_{t'_1, \ldots, t'_{n-1}}$ that lies in U. □

THEOREM 1.4. *Let* L/K *be a finite extension of n-dimensional complete fields. Then the topology of* K *coincide with the one induced by the topology of* L.

PROOF. Let $t_1, \ldots, t_{n-1}, \pi$ be local parameters of K. Then $t_1, \ldots, t_{n-1}$ are generalized local parameters of L. Let π_L be prime in L. If $e = e_n(L/K)$, put $\pi_{ae+b} = \pi^a \pi_L^b$ for $0 \leqslant b < e$. Finally, take arbitrary $e_1, \ldots, e_d \in \mathfrak{p}_L(0)$ such that $\{\overline{e_1}, \ldots, \overline{e_d}\}$ is a basis of $\overline{K}$-space $\overline{L}$.

By Lemma 1.5 we have the commutative diagram

$$\begin{array}{ccc} F_{0,K}((X_1)) \ldots ((X_{n-1})) & \xrightarrow{\text{in}} & F_{0,L}((X_1)) \ldots ((X_{n-1})) \\ {\scriptstyle h_{t_1, \ldots, t_{n-1}}} \downarrow & & {\scriptstyle h_{t_1, \ldots, t_{n-1}}} \downarrow \\ K & \xrightarrow{\text{in}} & L \end{array}$$

where $F_{0,K}$ and $F_{0,L}$ are the last residue fields of K and L respectively.

Now it is easy to see that

$$\{\mathcal{U}_{\{U_i\}} \cap K \mid \mathcal{U}_{\{U_i\}} \in B_{U,L}$$
$$= B_{t_1, \ldots, t_{n-1}, \{\pi_i\}, \{e_j\}}\} = B_{U,K} = B_{t_1, \ldots, t_{n-1}, \{\pi^i\}, \{1\}}. \quad \square$$

4°. Now let $\operatorname{char} K = \operatorname{char} \overline{K} = 0$. Fix a coefficient subfield in K. This means that we fix an embedding of fields $\alpha : \overline{K} \hookrightarrow K$ such that $\alpha(\overline{K}) \subset \mathfrak{p}(0)$, and $\overline{\alpha(a)} = a$ for $a \in \overline{K}$. Now we define a topology on K assuming that the topology on $\overline{K}$ is already defined. (On 1-dimensional fields we take the topology of discrete valuation.)

Let π be a prime in K. Consider a sequence of neighborhoods of zero $\{U_i\}_{i\in\mathbb{Z}}$ in $\overline{K}$, $U_i = \overline{K}$ for $i \gg -\infty$. Put

$$\mathcal{U}_{\{U_i\}} = \left\{ \sum_{i \gg -\infty} \alpha(a_i)\pi^i \mid a_i \in U_i \right\}.$$

All possible $\mathcal{U}_{\{U_i\}}$ form the set B_U. It is clear that B_U is a basis of neighborhoods of zero in some topology of the additive topological group on K. This is the required topology.

Notice that we can use this definition in the case $\operatorname{char} K = \operatorname{char} \overline{K} = p$, and we get the topology that was already defined in 1°.

Multiplication by $c \in K^*$ is a homeomorphism of K onto itself. Applying this assertion to $\overline{K}$, one easily deduces that the topology does not depend on the choice of π. All the properties stated in 2° also hold in the case $\operatorname{char} \overline{K} = 0$.

However, the considered topology depends on α. For example, let K be a complete discretely valuated field with residue field $\mathbb{Q}_p$. We have the general fact that $\alpha : \overline{K} \to \alpha(\overline{K})$ is a homeomorphism, and $\alpha(\overline{K})$ is closed in K. Thus, in our case $\alpha(\mathbb{Q}_p)$ is the closure of $\mathbb{Q}$ in K. Since the subfield $\alpha(\mathbb{Q}_p)$ actually depends on the choice of α (see [FV, Chapter II, Proposition 5.2]), we conclude that the topology of K depends on α as well.

§2. Arithmetic

In the present section we deal with the multiplicative group of an n-dimensional complete field. First of all, we have the expansion (2) in §0 for K^*. To study V_K, we must consider three essentially distinct cases. Here the last residue field of K is denoted by k.

1°. The equal characteristic case, $\operatorname{char} K = p$. This is the simplest case.

THEOREM 2.1. *Let $\operatorname{char} K = p$; $t_1, \ldots, t_n$ be the system of local parameters of K. Denote by B any subset of $\mathcal{R}_K$ such that the residues in k form a basis of k as an $\mathbb{F}_p$-space. For $\overline{r} \in \mathbb{Z}_+^n$, $\theta \in B$ choose an element $\varepsilon_{\overline{r},\theta} \equiv 1 + \theta t_1^{r^{(1)}} \ldots t_n^{r^{(n)}}$ $\bmod \mathfrak{p}(\overline{r}+)$. Then any element $\varepsilon \in V_K$ can be uniquely expanded as*

$$(1) \qquad \varepsilon = \prod_{\substack{\overline{r} \in \mathbb{Z}_+^n \\ p \nmid \overline{r}}} \prod_{\theta \in B} \varepsilon_{\overline{r},\theta}^{a_{\overline{r},\theta}},$$

where $a_{\overline{r},\theta} \in \mathbb{Z}_p$ satisfy the following condition:

(2) *for any $\overline{r} \in \mathbb{Z}_+^n$, $k \in \mathbb{N}$ we have $a_{\overline{r},\theta} \in p^k\mathbb{Z}_p$ for almost all θ,*
$\{\overline{r} \mid a_{\overline{r},\theta} \notin p^k\mathbb{Z}_p \text{ for some } \theta\}$ is an admissible set for any $k \in \mathbb{N}$.

REMARK. It is easy to show that the product (1) converges for any $a_{\bar{r},\theta}$ under conditions (2); a similar remark is valid for Theorem 2.2 below.

PROOF. Expand ε into a product of $\varepsilon_{\bar{r},\theta}$, $\bar{r}$ running over $\mathbb{Z}_+^n$, and notice that

$$\left(1 + \sum_{\bar{r}} \alpha_{\bar{r}} t_1^{r^{(1)}} \ldots t_n^{r^{(n)}}\right)^p = 1 + \sum_{\bar{r}} \alpha_{\bar{r}}^p t_1^{pr^{(1)}} \ldots t_n^{pr^{(n)}}. \quad \square$$

2°. The mixed characteristic case. Let $\operatorname{char} K = 0$, $\operatorname{char} \overline{K} = p$, $\bar{e} = \bar{v}(p)$.

THEOREM 2.2. *Let $t_1, \ldots, t_n$; B; $\varepsilon_{\bar{r},\theta}$ be as in Theorem 2.1. Suppose $(p-1)|\bar{e}$. Choose $a \in K$ so that $\bar{v}(a) = \frac{1}{p-1}\bar{e}$, and denote by $\wp_a$ the additive polynomial $X^p + \lambda X$, where λ is the residue of $\frac{p}{a^{p-1}}$ in k. Let B_1 be a subset of $\mathcal{R}_K$ such that the residues in k form a basis of the $\mathbb{F}_p$-space $k/\wp_a(k)$. For $\theta \in B_1$, choose an element $\eta_\theta \equiv 1 + \theta a^p \mod \mathfrak{p}(\frac{p}{p-1}\bar{e}+)$.*

Then any $\varepsilon \in V_K$ can be expanded as

$$\varepsilon = \prod_{\substack{\bar{r} \in \mathbb{Z}_+^n \\ p \nmid \bar{r} \\ \bar{r} < \frac{p}{p-1}\bar{e}}} \prod_{\theta \in B} \varepsilon_{\bar{r},\theta}^{a_{\bar{r},\theta}} \cdot \prod_{\theta \in B_1} \eta_\theta^{b_\theta}, \tag{3}$$

where the $a_{\bar{r},\theta}$ satisfy conditions (2), $b_\theta \in \mathbb{Z}_p$, and for any $k \in \mathbb{N}$ we have $b_\theta \in p^k \mathbb{Z}_p$ for almost all θ.

If $(p-1) \nmid \bar{e}$, then the product of $\eta_\theta^{b_\theta}$ is omitted.

REMARK 1. If K does not contain ζ_p, i.e., a primitive root of unity of degree p, then the expansion (3) is unique. Let p^m be the order of p-torsion in K^*. To get a complete description of K^* in this case, we must require

$$\varepsilon_{\bar{r}_0,\theta_0} = \zeta_{p^m} \quad \text{for } \bar{r}_0 = \frac{1}{p^{m-1}(p-1)}\bar{e}$$

and for the corresponding θ_0. Then $a_{\bar{r}_0,\theta_0}$ is determined modulo p^m, and all other $a_{\bar{r},\theta}$ are determined uniquely.

REMARK 2. Consider the case when the last residue field is finite or quasifinite. Then $\dim_{\mathbb{F}_p}(k/\wp_a(k))$ equals 1 or 0 depending on whether or not $\zeta_p \in K$. In the former case $B_1 = \{\theta\}$, and η_θ is a so-called primary element.

See [V] for an example of the system $\{\varepsilon_{\bar{r},\theta}, \eta_\theta\}$ (the Shafarevich basis), including an explicit construction of a primary element.

3°. The equal characteristic case, $\operatorname{char} K = 0$. Let F be a coefficient subfield in K and X a prime element, i.e., K is identified with $F((X))$. Then obviously

$$V_K = V_F \oplus U_K(1).$$

Furthermore, we have the following evident isomorphism:

$$\prod_{i=1}^{\infty} F \xrightarrow{\sim} U_K(1), \qquad (a_i)_{i=1}^{\infty} \mapsto \prod_{i=1}^{\infty}(1 + a_i X^i).$$

§3. Topological K-groups

Let $K_m^M(F)$ be the mth Milnor K-group of F. Recall that $K_m^M(F)$ is the abelian group generated by the symbols $\{x_1, \ldots, x_m\}$, where $x_i \in F$, which are subject to the relations

$$\{x_1, \ldots, x_k' x_k'', \ldots, x_m\} = \{x_1, \ldots, x_k', \ldots, x_m\} + \{x_1, \ldots, x_k'', \ldots, x_m\}$$
$$\{x_1, \ldots, x_k, 1 - x_k, \ldots, x_m\} \text{ for } k = 1, \ldots, m-1.$$

One has a canonical projection

$$\psi : F^* \times \cdots \times F^* \to K_m^M(F),$$
$$\psi(x_1, \ldots, x_m) = \{x_1, \ldots, x_m\}.$$

Now let F be an n-dimensional complete field. Consider the case when the characteristic of F is positive. In this case the expansion (2) from §0 takes the form

$$F^* \approx \mathbb{Z}t_n \oplus \cdots \oplus \mathbb{Z}t_1 \oplus F_0^* \oplus V_F,$$

where $t_1, \ldots, t_n$ is a system of local parameters of F, F_0 is the last residue field of F, and V_F is the group of principal units of F. The *Parshin topology* of F^* is defined to be the product of the discrete topology on $\mathbb{Z}t_n \oplus \cdots \oplus \mathbb{Z}t_1 \oplus F_0^*$ and the topology on V_F induced from F. The group F^* equipped with this topology is a topological group only in $\dim F \leqslant 2$ [K1, K2]. The Parshin topology is weaker than the topology defined by the discrete valuation of height 1 on F^*. The latter topology is compatible with the group structure on F^* but it does not take into account the topology of the residue field. The Parshin topology, obviously, does not depend on the choice of local parameters of F and on the embedding of the residue field in F (see §1).

Consider the set T of topologies on $K_m^M(F)$ satisfying the following conditions:
a) the projection ψ is sequentially continuous in each argument with respect to the topology $\tau \in T$ and the Parshin topology on F^*;
b) if $x_n \to x$ and $y_n \to y$ with respect to the topology $\tau \in T$, then $x_n + y_n \to x + y$ and $-x_n \to -x$ with respect to τ.

The set T is nonempty because it contains the weakest (discrete) topology.

Denote by τ_0 the least upper bound of the ordered set T. It is clear that $x_n \to x$ with respect to τ_0 if and only if $x_n \to x$ with respect to each $\tau \in T$. It follows immediately that the topology τ_0 satisfies conditions a) and b) above and, therefore, τ_0 is the strongest topology satisfying these conditions.

Supply the group $K_m^M(F)$ with the topology τ_0. Note that the condition b) for τ_0 means that the group operations on $K_m^M(F)$ are sequentially continuous with respect to τ_0. In particular, this implies that the intersection of all neighborhoods of 0 in $K_m^M(F)$ is a subgroup of $VK_m^M(F)$, where $VK_m^M(F)$ is generated by the symbols $\{x_1, x_2, \ldots, x_m\}$ with $x_1 \in V_F$ and $x_2, \ldots, x_m \in F^*$.

Denote by $\Lambda_m'(F)$ the subgroup of infinitely divisible elements in $K_m^M(F)$, and by $\Lambda_m''(F)$ the intersection of all neighborhoods of zero in $K_m^M(F)$.

If F is a local field, then F_0 is finite, and so $K_2^M(F_0) = 0$. This implies that $\Lambda_m'(F)$ is contained in $VK_m^M(F)$, and, moreover, it is contained in $\Lambda_m''(F)$.

Indeed, let U be a neighborhood of 0 in F^*. Then there exists an r such that $V_F^{p^r} \subset U$. In fact, Fesenko showed that in this case $\Lambda'_m(F) = \Lambda''_m(F)$.

In the case of an arbitrary residue field F_0, the group F_0^* may contain infinitely divisible elements, and $\Lambda'_m(F)$ need not lie in $\Lambda''_m(F)$. On the other hand, the classifying objects A_F of class field theory [F1] cannot contain infinitely divisible elements if one wants to obtain an inclusion $A_F \to \mathrm{Gal}(F^{\mathrm{ab}}/F)$. This justifies the following definition.

Let $\Lambda_m(F) = \Lambda'_m(F) + \Lambda''_m(F)$. We call the groups $K_m^M(F)/\Lambda_m(F)$ *Parshin topological K-groups of* F and denote them by $K_m^{\mathrm{top}}(F)$. These groups are described explicitly (see [P2] for the case of finite F_0 and [Be] for quasifinite F_0). The case of an n-dimensional local field of characteristic 0 with residue field of characteristic $p > 0$ was studied by Fesenko [F2]. In this case also $\Lambda'_m(F) = \Lambda''_m(F)$.

§4. Convergence of series over 2-dimensional complete fields

Now we deal with the problems of convergence of superpositions of series and infinite products of series over 2-dimensional complete fields. Let K be a 2-dimensional complete field, $\overline{K}$ be the first residue field of K, p the characteristic of the last residue field, $\overline{v} = (v_1, v_2)$ the valuation on K. We have the following classification of 2-dimensional complete fields by the structure theorem for n-dimensional complete fields.

1. If $\mathrm{char}\, K = \mathrm{char}\, \overline{K}$ (the equal characteristic case), then $K = \overline{K}((t))$.

2. If $\mathrm{char}\, K = 0$, $\mathrm{char}\, \overline{K} = p > 0$ (the mixed characteristic case), then there is a finite extension of K of the form $k\{\{t\}\}$, where k is a mixed characteristic complete field.

Let us recall the construction of the topology on K. If $K = \overline{K}((t))$, then the basis of neighborhoods of zero is the set of all subgroups

$$U = \left\{ \sum_j a^{(j)} t^j : a^{(j)} \in U_j \right\},$$

where U_j are open subgroups in the complete field $\overline{K}$ with the valuation v_1.

If $\mathrm{char}\, K \neq \mathrm{char}\, \overline{K}$, then we can assume that the topology of K is induced by the topology of $k\{\{t\}\}$, where $k\{\{t\}\}$ contains K, k is a mixed characteristic complete field.

The basis of neighborhoods of zero in $k\{\{t\}\}$ is a set of all subgroups

$$U = \left\{ \sum_j a^{(j)} t^j : a^{(j)} \in U_j \right\},$$

where U_j are open subgroups in the complete field k and
1) there is an i_0 such that $\mathfrak{p}_k(i_0) \subseteq U_j$ for all j,
2) any $\mathfrak{p}_k(i)$ lies in U_j with $j \geq j(i)$, where $\mathfrak{p}_k(i) = \{a \in k : v_1(a) \geq i\}$.
Moreover, one may require $U_{j-1} \subseteq U_j$ for all j.

1°. Convergence in equal characteristic complete fields.

In this subsection we assume K is an equal characteristic complete field. We identify K with $\overline{K}((t))$. Any $a \in K$ will be expanded as $\sum_{j \gg -\infty} a^{(j)} t^j$. We denote by v_1 and v_2 the discrete valuations of $\overline{K}$ and K respectively.

THEOREM 4.1. *Let* $\{a_n\}_{n\geqslant 1}$ *be a sequence,* $a_n \in K$. *We have the following necessary and sufficient conditions for convergence of* a_n *to* 0:
1. $\inf_n v_2(a_n) = j_0 > -\infty$;
2. *for any* j, $v_1(a_n^{(j)}) \xrightarrow[n\to+\infty]{} +\infty$.

PROOF. $\overline{K}$ is an ordinary complete field and admits a fundamental system of open subgroups of the form $U = \{\alpha \in \overline{K} : v_1(\alpha) \geqslant s\}$, where $s \in \mathbb{Z}$. Therefore, the basis of neighborhoods of zero in K is the set of subgroups

$$U = \left\{ \sum_{j \gg -\infty} a^{(j)} t^j : v_1(a^{(j)}) \geqslant s_j \right\}, \qquad s_j \in \mathbb{Z} \cup \{-\infty\};$$

$s_j = -\infty$ for $j \geqslant J$. It is clear that the second condition of our theorem is necessary, i.e., we can take any s_j for every j.

Let us prove that our first condition is necessary. This follows from Lemma 1.9, but we also give a direct proof for one case.

Let the first condition be not carried out and $\inf_n v_2(a_n) = -\infty$. Then there is n for every j_0 such that $v_2(a_n) < j_0$. We can take n_1: $v_2(a_{n_1}) = s_1 < -1$, $n_2 > n_1$: $v_2(a_{n_2}) = s_2 < s_1, \ldots, n_l < n_{l-1}$: $v_2(a_{n_l}) = s_l < s_{l-1}, \ldots$.
Let $U_{s_l} = \{\alpha \in \overline{K} : v_1(\alpha) \geqslant v_1(a_{n_l}^{(s_l)}) + 1\}$,

$$U = \left\{ \sum_{j \gg -\infty} a^{(j)} t^j : a^{(j)} \in U_j \right\},$$

and $U_j = U_{s_l}$ for $j = s_l$, $U_j = \overline{K}$ for j distinct from all the s_l.

U is a neighborhood of zero in K. But $v_1(a_{n_l}^{(s_l)}) < v_1(a_{n_l}^{(s_l)}) + 1$ for every a_{n_l}, i.e., $a_{n_l} \notin U$ for every n_l. Therefore, $a_n \not\xrightarrow[n\to+\infty]{} 0$.

The necessity of the first condition is proved. Now let us prove the sufficiency of our conditions.

Let U be a neighborhood of zero from our basis. $a_n^{(j)} = 0$ for every n, if $j < j_0$. But there is $J \in \mathbb{Z}$ such that $U_j = \overline{K}$ for $j > J$, and by our second condition we have $\inf_{j_0 \leqslant j \leqslant J} v_1(a_n^{(j)}) \xrightarrow[n\to+\infty]{} +\infty$. □

COROLLARY. *Let* $a(X) = \sum_{n\geqslant 1} a_n X^n$ *be a series over* K, $x \in K$. *Then we have the following sufficient conditions for the convergence of* $a(x)$:
1. $\inf_n (v_2(a_n) + n v_2(x)) > -\infty$;
2. *for any* j, $\inf\limits_{\substack{l \geqslant v_2(a_n) \\ i_1, \ldots, i_n \geqslant v_2(x) \\ l + i_1 + \cdots + i_n = j}} v_1(a_n^{(l)} x^{(i_1)} \ldots x^{(i_n)}) \xrightarrow[n\to+\infty]{} +\infty$.

THEOREM 4.2. *Let* $a(X) = \sum_{n\geqslant 1} a_n X^n$, $b(X) = \sum_{n\geqslant 1} b_n X^n$ *be series over* K, $c(X) = a \circ b(X) = \sum_{n\geqslant 1} c_n X^n$ *be the superposition of* $a(X)$ *and* $b(X)$, $x \in K$. *Then we have the following sufficient conditions for the convergence of* $c(x)$ *to* $a(b(x))$:
1. $\inf_n(j(n) + n v_2(x)) = j_1 > -\infty$, *where*

$$j(n) = \inf_{\substack{1 \leqslant r \leqslant n \\ s_1 + \cdots + s_n = n}} v_2(a_r b_{s_1} \ldots b_{s_r});$$

2. *for any* j, $M(j, n) \xrightarrow[n \to +\infty]{} +\infty$, *where*

$$M(j, n) = \inf_{\substack{1 \leqslant r \leqslant n \\ s_1 + \cdots + s_n = n \\ l \geqslant v_2(a_r) \\ l_1 \geqslant v_2(b_{s_1}), \ldots, l_r \geqslant v_2(b_{s_r}) \\ i_1, \ldots, i_n \geqslant v_2(x) \\ l + l_1 + \cdots + l_r = j}} v_1(a_r^{(l)} b_{s_1}^{(l_1)} \ldots b_{s_r}^{(l_r)} x^{(i_1)} \ldots x^{(i_n)}).$$

PROOF. We have

$$c_n = \sum_{j \geqslant j(n)} \left(\sum_{\substack{1 \leqslant r \leqslant n \\ s_1 + \cdots + s_n = n \\ l \geqslant v_2(a_r) \\ l_1 \geqslant v_2(b_{s_1}), \ldots, l_r \geqslant v_2(b_{s_r}) \\ l + l_1 + \cdots + l_r = j}} a_r^{(l)} b_{s_1}^{(l_1)} \ldots b_{s_r}^{(l_r)} \right) t^j.$$

We obtain the convergence of $c(x)$ by the Corollary to Theorem 4.1. We must prove that $c(x) = a(b(x))$.

We have $j(n) \leqslant v_2(a_1) + v_2(b_n)$. Therefore,

$$\inf_n (v_2(b_n) + n v_2(x)) \geqslant \inf_n (j(n) + n v_2(x)) - v_2(a_1) > -\infty.$$

$$M(j, n) \leqslant v_1(a_1^{v_2(a_1)}) + \inf_{\substack{l \geqslant v_2(b_n) \\ i_1, \ldots, i_n \geqslant v_2(x) \\ l + i_1 + \cdots + i_n = j - v_2(a)}} v_1(b_n^{(l)} x^{(i_1)} \ldots x^{(i_n)}) \xrightarrow[n \to +\infty]{} +\infty$$

by condition 2 of our theorem. Therefore, $b(x)$ converges by Corollary of Theorem 4.1.

$$a(b(x)) = \lim_{m \to +\infty} \sum_{i=1}^m a_i (b(x))^i.$$

Let

$$d_m(X) = \sum_{n \geqslant 1} d_{mn} X^n = \sum_{i=1}^m a_i (b(X))^i.$$

It is clear that $d_{mn} = c_n$ if $m \geqslant n$ and that $d_m(x)$ is finite under our conditions. Let $c(x) - d_m(x) = f_m$. We must prove that $f_m \xrightarrow[m \to +\infty]{} 0$.

$$f_m^{(j)} = \sum_{\substack{n \geqslant m+1 \\ m+1 \leqslant r \leqslant n \\ s_1 + \cdots + s_r = n \\ l + l_1 + \cdots + l_r + i_1 + \cdots + i_n = j \\ l \geqslant v_2(a_r) \\ l_1 \geqslant v_2(b_{s_1}), \ldots, l_r \geqslant v_2(b_{s_r}) \\ i_1, \ldots, i_n \geqslant v_2(x)}} a_r^{(l)} b_{s_1}^{(l_1)} \ldots b_{s_r}^{(l_r)} x^{(i_1)} \ldots x^{(i_n)}.$$

Furthermore, $\inf_m v_2(f_m) \geqslant \inf_n (n v_2(x) + j(n)) = j_1 > -\infty$ by our first condition, and

$$v_1(f_m^{(j)}) \geqslant \inf_{n \geqslant m+1} M(j, n) \xrightarrow[m \to +\infty]{} +\infty$$

by the second condition.

The result now follows from Theorem 4.1 for sequence $\{f_m\}_{m \geqslant 1}$. $\square$

LEMMA 4.1. *Let*
1. $\inf_n(v_2(a_1) + \cdots + v_2(a_n) + v_2(a_{n+1} - 1)) = \lambda > -\infty$;
2. *for any* j, $A(j, n) \xrightarrow[n \to +\infty]{} +\infty$, *where*

$$A(j, n) = \inf_{\substack{j_1 \geqslant v_2(a_1) \\ \cdots \\ j_n \geqslant v_2(a_n) \\ j_{n+1} \geqslant v_2(a_{n+1}-1) \\ j_1 + \cdots + j_{n+1} = j}} v_1(a_1^{(j_1)} \ldots a_n^{(j_n)}(a_{n+1} - 1)^{(j_{n+1})}).$$

Then $\prod_{i=1}^{+\infty} a_i$ *converges in* K.

The result follows from Theorem 4.1. □

COROLLARY. *Let*
1. *for any* n, $v_2(a_n) \geqslant 0$;
2. $a_n \xrightarrow[n \to +\infty]{} 1$ *in* K.

Then $\prod_{i=1}^{+\infty} a_i$ *converges in* K.

PROOF. We have $\inf_n(v_2(a_1) + \cdots + v_2(a_n) + v_2(a_{n+1} - 1)) \geqslant 0$ from our first condition.

Let $j \geqslant \lambda_1$. From our second condition we obtain $\inf_{0 \leqslant s \leqslant j} v_1(a_n^{(s)}) \geqslant 0$ for $n > N$, and $\inf_{0 \leqslant s \leqslant j} v_1((a_n - 1)^{(s)}) \xrightarrow[n \to +\infty]{} +\infty$. Therefore, for $n > N$,

$$A(j, n) \geqslant \inf_{\substack{0 \leqslant j_\alpha \leqslant j \\ 1 \leqslant \alpha \leqslant N}} v_1(a_1^{(j_1)} \ldots a_N^{(j_N)}) + \inf_{0 \leqslant s \leqslant j} v_1((a_n - 1)^{(s)}) \xrightarrow[n \to +\infty]{} +\infty. \quad \Box$$

THEOREM 4.3. *Suppose that*

$$h_i(X) = \sum_{s \geqslant 1} h_{is} X^s, \qquad i \in \mathbb{N};$$

$$1 + h(X) = 1 + \sum_{s \geqslant 1} c_s X^s = \prod_{i \geqslant 1}(1 + h_i(X))$$

are series over K; $x \in K$. *We have the following sufficient conditions for the convergence of* $h(x)$ *to* $\prod_{i \geqslant 1}(1 + h_i(x)) - 1$:
1. *for any* i, $\inf_{s \geqslant 1}(v_2(h_{is}) + sv_2(x)) \geqslant 0$;
2. *for all* i, j, $\Lambda(i, j, s) \xrightarrow[s \to +\infty]{} +\infty$, *where*

$$\Lambda(i, j, s) = \inf_{\substack{l \geqslant v_2(h_{is}) \\ j_1, \ldots, j_s \geqslant v_2(x) \\ l + j_1 + \cdots + j_s = j}} v_1(h_{is}^{(l)} x^{(j_1)} \ldots x^{(j_s)}),$$

3. *for any* j, $\Lambda(i, j) \xrightarrow[i \to +\infty]{} +\infty$, *where*

$$\Lambda(i, j) = \inf_{s \geqslant 1} \Lambda(i, j, s).$$

PROOF. The convergence of $h_i(x)$ follows from our conditions 1 and 2. Let us consider the series
$$d_m(X) = \prod_{i=1}^{m}(1 + h_i(X)) - 1$$
and prove that $h(x)$ is finite and $d_m(x) \xrightarrow[m \to +\infty]{} h(x)$. We have

$$((c_s - d_{ms})x^s)^{(j)}$$
$$= \sum_{\substack{n \geq 1 \\ i_1 \geq m+1 \\ i_2 \geq 1, \ldots, i_n \geq 1 \\ i_\alpha \text{ are different} \\ 1 \leq \alpha \leq n \\ s_1 + \cdots + s_n = s}} \sum_{\substack{j_1 + \cdots + j_n = j \\ j_1, \ldots, j_n \geq 0 \\ l_1 + \lambda_{11} + \cdots + \lambda_{1s_1} = j \\ \cdots \\ l_n + \lambda_{n1} + \cdots + \lambda_{ns_n} = j_n \\ l_\alpha \geq v_2(h_{i_\alpha s_\alpha}), \; \alpha \leq n \\ \lambda_{\alpha\beta} \geq v_2(x), \; \alpha \leq n, \; \beta \leq s}} h_{i_1 s_1}^{(l_1)} x^{(\lambda_{11})} \ldots x^{(\lambda_{1s_1})} \ldots h_{i_n s_n}^{(l_n)} x^{(\lambda_{n1})} \ldots x^{(\lambda_{ns_n})}.$$

We have the estimates

$$v_1(((c_s - d_{ms})x^s)^{(j)})$$
$$\geq \inf_{\substack{n \geq 1}} \inf_{\substack{i_1 \geq m+1 \\ i_2 \geq 1 \\ \vdots \\ i_n \geq 1 \\ i_\alpha \text{ are different}}} \inf_{\substack{s_1 + \cdots + s_n = s \\ j_1 + \cdots + j_n = j}} (\Lambda(i_1, j_1, s_1) + \cdots + \Lambda(i_n, j_n, s_n))$$

$$\geq \inf_{\substack{n \geq 1}} \inf_{\substack{i_1 \geq m+1 \\ i_2 \geq 1 \\ \vdots \\ i_n \geq 1 \\ i_\alpha \text{ are different}}} \inf_{\substack{0 \leq j_1 \leq j \\ \vdots \\ 0 \leq j_n \leq j}} (\Lambda(i_1, j_1) + \cdots + \Lambda(i_n, j_n))$$

$$\geq \inf_{n \geq 1} \Big[\inf_{0 \leq j_1 \leq j} \inf_{i_1 \geq m+1} \Lambda(i_1, j_1)$$
$$+ \inf_{i_2 < i_3 < \cdots < i_n} \Big(\inf_{0 \leq j_1 \leq j} \Lambda(i_2, j_1) + \cdots + \inf_{0 \leq j_1 \leq j} \Lambda(i_n, j_1) \Big) \Big]$$

$$\geq \inf_{i_1 \geq m+1} \inf_{0 \leq j_1 \leq j} \Lambda(i_1, j_1)$$
$$+ \inf_{n \geq 1} \inf_{i_2 < i_3 < \cdots < i_n} \Big(\inf(0, \inf_{0 \leq j_1 \leq j} \Lambda(i_2, j_1)) + \cdots + \inf(0, \inf_{0 \leq j_1 \leq j} \Lambda(i_n, j_1)) \Big)$$

$$\geq \inf_{i_1 \geq m+1} \inf_{0 \leq j_1 \leq j} \Lambda(i_1, j_1) + \sum_{i=1}^{\infty} \inf(0, \inf_{0 \leq j_1 \leq j} \Lambda(i, j_1)).$$

We have $\inf_{0 \leq j_1 \leq j} \Lambda(i, j_1) \xrightarrow[i \to +\infty]{} +\infty$ by condition 3 of our theorem. Therefore,
$$\inf(0, \inf_{0 \leq j_1 \leq j} \Lambda(i, j_1)) = 0$$
for $i > i_0$ and
$$\inf_s v_1(((c_s - d_{ms})x^s)^{(j)}) \xrightarrow[m \to +\infty]{} +\infty.$$

But $v_1((c_s x^s)^{(j)}) \geq \inf(v_1(((c_s - d_{ms})x^s)^{(j)}), v_1((d_{ms}x^s)^{(j)}))$ and for all m, j we have
$$v_1((d_{ms}x^s)^{(j)}) \xrightarrow[s \to +\infty]{} +\infty.$$

Let A be any number, then $\inf_s v_1(((c_s - d_{ms})x^s)^{(j)}) > A$ for $m \geq M$, and $v_1((d_{ms}x^s)^{(j)}) > A$ for $s \geq S$. Hence $v_1((c_s x^s)^{(j)}) > A$ for $s \geq S$, and

$$v_1((c_s x^s)^{(j)}) \xrightarrow[s \to +\infty]{} +\infty.$$

Now we have $v_2(c_s) \geq -s v_2(x)$ from condition 1 of our theorem. The series $h(x)$ is finite by the Corollary to Theorem 4.1.

$$v_1((h(x) - d_m(x))^{(j)}) \geq \inf_{s \geq 1} v_1(((c_s - d_{ms})x^s)^{(j)}) \xrightarrow[m \to +\infty]{} +\infty,$$

therefore, $d_m(x) \xrightarrow[m \to +\infty]{} h(x)$. □

2°. Convergence in mixed characteristic complete fields.

In this subsection we assume K to be a mixed characteristic 2-dimensional complete field. We fix an embedding of K into a field of the form $k\{\{t\}\}$, and expand any $a \in K$ as $\sum_{j=-\infty}^{\infty} a^{(j)} t^j$, $a^{(j)} \in k$. Denote by v_1 and v_2 the discrete valuations of k and K respectively.

THEOREM 4.4. *Let $\{a_n\}_{n \geq 1}$ be a sequence, $a_n \in K$. We have the following necessary and sufficient conditions for the convergence of a_n to 0:*
1. $\inf_n v_2(a_n) = j_0 > -\infty$;
2. *for all $J \in \mathbb{Z}$* $\inf_{j \leq J} v_1(a_n^{(j)}) \xrightarrow[n \to +\infty]{} +\infty$.

PROOF. k is an ordinary mixed characteristic complete field and admits a basis of open subgroups of the form $U = \{\alpha \in k : v_1(\alpha) \geq s\}$, where $s \in \mathbb{Z}$. Therefore, the basis of neighborhoods of zero in K is the set of subgroups

$$U = \left\{ \sum_{j=-\infty}^{+\infty} a^{(j)} t^j : v_1(a^{(j)}) \geq s_j \right\}, \qquad s_j \in \mathbb{Z} \cup \{-\infty\},$$

s_j are bounded from above, $s_j \xrightarrow[j \to +\infty]{} -\infty$, $s_{j+1} \leq s_j$ for all j.

It is clear that the second condition of our theorem is necessary, i.e., we can take any s_j for $j \leq J$. Now we shall prove the necessity of the first condition.

Let the first condition break down, and $\inf_n v_2(a_n) = -\infty$. Then there are $n_1 < n_2 < \ldots$ such that $-1 > v_2(a_{n_1}) > v_2(a_{n_2}) > \ldots$. For convenience we shall write a_n instead of a_{n_l}. Clearly,

$$v_2(a_n) = \inf_j v_1(a_n^{(j)}) = v_1(a_n^{(j_n)}).$$

If $j_n \leq J_1$ for all n, then $\inf_{j \leq J_1} v_1(a_n^{(j)}) < -1$ for all n. We get a contradiction with our condition $a_n \xrightarrow[n \to +\infty]{} 0$.

Therefore, we can choose a subsequence j_{n_l} such that for all l $j_{n_l} \geq 1$ and $j_{n_{l+1}} > j_{n_l}$. Once again, we shall write a_n instead a_{n_l}. We have a sequence $\{a_n\}_{n \geq 1}$ (it is a subsequence of our first sequence) such that for all n:
- $v_2(a_n) \leq -1$;

- $v_2(a_{n+1}) < v_2(a_n)$;
- $v_2(a_n) = v_1(a_n^{(j_n)})$; $j_n \geq 1$; $j_{n+1} > j_n$.

We can take an s_j such that $s_j = 0$ for $j \leq 0$, $s_{j_n} = v_2(a_n) + 1$ for all n, $s_j = s_{j_n}$ for $j_n \leq j < j_{n+1}$. But $v_1(a_n^{(j_n)}) = v_2(a_n) < s_{j_n}$ for every n. Therefore, $a_n \not\xrightarrow[n\to+\infty]{} 0$. We obtain a contradiction with our assumptions.

The proof of the necessity is complete. Next, we prove the sufficiency. Let $\{s_j\}_{j=-\infty}^{+\infty}$ be a sequence with properties:

$$s_j \text{ are bounded from above}, \quad s_j \xrightarrow[j\to+\infty]{} -\infty, \quad \text{for all } j \ s_{j+1} \leq s_j.$$

Since $s_j \xrightarrow[j\to+\infty]{} -\infty$, we have $s_j \leq j_0 = \inf_n v_2(a_n)$ for $j > J_1$. Therefore, $v_1(a_n^{(j)}) \geq s_j$ for $j > J_1$. But $\inf_{j \leq J_1} v_1(a_n^{(j)}) \geq s_0 \geq s_j$ for $n > N$ from condition 2 of our theorem. Sufficiency is proved. □

COROLLARY. Let $a(X) = \sum_{n \geq 1} a_n X^n$ be a series over K, $x \in K$. Then we have the following sufficient conditions for convergence of $a(x)$:
1. $\inf_n(v_2(a_n) + nv_2(x)) > -\infty$;
2. for any $J \in \mathbb{Z}$, $\inf_{j \leq J} \inf_{l+i_1+\cdots+i_n = j} v_1(a_n^{(l)} x^{(i_1)} \ldots x^{(i_n)}) \xrightarrow[n\to+\infty]{} +\infty$. □

THEOREM 4.5. Let $a(X) = \sum_{n \geq 1} a_n X^n$, $b(X) = \sum_{n \geq 1} b_n X^n$ be series over K, $c(X) = a \circ b(X) = \sum_{n \geq 1} c_n X^n$ be a superposition of $a(X)$ and $b(X)$, $x \in K$. Then we have the following sufficient conditions for the convergence of $c(x)$ to $a(b(x))$:
1. $\inf_n(j(n) + nv_2(x)) = j_1 > -\infty$, where

$$j(n) = \inf_j \inf_{\substack{1 \leq r \leq n \\ s_1+\cdots+s_r=n \\ l+l_1+\cdots+l_r=j}} v_1(a_r^{(l)} b_{s_1}^{(l_1)} \ldots b_{s_r}^{(l_r)});$$

2. for any $J \in \mathbb{Z}$, $\inf_{j \leq J} M(j, n) \xrightarrow[n\to+\infty]{} +\infty$, where

$$M(j, n) = \inf_{\substack{1 \leq r \leq n \\ s_1+\cdots+s_r=n \\ l+l_1+\cdots+l_r+i_1+\cdots+i_n=j}} v_1(a_r^{(l)} b_{s_1}^{(l_1)} \ldots b_{s_r}^{(l_r)} x^{(i_1)} \ldots x^{(i_n)}).$$

PROOF. We have

$$c_n = \sum_{j=-\infty}^{+\infty} \left(\sum_{\substack{1 \leq r \leq n \\ s_1+\cdots+s_r=n \\ l+l_1+\cdots+l_r=j}} a_r^{(l)} b_{s_1}^{(l_1)} \ldots b_{s_r}^{(l_r)} \right) t^j.$$

We can assert the convergence of $c(x)$ by the Corollary to Theorem 4.4. We must show that $c(x) = a(b(x))$. The inequality $j(n) \leq v_1(a_1^{(0)}) + v_2(b_n)$ implies

$$\inf_n(v_2(b_n) + nv_2(x)) > -\infty.$$

Further,

$$\inf_{j \leqslant J} M(j, n) \leqslant v_1(a_1^{(0)}) + \inf_{j \leqslant J} \inf_{l+i_1+\cdots+i_n=j} v_1(b_n^{(l)} x^{(i_1)} \ldots x^{(i_n)}) \xrightarrow[n \to +\infty]{} +\infty$$

by condition 2 of our theorem. Therefore, $b(x)$ converges by the Corollary to Theorem 4.4. Let

$$d_m(X) = \sum_{n \geqslant 1} d_{mn} X^n = \sum_{i=1}^{m} a_i (b(X))^i.$$

It is clear that $d_{mn} = c_n$ if $m \geqslant n$ and that $d_m(x)$ is finite under our conditions. Let $c(x) - d_m(x) = f_m$; we must prove that $f_m \xrightarrow[m \to +\infty]{} 0$. In fact, we have

$$f_m^{(j)} = \sum_{\substack{n \geqslant m+1 \\ m+1 \leqslant r \leqslant n \\ s_1+\cdots+s_r=n \\ l+l_1+\cdots+l_r+i_1+\cdots+i_n=j}} a_r^{(l)} b_{s_1}^{(l_1)} \ldots b_{s_r}^{(l_r)} x^{(i_1)} \ldots x^{(i_n)},$$

whence $\inf_m v_2(f_m) \geqslant \inf_n (nv_2(x) + j(n)) > -\infty$, and for any $J \in \mathbb{Z}$,

$$\inf_{j \leqslant J} v_1(f_m^{(j)}) \geqslant \inf_{j \leqslant J} \inf_{n \geqslant m+1} M(j, n) \xrightarrow[m \to +\infty]{} +\infty.$$

The result now follows from Theorem 4.4 for the sequence $\{f_m\}_{m \geqslant 1}$. □

LEMMA 4.2. *Let*
1. $\inf_n (v_2(a_1) + \cdots + v_2(a_n) + v_2(a_{n+1} - 1)) = \lambda_1 > -\infty$;
2. *for any* $J \in \mathbb{Z}$, $\inf_{j \leqslant J} A(j, n) \xrightarrow[n \to +\infty]{} +\infty$, *where*

$$A(j, n) = \inf_{j_1+\cdots+j_{n+1}=j} v_1(a_1^{(j_1)} \ldots a_n^{(j_n)} (a_{n+1} - 1)^{(j_{n+1})}).$$

Then $\prod_{i=1}^{+\infty} a_i$ *converges in* K.

This fact follows from Theorem 4.4. □

COROLLARY. *Let*
1. *for any* $j \in \mathbb{Z}$ *and any* $n \in \mathbb{N}$, $v_1(a_n^{(j)}) \geqslant 0$;
2. $a_n \xrightarrow[n \to +\infty]{} 1$ *in* K.

Then $\prod_{i=1}^{+\infty} a_1$ *converges in* K.

PROOF. We have $\lambda_1 \geqslant 0$ from the first condition. Let $J \in \mathbb{Z}$. We take any M and we must prove that

$$\inf_{j \leqslant J} \inf_{j_1+\cdots+j_{n+1}=j} v_1(a_1^{(j_1)} \ldots a_n^{(j_n)} (a_{n+1} - 1)^{(j_{n+1})}) > M$$

for $n > N$. We have $v_1(a_n^{(j)}) \xrightarrow[j \to -\infty]{} +\infty$ for all n by the definition of $k\{\{t\}\}$. Therefore, $v_1(a_n^{(j)}) > M$ for $j < J(n)$.

If there is an n_1 such that $v_1(a_{n_1}^{(j)}) > M$ for every j, then $\inf_{j \leqslant J} A(j, n) > M$ for $n > n_1$. Otherwise we can choose $J(n)$ so that $v_1(a_n^{(J(n))}) \leqslant M$ for every n.

Now let $j_\alpha \geqslant J(\alpha)$ now. Therefore,

$$j_{n+1} = j - (j_1 + \cdots + j_n) \geqslant J - (J(1) + \cdots + J(n)).$$

But $\inf_{j<0} v(a_n^{(j)}) > M$ for $n > N$ by our condition 2. Hence $J(n) \geqslant 0$ for $n > N$, and $\sum_{n \geqslant 1} J(n) = L > -\infty$. For all n we have

$$J - (J(1) + \cdots + J(n)) \leqslant J - L > +\infty.$$

Finally

$$\inf_{\substack{j \leqslant J \\ }} \inf_{\substack{j_1 + \cdots + j_{n+1} = j \\ j_\alpha \geqslant J(\alpha) \\ 1 \leqslant \alpha \leqslant n}} v_1(a_1^{(j_1)} \ldots a_n^{(j_n)} (a_{n+1} - 1)^{(j_{n+1})})$$

$$\geqslant \inf_{j_{n+1} \leqslant J - L} v((a_{n+1} - 1)^{(j_{n+1})}) \xrightarrow[n \to +\infty]{} +\infty$$

by condition 2. □

THEOREM 4.6. *Suppose that* $h_i(X) = \sum_{s \geqslant 1} h_{is} X^s$ *be series over K for all natural numbers i, and*

$$1 + h(X) = 1 + \sum_{s \geqslant 1} c_s X^s = \prod_{i \geqslant 1} (1 + h_i(X)),$$

$x \in K$. *We have the following sufficient conditions for the convergence of $h(x)$ to $\prod_{i \geqslant 1} (1 + h_i(x)) - 1$:*

1. *for all i, j, s we have $\Lambda(i, j, s) \geqslant 0$, where*

$$\Lambda(i, j, s) = \inf_{l + j_1 + \cdots + j_s = j} v_1(h_{is}^{(l)} x^{(j_1)} \ldots x^{(j_s)});$$

2. *for all $i \in \mathbb{N}$ and all $J \in \mathbb{Z}$ we have $\inf_{j \leqslant J} \Lambda(i, j, s) \xrightarrow[s \to +\infty]{} +\infty$;*

3. *for all $J \in \mathbb{Z}$ we have $\inf_{j \leqslant J} \Lambda(i, j) \xrightarrow[i \to +\infty]{} +\infty$, where*

$$\Lambda(i, j) = \inf_{s \geqslant 1} \Lambda(i, j, s).$$

PROOF. We have $\inf_{s \geqslant 1}(v_2(h_{is}) + sv_2(x)) \geqslant 0$ for all i by the first condition. Therefore, any $h_i(x)$ converges.

Consider the series $d_m(X) = \prod_{i=1}^m (1 + h_i(X)) - 1$; let us prove that $h(x)$ is finite and $d_m(x) \xrightarrow[m \to +\infty]{} h(x)$. We have

$$((c_s - d_{ms})x^s)^{(j)}$$

$$= \sum_{\substack{n \geqslant 1 \\ i_1 \geqslant m+1 \\ i_2 \geqslant 1 \\ i_3 \geqslant 2 \\ \vdots \\ i_n \geqslant n-1 \\ i_\alpha \text{ are different} \\ s_1 + \cdots + s_n = s}} \sum_{\substack{j_1 + \cdots + j_n = j \\ l_1 + \lambda_{11} + \cdots + \lambda_{1s_1} = j_1 \\ \vdots \\ l_n + \lambda_{n1} + \cdots + \lambda_{ns_n} = j_n}} h_{i_1 s_1}^{(l_1)} x^{(\lambda_{11})} \ldots x^{(\lambda_{1s_1})} \ldots h_{i_n s_n}^{(l_n)} x^{(\lambda_{n1})} \ldots x^{(\lambda_{ns_n})}.$$

Hence

$$\inf_{s\geqslant 1} v_1(((c_s - d_{ms})x^s)^{(j)}) \geqslant \inf_{\substack{s\geqslant 1 \\ n\geqslant 1}} \inf_{\substack{s_1+\cdots+s_n=s \\ j_1+\cdots+j_n=j \\ i_1\geqslant m+1 \\ i_2\geqslant 1,\ldots,i_n\geqslant n-1}} (\Lambda(i_1, j_1, s_1) + \cdots + \Lambda(i_n, j_n, s_n))$$

$$\geqslant \inf_{n\geqslant 1} \inf_{\substack{j_1+\cdots+j_n=j \\ i_1\geqslant m+1 \\ i_2\geqslant 1,\ldots,i_n\geqslant n-1}} (\Lambda(i_1, j_1) + \cdots + \Lambda(i_n, j_n)),$$

$$\inf_{s\geqslant 1} \inf_{j\leqslant J} v_1(((c_s - d_{ms})x^s)^{(j)}) \geqslant \inf_{j\leqslant J} \inf_{\substack{n\geqslant 1 \\ j_1+\cdots+j_n=j \\ i_1\geqslant m+1 \\ i_2\geqslant 1 \\ \vdots \\ i_n\geqslant n-1}} (\Lambda(i_1, j_1) + \cdots + \Lambda(i_n, j_n)).$$

Let A be any number. We want to show that

$$\inf_{s\geqslant 1} \inf_{j\leqslant J} v_1(((c_s - d_{ms})x^s)^{(j)}) > A \quad \text{for } m > M.$$

Let us estimate $\inf_{i\geqslant i_0} \Lambda(i, j)$. If $j \leqslant 0$ then $\inf_s \Lambda(i, j, s) > A$ for $i > I$ by the third condition of our theorem. Let $i_0 \leqslant i \leqslant I$. But $\Lambda(i, j) \xrightarrow[j\to -\infty]{} +\infty$ by the definition of $k\{\{t\}\}$. Therefore, $\inf_{i_0\leqslant i\leqslant I} \Lambda(i, j) > A$ for $j \leqslant J(i_0, I)$.

We have $\inf_{i\geqslant i_0} \inf_s \Lambda(i, j, s) > A$ for $j \leqslant \inf(0, J(i_0, I)) = J(i_0)$, where $J(i_0) = 0$ for $i_0 > I$. If $j_\alpha \leqslant J(\alpha - 1)$ for some α ($2 \leqslant \alpha \leqslant n$), then $\Lambda(i_1, j_1) + \cdots + \Lambda(i_n, j_n) > A$ in the case of $i_1 \geqslant m+1$, $i_2 \geqslant 1, \ldots, i_n \geqslant n-1$.

Let $j_2 > J(1), \ldots, j_n > J(n-1)$. We have $j_1 + \cdots + j_n = j \leqslant J$, i.e., $j_1 \leqslant J - (J(1) + \cdots + J(n-1))$. Since $J(\alpha) \leqslant 0$ for any α, we have

$$\sum_{\alpha=1}^{n} J(\alpha) \geqslant \sum_{\alpha=1}^{+\infty} J(\alpha).$$

Hence $j_1 \leqslant J - \sum_{\alpha=1}^{+\infty} J(\alpha)$. But $J(\alpha) = 0$ for $\alpha > I$. We have $j_1 \leqslant J - \sum_{\alpha=1}^{I} J(\alpha)$. If we take $j \leqslant J - \sum_{\alpha=1}^{I} J(\alpha)$, then

$$\inf_j \inf_{i\geqslant m+1} \Lambda(i, j) > A \quad \text{for } m > M$$

by the third condition of our theorem. Therefore,

$$\inf_s \inf_{j\leqslant J} v_1(((c_s - d_{ms})x^s)^{(j)}) \xrightarrow[m\to +\infty]{} +\infty.$$

But $v_1((c_s x^s)^{(j)}) \geqslant \inf(v_1(((c_s - d_{ms})x^s)^{(j)}), v_1((d_{ms}x^s)^{(j)}))$. It is clear that for all m

$$\inf_{j\leqslant J} v_1((d_{ms}x^s)^{(j)}) \xrightarrow[s\to +\infty]{} +\infty.$$

Let $\inf_{j\leqslant J} v_1(((c_s - d_{ms})x^s)^{(j)}) > A$ for $m > M$. Put $m_0 = M + 1$. We have

$$\inf_{j\leqslant J} v_1((d_{m_0 s}x^s)^{(j)}) > A$$

for $s > S$. Hence $\inf_{j\leqslant J} v_1((c_s x^s)^{(j)}) > A$ for $s > S$.

Therefore, $h(x)$ is finite, and $d_m(x) \xrightarrow[m\to +\infty]{} h(x)$. □

References

[Be] B. M. Bekker, *Abelian extensions of a complete discrete vauation field of finite height*, Algebra i Analiz **3** (1991), no. 6, 76–84; English transl. in St. Petersburg Math. J. **3** (1992), no. 6.

[Bou] N. Bourbaki, *Algèbre commutative. Ch. 5. Entiers. Ch. 6. Valuations.*, Hermann, Paris, 1964.

[F1] I. B. Fesenko, *On class field theory of multidimensional local fields of positive characteristic*, Advances in Soviet Math., vol. 4, Amer. Math. Soc., Providence, RI, 1991, pp. 103–127.

[F2] _____, *Class field theory of multidimensional local fields of characteristic zero with residue field of positive characteristic*, Algebra i Analiz **3** (1991), no. 3, 649–677; English transl. in St. Petersburg Math. J. **3** (1992), no. 3.

[FV] I. B. Fesenko and S. V. Vostokov, *Local fields and their extensions: a constructive approach*, Amer. Math. Soc., Providence, RI, 1993.

[K1] K. Kato, *A generalization of local class field theory by using K-groups*, I, J. Fac. Sci. Univ. Tokyo. Sect. 1A Math. **26** (1979), 303–376.

[K2] _____, *A generalization of local class field theory by using K-groups*, II, J. Fac. Sci. Univ. Tokyo. Sect. 1A Math. **27** (1980), 603–685.

[P1] A. N. Parshin, *Abelian coverings of arithmetic schemes*, Dokl. Akad. Nauk SSSR **243** (1978), no. 4, 855–858; English transl., Soviet Math. Dokl. **19** (1978), 1438–1442.

[P2] _____, *Local class field theory*, Trudy Mat. Inst. Steklov **165** (1984), 143–170; Englsh transl., Proc Steklov Inst. Math. **165** (1985), no. 3, 157–185.

[Z] I. B. Zhukov, *Structure theorems for complete fields*, this volume.

Translated by THE AUTHORS

St. Petersburg State University, Department of Mathematics and Mechanics, Chair of Higher Algebra and Number Theory, St. Petersburg, 198904 Russia

E-mail address: zhukov@math.lgu.spb.su

A Ramification Filtration of the Galois Group of a Local Field

V. A. ABRASHKIN

§0. Introduction

Let K be a local complete discrete valuation field with perfect residue field k of characteristic p. We denote by K_{sep} a separable closure of K and by $\Gamma = \text{Gal}(K_{\text{sep}}/K)$ the absolute Galois group of K. Consider the standard tower of algebraic extensions
$$K \subset K_{\text{ur}} \subset K_{\text{tr}} \subset K_{\text{sep}},$$
where K_{ur} (resp., K_{tr}) is a maximal unramified (resp., tamely ramified) subfield of K_{sep}. This gives a filtration of Γ:
$$\Gamma \supset \Gamma^{(0)} \supset I,$$
where $K_{\text{ur}} = K_{\text{sep}}^{\Gamma^{(0)}}$ and $K_{\text{tr}} = K_{\text{sep}}^{I}$. It is known that the subquotient $\Gamma^{(0)}/I$ of Γ is a procyclic group of order relatively prime to p and I is a pro-p-group, which is profree if $\text{char}\, K = p$. The group-theoretic structure of Γ/I, as well as the arithmetic nature of the above filtration, are well known. A further decomposition of Γ is related to the decreasing filtration $\{\Gamma^{(v)}\}_{v>0}$ of normal subgroups of I. These $\Gamma^{(v)}$ are said to be the *ramification subgroups* of I in the upper numbering. This filtration plays a very important role in many arithmetic topics:

(a) Let L be a finite Galois extension of K with Galois group $\Gamma_{L/K}$. The knowledge of the image of the filtration $\{\Gamma^{(v)}\}_{v \geq 0}$ in $\Gamma_{L/K}$ gives full information about the values of the different and the discriminant of L/K.

(b) The majority of applications of the local class field reciprocity map $\psi \colon K^* \to \Gamma^{ab}$ use the arithmetic structure of K^*. This structure is related to a filtration $\{U_n\}_{n \geq 0}$ of K^*, where $U_n = \{u \in O_K \mid u \equiv 1(\pi_K^n)\}$ (π_K is any uniformizer of the valuation ring O_K of K), which corresponds under ψ to the image of $\{\Gamma^{(v)}\}_{v \geq 0}$ in Γ^{ab}.

(c) Though an explicit description of Γ in purely group-theoretic terms is well known, see [J-W, Jak], the area of its applicability is not large enough. The reason is the absence of any information about the arithmetic nature of the known generators and relations. A description of the ramification filtration in terms of these generators would certainly fill this gap.

© 1995, American Mathematical Society

(d) Let $\operatorname{char} K = 0$. A description of finite commutative group schemes G_K over K is not "arithmetic". They can be identified with representations of Γ in a module $G(K_{\text{sep}})$ and, therefore, can be described in purely group-theoretic terms. But the question on "which of these representations arise from group schemes G over a valuation ring O of K" is already arithmetic. For example, the property "an abelian variety A over K has a good reduction over K" closely related to the property "group schemes $\operatorname{Ker}(p^n \operatorname{id}_A)$, $n \geqslant 1$, can be defined over O". All known properties of Γ-modules H that arise from a finite flat commutative group scheme over O can be expressed in terms of the ramification filtration. They are:

(d_1) "Serre's conjecture" about the action of the tamely ramified part of Γ on a semisimple envelope of H, [Se1, R];

(d_2) the condition for the action of $\Gamma^{(v)}$ to be trivial on H for any $v > v_0(H)$, where $v_0(H)$ depends on certain invariants of K and H, [F1].

In some cases these conditions are also sufficient for H to be realized in the form $G(K_{\text{sep}})$, where G is a group scheme over O, [A1, A2].

(e) A similar problem about representations of Γ in the étale cohomology of proper schemes over K having a good reduction gives the same picture, [F2, A3, A4].

(f) The study of the ramification filtration is interesting by itself. This question was posed by E. Maus [Ma] and I. R. Shafarevich, see the Introduction to [Ko]. Some information about the nontrivial character of this filtration was obtained by E. Maus [Ma] and N. Gordeev [Go].

The main purpose of this paper is to give an explicit description of the ramification filtration $\{\Gamma^{(v)}\}_{v \geqslant 0}$ in group-theoretic terms. We consider the simplest case of the problem: $\operatorname{char} K = p$, $k = \overline{\mathbb{F}_p}$. The reasons are:

1) the subgroup $I = \bigcup_{v > 0} \Gamma^{(v)}$ of Γ is a pro-p-free group, so we have no additional problems with the abstract group-theoretic structure of I;

2) the requirement that the residue field k of K be algebraically closed does not affect the ramification filtration and allows to identify Γ and $\Gamma^{(0)}$;

3) the arithmetic properties of K depend on its cohomological dimension n, so we treat the case $n = 1$.

Our main result gives an explicit description of the ramification filtration modulo the subgroup $I^p C_p(I)$, where $C_p(I)$ is the subgroup of I generated by all commutators of order p.

Our considerations are based on the nonabelian version of the Artin-Schreier theory.

Let K be an arbitrary field of characteristic p. Classical Artin-Schreier theory gives an explicit description of any p-elementary extension of K with an action of the Galois group. E. Witt gave an extension of this theory to the case of p-cyclic extensions, [Wtt]. A matrix form of this theory was developed by H. Inaba [In1-3]. Under this approach, it is possible to treat arbitrary extensions of K, but as a matter of fact it gives a theory of representations of Γ in vector spaces over $\mathbb{F}_p$. The invariant form of this theory appeared in the study of crystalline representations, [F3, A3]. But this generalisation is not very convenient if we want to study the Galois group itself (rather than its image under some modular representation).

We construct our version of the Artin-Schreier theory in §1. This construction depends on the choice of some filtered associative bialgebra (f.a.b.) over $\mathbb{F}_p$ and its area of applicability depends on a certain condition (C_{s_0}), where $s_0 \in \mathbb{N}$, see 1.1. We construct a f.a.b. that satisfies condition (C_{s_0}) for any $s_0 < p$ and apply this construction to the study of arbitrary extensions of K with Galois group of exponent p and class of nilpotency $< p$. The Galois group of such an extension may be related in a very natural way to some Lie algebra over $\mathbb{F}_p$ (having class of nilpotency $< p$) and in its terms the action of the Galois group can be described explicitly, see §2.

We specify our arguments in §3 in the case of a local complete discrete valuation field K of characteristic p. So, we have an explicit description of the extension $\widetilde{K} = K_{\text{sep}}^{I^p C_p(I)}$ over K in terms of a profree Lie $\mathbb{F}_p$-algebra $\mathcal{L}$. Under certain identifications, the Galois group $\text{Gal}(\widetilde{K}/K_{\text{tr}})$ and the Lie algebra $\widetilde{\mathcal{L}} = \mathcal{L}/C_p(\mathcal{L})$ (where $C_p(\mathcal{L})$ is the ideal of $\mathcal{L}$ generated by all commutators of order $\geq p$) are related by the truncated exponent.

In §4 we define a decreasing filtration $\{\widetilde{\mathcal{L}}^{(v)}\}_{v>0}$ of $\widetilde{\mathcal{L}}$ by its ideals $\widetilde{\mathcal{L}}^{(v)}$, $v > 0$. The description of this filtration becomes clearer when considered over $k = \overline{\mathbb{F}}_p$ (see §5).

The main theorem (see §7) shows that if the residue field of K is $k = \overline{\mathbb{F}}_p$, then the truncated exponent transfers the above filtration $\{\widetilde{\mathcal{L}}^{(v)}\}_{v>0}$ to the image of the ramification filtration $\{\Gamma^{(v)}\}_{v>0}$ of the Galois group $\Gamma = \text{Gal}(K_{\text{sep}}/K)$ in $\text{Gal}(\widetilde{K}/K)$.

There is evidence that this approach must work without the "mod $I^p C_p(I)$" and "char $K = p$" assumptions. One can try to apply the characteristic p version of the Campbell-Hausdorff formula for the construction of the f.a.b. satisfying condition (C_{s_0}) for $s_0 > p$, [Di]. The Fontaine-Wintenberger functor "les corps des normes", [Wnt], should give an extension of our description to the case of local fields of characteristic zero. We hope to consider these questions in forthcoming papers.

The main results of this paper were obtained during my stay at Utrecht University (Holland, February–May 91) and in the Max-Planck-Institut für Mathematik (Bonn, February 92–January 93). I would like to express gratitude to these organisations (and especially to the organisers of these visits, Professors G. van der Geer, F. Oort, and F. Hirzebruch) for their hospitality.

§1. A generalisation of the Artin-Schreier theory

1.1. Statement of the main theorem. Let k be any field.

1.1.1. DEFINITION. A is said to be a *filtered associative bialgebra* (f.a.b.) over k, if

(a) A is an associative k-algebra with unit element $1_A = 1$;
(b) there is a decreasing filtration in A:

$$A = J_0(A) \supset J_1(A) \supset \cdots \supset J_n(A) \supset \ldots,$$

where all $J_n(A)$ are two-sided ideals, $J_{n_1}(A)J_{n_2}(A) \subset J_{n_1+n_2}(A)$ for all n_1, $n_2 \geq 0$ and $A = k 1_A \oplus J_1(A)$ as a k-module;

(c) there is a structure of a coassociative coalgebra over k on A given by the k-algebra morphisms: $\Delta: A \to A \otimes A$ (comultiplication) and $\varepsilon: A \to k$ (counit);

(d) for every $n \geq 1$ we have
$$\Delta(J_n(A)) \subset \bigoplus_{n_1+n_2=n} J_{n_1}(A) \otimes J_{n_2}(A) \quad \text{and} \quad \varepsilon(J_n(A)) = 0.$$

If A, B are f.a.b.'s over k, then $A \otimes_k B$ is equipped with the natural structure of an f.a.b. over k. We note that for $n \geq 0$,
$$J_n(A \otimes B) = \sum_{n_1+n_2=n} J_{n_1}(A) \otimes J_{n_2}(B).$$

If K is an extension of k and A is an f.a.b. over k, then $A \otimes_k K$ has the natural structure of an f.a.b. over K.

1.1.2. Let A be an f.a.b. over k and s be some nonnegative integer.

DEFINITION. $a \in A/J_{s+1}(A)$ is called *s-diagonal* if for some (and hence for any) $\hat{a} \in A$ such that $a = \hat{a} \mod J_{s+1}(A)$ we have $\Delta(\hat{a}) \equiv \hat{a} \otimes \hat{a} \mod J_{s+1}(A \otimes A)$ and $\varepsilon(\hat{a}) = 1$.

Denote the set of all s-diagonal elements in the f.a.b. A by $G_A(s)$. Obviously, $G_A(s)$ is a group with respect to the operation induced by the multiplication in A. For $s_1 \geq s_2$ the quotient morphisms $A/J_{s_1+1}(A) \to A/J_{s_2+1}(A)$ induce the reduction morphisms $r_{s_1, s_2} \colon G_A(s_1) \to G_A(s_2)$.

DEFINITION. An f.a.b. A defined over $\mathbb{F}_p$ *satisfies condition* (C_{s_0}) if for all natural numbers s_1, s_2 such that $s_2 \leq s_1 \leq s_0$ and for all fields K of characteristic p, the map
$$r_{s_1, s_2} \colon G_{A_K}(s_1) \to G_{A_K}(s_2)$$
is an epimorphism.

1.1.3. Let A be an f.a.b. over the field K and let L/K be some Galois extension with Galois group $\mathrm{Gal}(L/K)$. For any natural number s consider the groups $G_A(s)$ and $G_{A_L}(s)$ of s-diagonal elements in A and $A_L = A \otimes L$, respectively. Obviously,
$$\{\, a \in G_{A_L}(s) \mid \tau a = a \text{ for all } \tau \in \mathrm{Gal}(L/K) \,\} = G_A(s).$$

1.1.4. Let p be any prime, let A be an f.a.b. over $\mathbb{F}_p$ and let K be a field of characteristic p. For $s \geq 1$ we use the notation $G_{\mathbb{F}_p}(s)$ and $G_K(s)$ for the groups of s-diagonal elements in A and $A \otimes K$, respectively. The absolute Frobenius morphism of K acts on $G_K(s)$; we shall use the notation $a^{(p)}$ for the image of $a \in G_K(s)$ under this action.

We have
$$\{\, a \in G_K(s) \mid a = a^{(p)} \,\} = G_{\mathbb{F}_p}(s).$$

Let us introduce an equivalence relation R_s on $G_K(s)$:

for all a_1, $a_2 \in G_K(s)$, $a_1 \equiv a_2 \mod R_s$

$\iff$ there exists a $b \in G_K(s)$ such that $a_1 = b^{-1} a_2 b^{(p)}$.

Let K_{sep} be a separable closure of K, $\Gamma = \mathrm{Gal}(K_{\mathrm{sep}}/K)$. By definition, $f_1, f_2 \in \mathrm{Hom}(\Gamma, G_{\mathbb{F}_p}(s))$ are in the same conjugation class if and only if there exists $c \in G_{\mathbb{F}_p}(s)$ such that $f_1(\tau) = c^{-1} f_2(\tau) c$ for any $\tau \in \Gamma$.

THEOREM. *Let K be a field of characteristic $p > 0$ and A be an f.a.b. over $\mathbb{F}_p$ satisfying condition (C_{s_0}) for some $s_0 \geq 1$. Then for any $s \leq s_0$ there exists a one-to-one correspondence*

$$\tilde{\pi}_s \colon \{G_K(s)/R_s\} \to \{\text{conj. cl. of } \operatorname{Hom}(\Gamma, G_{\mathbb{F}_p}(s))\}.$$

REMARK. It follows from the proof below (see 1.2), that these correspondences agree on s, i.e., for any $s_2 \leq s_1 \leq s_0$ the following diagram is commutative:

$$\begin{array}{ccc} \{G_K(s_1)/R_{s_1}\} & \xrightarrow{\tilde{\pi}_{s_1}} & \{\text{conj. cl. of } \operatorname{Hom}(\Gamma, G_{\mathbb{F}_p}(s_1))\} \\ {\scriptstyle r_{s_1, s_2}} \downarrow & & \downarrow {\scriptstyle r_{s_1, s_2*}} \\ \{G_K(s_2)/R_{s_2}\} & \xrightarrow{\tilde{\pi}_{s_2}} & \{\text{conj. cl. of } \operatorname{Hom}(\Gamma, G_{\mathbb{F}_p}(s_2))\} \end{array}$$

1.2. PROOF OF THE THEOREM. Let L be an algebraic extension of K and $e \in G_K(s)$ for some $s \geq 0$. Consider the set

$$M_s(L, e) = \{ f \in A_L/J_{s+1}(A_L) \mid f \in G_L(s), \ f^{(p)} = fe \}$$

(here the f.a.b. $A_L = A \otimes L$ is obtained from A by extension of scalars).

1.2.1. LEMMA. *For any $s \leq s_0$ and $e \in G_K(s)$ there exists a separable extension L of K such that $M_s(L, e) \neq \varnothing$.*

PROOF. For any $a \in A$ define its degree by $d(a) = \min\{n \mid a \in J_n(A)\}$. Choose a family $\{c_\alpha\}_{\alpha \in I} \subset J_1(A)$ such that for any natural number s the set

$$\{ c_\alpha \mid \alpha \in I, \ c_\alpha \in J_s(A)\}$$

is an $\mathbb{F}_p$-basis of $J_s(A)$. This means also that for any integer $s \geq 0$ the elements of the set

$$\{ c_\alpha \mid \alpha \in I, \ c_\alpha \notin J_{s+1}(A)\} \cup \{1\}$$

taken modulo $J_{s+1}(A)$ give an $\mathbb{F}_p$-basis of $A/J_{s+1}(A)$.

Now we can choose (uniquely) $E \in A$ so that $e = E \bmod J_{s+1}(A)$ and $E = 1 + \sum_\alpha \eta(\alpha) c_\alpha$, where each $\eta(\alpha) \in K$ and $\eta(\alpha) = 0$ for $d(c_\alpha) > s$. We must prove that there exists an $F = 1 + \sum_\alpha T(\alpha) c_\alpha \in A_{K_{\text{sep}}}$ such that

(1) $F^p \equiv FE \bmod J_{s+1}(A_{K_{\text{sep}}})$;
(2) $\Delta F \equiv F \otimes F \bmod J_{s+1}(A_{K_{\text{sep}}} \otimes A_{K_{\text{sep}}})$.

One can assume that $s \geq 1$ and that all $T(\alpha)$ are defined for $d(c_\alpha) < s$ in such a manner that the equivalences (1) and (2) are valid modulo $J_s(A_{K_{\text{sep}}})$ and $J_s(A_{K_{\text{sep}}} \otimes A_{K_{\text{sep}}})$, respectively.

Let $L_1 = K(\{T(\alpha) \mid d(c_\alpha) < s\})$. By inductive assumption, $L_1 \subset K_{\text{sep}}$. It follows from part b) of the definition of f.a.b. that for any $\alpha_1, \alpha_2 \in I$ we have

$$c_{\alpha_1} c_{\alpha_2} = \sum_{\alpha \in I} A(\alpha_1, \alpha_2; \alpha) c_\alpha,$$

where $A(\alpha_1, \alpha_2; \alpha) \in \mathbb{F}_p$ and $A(\alpha_1, \alpha_2; \alpha) = 0$ for $d(\alpha_1) + d(\alpha_2) > d(\alpha)$.

If $\alpha \in I$ and $d(c_\alpha) = s$, then the expression

$$\sum_{\alpha_1, \alpha_2 \in I} T(\alpha_1) A(\alpha_1, \alpha_2; \alpha) \eta(\alpha_2)$$

is well defined and gives an element of L_1. Indeed, if $A(\alpha_1, \alpha_2; \alpha) \eta(\alpha_2) \neq 0$, then $d(c_{\alpha_1}) + d(c_{\alpha_2}) \leqslant d(c_\alpha) = s$ and $d(c_{\alpha_1}) < s$, so $T(\alpha_1)$ defines an element of L_1 by the inductive assumption.

For any $\alpha \in I$ satisfying $d(c_\alpha) = s$, let us consider the extension $L = L_1(\{T(\alpha) \mid d(c_\alpha) = s\})$, where

$$T(\alpha)^p - T(\alpha) = \sum_{\alpha_1, \alpha_2 \in I} T(\alpha_1) A(\alpha_1, \alpha_2; \alpha) \eta(\alpha_2) + \eta(\alpha).$$

Obviously, L is separable over L_1 and one can assume that $L \subset K_{\text{sep}}$. Let

$$F_1 = 1 + \sum_{d(c_\alpha) \leqslant s} T(\alpha) c_\alpha \in A_L.$$

By the choice of $T(\alpha)$ for $\alpha \in I$ with $d(c_\alpha) = s$, we have $F_1^{(p)} \equiv F_1 E$ mod $J_{s+1}(A_L)$. By the inductive assumption

$$\Delta F_1 \equiv C_0(F_1 \otimes F_1) \mod J_{s+1}(A_L \otimes A_L),$$

where

$$C_0 = 1 + \sum_{\alpha_1, \alpha_2} \gamma(\alpha_1, \alpha_2) c_{\alpha_1} \otimes c_{\alpha_2},$$

$\gamma(\alpha_1, \alpha_2) \in L$ and $\gamma(\alpha_1, \alpha_2) = 0$ for $d(c_{\alpha_1}) + d(c_{\alpha_2}) \neq s$.

From the equivalences

$$\Delta F_1^{(p)} \equiv C_0^{(p)}(F_1^{(p)} \otimes F_1^{(p)}) \equiv C_0^{(p)}(F_1 \otimes F_1)(E \otimes E) \mod J_{s+1}(A_L \otimes A_L),$$
$$\Delta F_1^{(p)} \equiv \Delta F_1 \Delta E \equiv C_0(F_1 \otimes F_1)(E \otimes E) \mod J_{s+1}(A_L \otimes A_L),$$

it follows that $C_0 = C_0^{(p)}$, i.e., all $\gamma(\alpha_1, \alpha_2)$ belong to $\mathbb{F}_p$.

Let us prove the existence of $C \in A$ such that $C \equiv 1 \mod J_s(A)$ and

$$\Delta C \equiv C_0(C \otimes C) \mod J_{s+1}(A \otimes A).$$

Let us put

$$C = 1 + \sum_{d(c_\alpha) = s} \mu(\alpha) c_\alpha,$$

where $\mu(\alpha) \in \mathbb{F}_p$. We can assume that

$$\Delta c_\alpha = c_\alpha \otimes 1 + 1 \otimes c_\alpha + \sum_{\alpha_1, \alpha_2 \in I} B(\alpha; \alpha_1, \alpha_2) c_{\alpha_1} \otimes c_{\alpha_2}$$

for some $B(\alpha; \alpha_1, \alpha_2) \in \mathbb{F}_p$. We have $B(\alpha; \alpha_1, \alpha_2) = 0$ for $d(c_{\alpha_1}) + d(c_{\alpha_2}) < s$ by the definition of f.a.b. It is clear that the existence of C is equivalent to the existence of $\mu(\alpha) \in \mathbb{F}_p$, where $\alpha \in I$ and $d(c_\alpha) = s$, satisfying the following equations

$$(*) \qquad \sum_\alpha B(\alpha; \alpha_1, \alpha_2)\mu(\alpha) = \gamma(\alpha_1, \alpha_2),$$

where $\alpha_1, \alpha_2 \in I$ and $d(c_{\alpha_1}) + d(c_{\alpha_2}) = s$.

The coefficients and constant terms of this system are in $\mathbb{F}_p$, so it is sufficient to prove the solvability of the system $(*)$ in some field of characteristic p.

By condition (C_{s_0}), the map $G_L(s) \to G_L(s-1)$ is an epimorphism, therefore there exists $F_2 \in A_L$ such that $F_2 \mod J_{s+1}(A_L) \in G_L(s)$ and $F_2 \equiv F_1 \mod J_s(A_L)$. Let $\widetilde{C} = F_1 F_2^{-1}$; then $\widetilde{C} \in A_L$, $\widetilde{C} \equiv 1 \mod J_s(A_L)$ and

$$\Delta \widetilde{C} \equiv (\Delta F_1)(\Delta F_2)^{-1} \equiv C_0(F_1 \otimes F_1)(F_2 \otimes F_2)^{-1}$$
$$\equiv C_0(\widetilde{C} \otimes \widetilde{C}) \mod J_{s+1}(A_L \otimes A_L).$$

If we have

$$\widetilde{C} \equiv 1 + \sum_{d(c_\alpha)=s} \widetilde{\mu}(\alpha) c_\alpha \mod J_{s+1}(A_L),$$

then the equivalence above means that the collection $\{\widetilde{\mu}(\alpha) \mid d(c_\alpha) = s\}$ gives the solution of system $(*)$ in L. So system $(*)$ is solvable in $\mathbb{F}_p$ and the required element C exists.

Now for $F = C^{-1} F_1 \in A_L$ we have

$$F^{(p)} = C^{(p)-1} F_1^{(p)} \equiv C^{-1} F_1 E \equiv FE \mod J_{s+1}(A_L),$$
$$\Delta F \equiv (\Delta C)^{-1} \Delta F_1 \equiv C_0^{-1}(C \otimes C)^{-1} C_0(F_1 \otimes F_1) \equiv F \otimes F \mod J_{s+1}(A_L).$$

The lemma is proved.

1.2.2. PROPOSITION. *Let $e \in G_K(s)$ and L be an extension of K such that $M_s(L, e) \neq \varnothing$. If $f_1, f_2 \in M_s(L, e)$, then $f_1 f_2^{-1} \in G_{\mathbb{F}_p}(s)$.*

PROOF. $f_1^{(p)} = f_1 e$, $f_2^{(p)} = f_2 e$, therefore $(f_1 f_2^{-1})^{(p)} = f_1 f_2^{-1}$, i.e., we have $f_1 f_2^{-1} \in G_{\mathbb{F}_p}(s)$.

1.2.3. COROLLARY. *Fix some separable closure K_{sep} of K. Then for $s \leq s_0$ and $e \in G_K(s)$, there exists a uniquely determined Galois extension $K_s(e) \subset K_{\text{sep}}$ of K such that*
  (1) $M_s(K_s(e), e) \neq \varnothing$;
  (2) *if $M_s(L, e) \neq \varnothing$ for some subfield $L \subset K_{\text{sep}}$, then $L \supset K_s(e)$ and $M_s(L, e) = M_s(K_s(e), e)$.*

PROOF. Let $L_0 \subset K_{\text{sep}}$ be some minimal element in the partially ordered (by inclusion) set of subfields L in K_{sep} such that $M_s(L, e) \neq \varnothing$. Choose some $f_0 \in M_s(L_0, e)$. If $L \subset K_{\text{sep}}$ satisfies $M_s(L, e) \neq \varnothing$ and $f \in M_s(L, e)$, then $f_0, f_1 \in M_s(LL_0, e)$. It follows from the above proposition that $f_0, f_1 \in M_s(L \cap L_0, e)$. L_0 is minimal, so $L \cap L_0 = L_0$, i.e., $L \supset L_0$. Similarly, L_0 is the Galois extension of K. So we can take $K_s(e) = L_0$. □

1.2.4. Now we can use the new notation $M_s(e)$ for the set $M_s(K_s(e), e)$.

PROPOSITION. *Let* $s_2 \leq s_1 \leq s_0$, $e_1 \in G_K(s_1)$, $e_2 \in G_K(s_2)$ *and* $r_{s_1,s_2}(e_1) = e_2$, *where* $r_{s_1,s_2}\colon G_K(s_1) \to G_K(s_2)$ *is the reduction morphism from* 1.2. *Then* $K_{s_1}(e_1) \supset K_{s_2}(e_2)$ *and* r_{s_1,s_2} *induces an epimorphic mapping* $M_{s_1}(e_1) \to M_{s_2}(e_2)$.

The proof follows immediately from 2.2.

1.2.5. Let $s \leq s_0$, $e \in G_K(s)$, $f \in M_s(e)$. For any $\tau \in \Gamma = \mathrm{Gal}(K_{\mathrm{sep}}/K)$ one has $\tau f = c(\tau)f$, where $c(\tau) \in G_{\mathbb{F}_p}(s)$. Obviously, the correspondence $\tau \mapsto c(\tau)$ defines an element of $\mathrm{Hom}(\Gamma, G_{\mathbb{F}_p}(s))$, which we denote by $\pi_{e,f,s}$. The following proposition follows immediately from definitions.

PROPOSITION.
(1) $\mathrm{Ker}(\pi_{e,f,s}) = \mathrm{Gal}(K_{\mathrm{sep}}/K_s(e))$;
(2) *if* $f_1, f_2 \in M_s(e)$, *then* $\pi_{e,f_1,s}$ *and* $\pi_{e,f_2,s}$ *are conjugate under some inner automorphism of the group* $G_{\mathbb{F}_p}(s)$;
(3) *if some homomorphism* $\pi\colon \Gamma \to G_{\mathbb{F}_p}(s)$ *is conjugate to* $\pi_{e,f,s}$ *then there exists an* $f' \in M_s(e)$ *such that* $\pi_{e,f',s} = \pi$.

1.2.6. So, for $1 \leq s \leq s_0$, the correspondence defined by $e \mapsto \{\mathrm{conj.~cl.~of~} \mathrm{Hom}(\Gamma, G_{\mathbb{F}_p}(s))\}$ gives the mappings

$$\widetilde{\pi}_s\colon G_K(s) \to \{\mathrm{conj.~cl.~of~} \mathrm{Hom}(\Gamma, G_{\mathbb{F}_p}(s))\}.$$

PROPOSITION. *Let* $e_1, e_2 \in G_K(s)$, $s \leq s_0$. *Then* $\widetilde{\pi}_s(e_1) = \widetilde{\pi}_s(e_2) \iff e_1 \equiv e_2$ mod R_s, *i.e., there exists some* $h \in G_K(s)$ *such that* $e_2 = h^{-1}e_1 h^{(p)}$.

PROOF. Let $f_1 \in M_s(e_1)$, $f_2 \in M_s(e_2)$. Then $\widetilde{\pi}_s(e_1) = \widetilde{\pi}_s(e_2) \iff$ there exists $a \in G_{\mathbb{F}_p}(s)$ such that for any $\tau \in \Gamma$ one has the relation $c_1(\tau) = a^{-1}c_2(\tau)a$, where $\tau f_1 = c_1(\tau)f_1$, $\tau f_2 = c_2(\tau)f_2$.

Let $h = f_1^{-1}a^{-1}f_2 \in G_{K_{\mathrm{sep}}}(s)$. Then

$$h^{(p)} = f_1^{(p)-1}a^{-1}f_2^{(p)} = e_1^{-1}f_1^{-1}a^{-1}f_2 e_2 = e_1^{-1}h e_2, \quad \text{i.e.,} \quad e_2 = h^{-1}e_1 h^{(p)}.$$

But for any $\tau \in \Gamma$ we have

$$\tau h = (\tau f_1)^{-1}a^{-1}(\tau f_2) = f_1^{-1}c_1(\tau)^{-1}a^{-1}c_2(\tau)f_2 = f_1^{-1}a^{-1}f_2 = h,$$

i.e., $h \in G_K(s)$.

1.2.7. It follows from the previous proposition that $\widetilde{\pi}_s$ defines an injective mapping

$$\pi_s\colon G_K(s)/R_s \to \{\mathrm{conj.~cl.~of~} \mathrm{Hom}(\Gamma, G_{\mathbb{F}_p}(s))\}.$$

The surjectivity of π_s follows from the next proposition.

PROPOSITION. *Let* $s \leq s_0$ *and* $\pi \in \mathrm{Hom}(\Gamma, G_{\mathbb{F}_p}(s))$. *Then there exist* $e \in G_K(s)$ *and* $f \in M_s(e)$ *such that* $\pi = \pi_{e,f,s}$.

PROOF. We can assume that $s > 1$ and apply induction on s. Then there exist $e_1 \in G_K(s-1)$, $f_1 \in M_{s-1}(e_1)$ such that for any $\tau \in \Gamma$, $\tau f_1 = \pi'(\tau)f_1$, where $\pi'(\tau) = \pi(\tau) \bmod J_s(A)$. Choose $e_2 \in G_K(s)$ such that $r_{s,s-1}(e_2) = e_1$ and choose $f_2 \in M_s(e_2)$ such that $r_{s,s-1}(f_2) = f_1$.

Now we use the special $\mathbb{F}_p$-basis $\{c_\alpha\}_{\alpha \in I}$ constructed in the proof of Lemma 1.2.1. By means of this basis we can take liftings

$$E_2 = 1 + \sum_{\alpha \in I} \eta(\alpha) c_\alpha \in A_K, \qquad F_2 = 1 + \sum_{\alpha \in I} \mu(\alpha) c_\alpha \in A_{K_{\text{sep}}}$$

of e_2 and f_2, which are uniquely determined by the conditions $\eta(\alpha) = 0$, $\mu(\alpha) = 0$ if $d(c_\alpha) > s$. We have:

$$F_2^{(p)} \equiv F_2 E_2 \mod J_{s+1}(A_{K_{\text{sep}}}),$$
$$\Delta F_2 \equiv F_2 \otimes F_2 \mod J_{s+1}(A_{K_{\text{sep}}} \otimes A_{K_{\text{sep}}}),$$
$$\tau F_2 \equiv \pi_1(\tau) F_2 \mod J_{s+1}(A_{K_{\text{sep}}}),$$

where $\tau \in \Gamma$, $\pi_1 \in \text{Hom}(\Gamma, G_{\mathbb{F}_p}(s))$ and $\pi_1 \equiv \pi \mod J_s(A)$.

Let $\pi_1(\tau) = c(\tau) \pi(\tau)$ for any $\tau \in \Gamma$. Then $c(\tau) \in G_{\mathbb{F}_p}(s)$, $r_{s,s-1}(c(\tau)) = 1$ and $c(\tau_1 \tau_2) = c(\tau_1) c(\tau_2)$ for any $\tau_1, \tau_2 \in \Gamma$. Let

$$C(\tau) \in 1 + J_s(A) \subset A$$

be the liftings of $c(\tau)$ defined by the above choice of special $\mathbb{F}_p$-basis. By the cohomological triviality of the Galois module K_{sep}, there exists a $C_1 \in A_{K_{\text{sep}}}$ such that $C_1 \equiv 1 \mod J_s(A_{K_{\text{sep}}})$ and

$$\tau C_1 \equiv C(\tau) C_1 \mod J_{s+1}(A_{K_{\text{sep}}}).$$

Now for $F_3 = C_1 F_2$ we have

$$\tau F_3 \equiv \pi(\tau) F_3 \mod J_{s+1}(A_{K_{\text{sep}}}),$$
$$\Delta F_3 \equiv (F_3 \otimes F_3) C_2 \mod J_{s+1}(A_{K_{\text{sep}}} \otimes A_{K_{\text{sep}}}),$$

where $C_2 \equiv 1 \mod J_s(A_{K_{\text{sep}}} \otimes A_{K_{\text{sep}}})$ and

$$\Delta C_1 \equiv (C_1 \otimes C_1) C_2 \mod J_{s+1}(A_{K_{\text{sep}}} \otimes A_{K_{\text{sep}}}).$$

For every $\tau \in \Gamma$, $(\tau C_1) C_1^{-1} \mod J_{s+1}(A_{K_{\text{sep}}})$ is an s-diagonal element, therefore, $\tau C_2 = C_2$, i.e., $C_2 \in (A \otimes A) \mod J_{s+1}(A_{K_{\text{sep}}} \otimes A_{K_{\text{sep}}})$. Similarly, as in the proof of Lemma 2.1, we can obtain the existence of a $C_3 \in A$ such that $C_3 \equiv 1 \mod J_s(A)$ and

$$\Delta C_3 \equiv (C_3 \otimes C_3) C_2 \mod J_{s+1}(A \otimes A).$$

So, for $F = C_3^{-1} F_3$ we have:

$$\tau F \equiv \pi(\tau) F \mod J_{s+1}(A_{K_{\text{sep}}}), \qquad \Delta F \equiv F \otimes F \mod J_{s+1}(A_{K_{\text{sep}}} \otimes A_{K_{\text{sep}}}).$$

It now follows that $E = F^{(p)} F^{-1} \mod J_{s+1}(A_{K_{\text{sep}}})$ is a Γ-invariant element of $G_{K_{\text{sep}}}(s)$, hence $E \mod J_{s+1}(A_{K_{\text{sep}}}) = e \in G_K(s)$. If we set $f = F \mod J_{s+1}(A_{K_{\text{sep}}})$, then $\pi = \pi_{e,f,s}$.

So, the proposition and the theorem of 1.2 are proved.

1.3. Examples and applications.

1.3.1. The theorem in 1.1.4 gives nothing in the case $s = 0$ because for every f.a.b. A over $\mathbb{F}_p$ one has $G_{\mathbb{F}_p}(0) = G_K(0) = 1$.

Consider the first nontrivial case where one can apply our theorem. Let $A = \mathbb{F}_p[D]$ (here D is an indeterminate) with a filtration by the ideals $J_s(A) = (D^s)$, $s \geqslant 0$, and coalgebra structure given by relations $\Delta D = D \otimes 1 + 1 \otimes D$, $\varepsilon(D) = 0$. It is easy to verify that for $s = 1$ and this choice of f.a.b. our theorem gives the usual Artin-Schreier theory. Indeed we have the identifications $G_{\mathbb{F}_p}(s) = \mathbb{F}_p$, $G_K(1) = K$ given by correspondences

$$1 + aD \bmod (D^2) \mapsto a,$$

where $a \in \mathbb{F}_p$ or $a \in K$. The equivalence relation R_1 on $G_K(1)$ is transformed here to the relation R on K:

$$a_1 \equiv a_2 \bmod R \quad \text{iff} \quad a_1 = a_2 + b^p - b \quad \text{for some } b \in K.$$

So π_1 can be regarded as a one-to-one correspondence

$$\pi'_1: K/R \to \operatorname{Hom}(\Gamma, \mathbb{F}_p),$$

where $\Gamma = \operatorname{Gal}(K_{\text{sep}}/K)$. It follows from the construction of π_1 that for any $a \in K$ the homomorphism $\chi = \pi'_1(a \bmod R)$ maps any $\tau \in \Gamma$ to $\chi(\tau) = \tau T - T \in \mathbb{F}_p$, where $T^p - T = a$. Of course, π'_1 is also an isomorphism of groups.

Let $1 \leqslant s < p$. It is easy to show that for these s we have

$$G_K(s) = \{\widetilde{\exp}(aD) \bmod J_s(A) \mid a \in K\},$$

where $\widetilde{\exp}(l) = \sum_{0 \leqslant n < p} l^n/n!$ is the truncated exponent. This means that the reduction maps $r_{s_1, s_2}: G_K(s_1) \to G_K(s_2)$, for $1 \leqslant s_2 \leqslant s_1 < p$, are one-to-one mappings and therefore the f.a.b. A satisfies condition (C_{p-1}). But this means also that the theorem in 1.1.4 for the f.a.b. A and arbitrary $1 \leqslant s < p$ gives nothing more than the Artin-Schreier theory.

It may be shown that

$$G_K(p) = \{\widetilde{\exp}(aD^p) \bmod J_{p+1}(A) \mid a \in K\}.$$

Therefore, $r_{p,1}: G_K(p) \to G_K(1)$ is the zero mapping, so the f.a.b. A does not satisfy condition (C_p).

1.3.2. Let $W = \operatorname{Spec} B$ be the scheme of Witt vectors. One can assume that $B = \mathbb{Z}_p[Y_0, Y_1, \ldots, Y_n, \ldots]$ and the operations on W are given by means of the Witt polynomials

$$w_n(Y_0, \ldots, Y_n) = Y_0^{p^n} + pY_1^{p^{n-1}} + \cdots + p^n Y_n, \quad n \geqslant 0.$$

Let $\widetilde{W} = W \otimes \mathbb{F}_p$ be the reduction of W modulo p. Then $\widetilde{W} = \operatorname{Spec} A$, where $A = \mathbb{F}_p[X_0, X_1, \ldots, X_n, \ldots]$ and $X_n = Y_n \otimes 1$ for $n \geqslant 0$. The bialgebra structure on A is induced by the bialgebra structure on B. Introduce the grading of A by the conditions $d(X_n) = p^n$ for all $n \geqslant 0$. Then the ideals

$$J_s(A) = \{a \in A \mid d(a) \geqslant s\}$$

for $s \geqslant 0$ define a decreasing filtration of A. So we have the structure of an f.a.b. on A.

Let $E \in \mathbb{Z}_p[[Y]]$ be the power series equal to

$$\exp(Y + Y^p/p + \cdots + Y^{p^n}/p^n + \ldots)$$

(the Artin-Hasse exponential). We can consider the collection of variables

$$\overline{X} = (X_0, X_1, \ldots, X_n, \ldots)$$

as an element of the ring $\widetilde{W}(A)$ and define

$$\overline{E}(\overline{X}) = \prod_{n \geqslant 0} E(X_n).$$

$\overline{E}(\overline{X})$ is the element of the completion of A in the topology induced by the grading d. If K is a field of characteristic p, then the collection of coordinates of the product of Witt vectors $\overline{w}$ and $\overline{X}$ will be denoted by $\overline{w}\overline{X}$.

PROPOSITION. *Let s be a natural number, $G_{A,K}(s)$ be the group of s-diagonal elements of the f.a.b. $A_K = A \otimes K$. Then*

$$G_{A,K}(s) = \{\overline{E}(\overline{w}\overline{X}) \bmod J_{s+1}(A_K) \mid \overline{w} \in \widetilde{W}(K)\}.$$

PROOF. Note that elements of the $\mathbb{F}_p$-module

$$M_s = \{m \in K[X_0, \ldots, X_n, \ldots] \mid d(m) \leqslant s\}$$

give the full set of representatives of the elements of a $A_K/J_{s+1}(A_K)$ in A_K. Let

$$e \in A_K/J_{s+1}(A_K), \qquad E = \sum_{i=0}^{s} E_i \in M_s,$$

be a representative of e, where E_i for $0 \leqslant i \leqslant s$ are isobaric polynomials of X_0, $X_1, \ldots, X_n, \ldots$ of weight $d = i$. The coaddition Δ in A is given by the isobaric polynomials, therefore $e \in G_{A,K}(s)$ if and only if $E_0 = 1$ and

$$\Delta E_i = \sum_{i_1 + i_2 = i} E_{i_1} \otimes E_{i_2}$$

for $0 \leqslant i \leqslant s$. This means that $E_0 + E_1 t + \cdots + E_s t^s$ is a "curve" for $\widetilde{W}$ modulo $\deg(s+1)$, see [Di]. Now our proposition follows from the explicit description of all curves for the scheme of Witt vectors, [Di, §7].

COROLLARY. *The f.a.b.* A *satisfies the condition* (C_{s_0}) *for any natural number* s_0.

COROLLARY. *For any field* K *of characteristic* p *and any natural number* s, *the correspondence* $\overline{E(\overline{w}X)} \mapsto \overline{w}$ *gives an isomorphism*

$$G_{A,K}(s) \to \widetilde{W}_s(K),$$

where $\widetilde{W}_s$ *is the group scheme of Witt vectors of finite length* s *over* $\mathbb{F}_p$.

So this choice of f.a.b. A yields the following result of Witt [Wtt]:

If $s \geqslant 1$ *and* K *is a field of characteristic* $p > 0$, *then there exists a one-to-one correspondence*

$$W_s(K)/(F - \mathrm{id})W_s(K) \to \mathrm{Hom}(\Gamma, \mathbb{Z}/p^s\mathbb{Z})$$

(here $F: \widetilde{W}_s(K) \to \widetilde{W}_s(K)$ is the Frobenius morphism and, of course, this correspondence is a group isomorphism).

Taking the projective limit over s, we obtain the isomorphism

$$\widetilde{W}(K)/(F - \mathrm{id})\widetilde{W}(K) \to \mathrm{Hom}_{\mathrm{cont}}(\Gamma, \mathbb{Z}_p).$$

So, we have a complete description of all $\mathbb{Z}_p$-extensions of K with an explicitly described action of the Galois group.

1.3.3. Let p be a fixed prime, $\mathcal{L}$ be a nilpotent finite-dimensional Lie algebra over $\mathbb{F}_p$. We assume that the nilpotency class of $\mathcal{L}$ is less than p.

Let $A = A_{\mathcal{L}}$ be the enveloping algebra of $\mathcal{L}$. Note that there exists a canonical embedding $\mathcal{L} \subset A$. For any field K of characteristic p the coalgebra structure on $A_K = A \otimes K$ is uniquely given by the conditions: $\Delta(l) = l \otimes 1 + 1 \otimes l$ and $\varepsilon(l) = 0$ for $l \in \mathcal{L}$. If $J(A_K) = \mathrm{Ker}\,\varepsilon$, then we define the decreasing filtration $\{J_s(A_K)\}$ of A_K for all $s \geqslant 0$ by $J_s(A_K) := J^s(A_K)$. It is easy to see that A is an f.a.b.

Let

$$\widetilde{\log}(T) = \sum_{1 \leqslant i \leqslant p-1} (-1)^{i-1}(T-1)^i/i$$

be the truncated logarithm. It is clear that for $s < p$ the correspondence $a \mapsto \widetilde{\log}\,a$ defines a one-to-one correspondence between the sets

$$1 + J_1(A_K) \bmod J_{s+1}(A_K) \quad \text{and} \quad J_1(A_K) \bmod J_{s+1}(A_K).$$

PROPOSITION. *For* $s < p$, *the correspondence* $a \mapsto \widetilde{\log}\,a$ *defines a one-to-one mapping between* $G_K(s)$ *and* $\mathcal{L} \otimes K \bmod J_{s+1}(A_K)$.

PROOF. Let $\mathcal{L}_K = \mathcal{L} \otimes K$, $C_1(\mathcal{L}_K) = \mathcal{L}_K$, and, for $s > 1$, $C_s(\mathcal{L}_K) = [C_{s-1}(\mathcal{L}_K), \mathcal{L}_K]$. We have: $C_p(\mathcal{L}) = 0$ and, for any $s_1, s_2 \geqslant 1$, $[C_{s_1}(\mathcal{L}), C_{s_2}(\mathcal{L})] \subset C_{s_1+s_2}(\mathcal{L})$.

For any $l \in \mathcal{L}$ set $w(l) = \max\{i \mid l \in C_i(\mathcal{L})\}$. Choose a special basis $l_1, l_2, \ldots, l_N$ of $\mathcal{L}_K$ over K, $N = \dim_K \mathcal{L}_K$, satisfying the following condition:

$\{l_i \mid l_i \in C_s(\mathcal{L})\}$ is a K-basis of $C_s(\mathcal{L})$ for all $s < p$.

This condition is equivalent to the condition:

$\{l_i \mid w(l_i) = s\}$ is a basis of the supplementary vector space for $C_{s+1}(\mathcal{L})$ in $C_s(\mathcal{L})$ for all $s < p$.

By the Birkhoff-Witt theorem, the monomials $l_1^{a_1} l_2^{a_2} \ldots l_N^{a_N}$, where $a_i \in \mathbb{N} \cup \{0\}$ for $1 \leqslant i \leqslant N$, give a K-basis of A_K. We set

$$w(l_1^{a_1} l_2^{a_2} \ldots l_N^{a_N}) = \sum_{i=1}^{N} a_i w(l_i).$$

LEMMA. *For any s, the set $\{l_1^{a_1} l_2^{a_2} \ldots l_N^{a_N} \mid w(l_1^{a_1} l_2^{a_2} \ldots l_N^{a_N}) \geqslant s\}$ is a basis of $J_s(A_K)$ over K.*

PROOF. By definition, we have $l \in J_{w(l)}(A_K)$ for any $l \in \mathcal{L}_K$. Hence, any monomial $l_1^{a_1} l_2^{a_2} \ldots l_N^{a_N}$ of w-weight greater than or equal to s is in $J_s(A_K)$.

Conversely, the ideal $J(A_K) = \operatorname{Ker}(\varepsilon)$ is generated by the set $\{l_i \mid w(l_i) = 1\}$. Therefore, it is sufficient to prove that every product $l_{i_1} \ldots l_{i_{s_1}}$, where $s_1 \geqslant s$, $1 \leqslant i_1, \ldots, i_{s_1} \leqslant N$, $w(l_{i_1}) = \cdots = w(l_{i_{s_1}}) = 1$, can be expressed as a sum of monomials $l_1^{a_1} l_2^{a_2} \ldots l_N^{a_N}$ of w-weight greater than or equal to s_1.

If $i_1 \leqslant i_2 \leqslant \ldots \leqslant i_{s_1}$, then $l_{i_1} \ldots l_{i_{s_1}}$ is one of these monomials and it has w-weight $s_1 \geqslant s$. So, here is nothing to prove.

If the sequence of indices $i_1, i_2, \ldots, i_{s_1}$ does not grow, we must use the commutator relations for presenting $l_{i_1} \ldots l_{i_{s_1}}$ as a sum of monomials from the Birkhoff-Witt basis. These relations are of the following kind: $l'l'' = l''l' + \sum_i \alpha_i l_i$, where $l', l'' \in \{l_1, l_2, \ldots, l_N\}$, all $\alpha_i \in K$ and $\alpha_i = 0$ if $w(l_i) < w(l') + w(l'')$ (this follows from the special choice of the basis $l_1, l_2, \ldots, l_N$). It is clear that these relations can give only monomials of weight greater than or equal to s_1. The lemma is proved.

We continue the proof of the proposition. Let $a \in G_K(s)$ and $\widehat{a} \in A_K$ satisfy $a = \widehat{a} \mod J_{s+1}(A_K)$. Consider $b = \widetilde{\log}(\widehat{a})$. Then

$$\Delta b \equiv b \otimes 1 + 1 \otimes b \mod J_{s+1}(A_K \otimes A_K).$$

Let $b = \sum_{i \geqslant 1} b_i$, where every b_i is a linear combination of monomials from the Birkhoff-Witt basis with w-weight equal to i. We note that the elements

$$l_1^{a_1} l_2^{a_2} \ldots l_N^{a_N} \otimes l_1^{b_1} l_2^{b_2} \ldots l_N^{b_N},$$

where $a_i, b_i \in \mathbb{N} \cup \{0\}$ for $1 \leqslant i \leqslant N$, give the Birkhoff-Witt basis of the enveloping algebra of $\mathcal{L}_K \oplus \mathcal{L}_K$. Obviously, for any i, $\Delta(b_i) - (b_i \otimes 1 + 1 \otimes b_i)$ can be expressed as a linear combination of such monomials with w-weight equal to i. So, $\Delta(b_i) - (b_i \otimes 1 + 1 \otimes b_i)$ is equal to 0 or has w-weight equal to i. By assumption we have

$$\Delta(b) - (b \otimes 1 + 1 \otimes b) \in J_{s+1}(A_K \otimes A_K).$$

It follows now from the above lemma that

$$w(\Delta b - (b \otimes 1 + 1 \otimes b)) = w\left(\sum_{i \geqslant 1}(\Delta b_i - (b_i \otimes 1 + 1 \otimes b_i))\right) \geqslant s + 1.$$

Hence, for $i \leqslant s$ we have $\Delta b_i = b_i \otimes 1 + 1 \otimes b_i$. This means that $b_i \in \mathcal{L}_K$ for $i \leqslant s$ (see [B]) and the proposition is proved.

COROLLARY. *The f.a.b. A satisfies condition* (C_{p-1}).

By the Campbell-Hausdorff formula, we can now deduce the following well-known fact (see [B, Chapter 2, Exercises 4–8]):

COROLLARY. *The correspondence* $\mathcal{L} \mapsto G_{\mathbb{F}_p}(\mathcal{L})$ *gives an equivalence of the category of Lie* $\mathbb{F}_p$-*algebras of class of nilpotency* $< p$ *and the category of p-periodic groups of nilpotency class less than* p.

After these preparations, we are able to apply the theorem in 1.1.4 to the explicit description of the Galois extensions of K with arbitrary p-periodic Galois group of nilpotency class less than p.

Let $e \in G_K(s)$, $s < p$, and $\widetilde{\pi}_s(e)$ be the corresponding conjugacy class in $\mathrm{Hom}(\Gamma, G_{\mathbb{F}_p}(s))$. Let $l_1, \ldots, l_n$ be the part of the special basis of $\mathcal{L}$ (from the above proposition) that consists of elements with w-weight equal to 1. Then for the reduction $r_{s,1}(e)$ of e, we have

$$r_{s,1}(e) = 1 + w_1 l_1 + \cdots + w_n l_n \bmod J_2(A_K),$$

where $w_1, \ldots, w_n \in K$.

PROPOSITION. *The conjugacy class* $\widetilde{\pi}_s(e)$ *consists of epimorphisms if and only if the images of* $w_1, \ldots, w_n$ *in* $K/(F - \mathrm{id})K$ *are linearly independent (here F is the Frobenius morphism of K).*

PROOF. For $s = 1$ this can be easily checked by ordinary Artin-Schreier theory. Let $s > 1$, $f: \Gamma \to G_{\mathbb{F}_p}(s)$ be any homomorphism from the class $\widetilde{\pi}_s(e)$. $G_{\mathbb{F}_p}(s)$ is a p-group, hence f factors through the quotient $\Gamma \to \Gamma(p)$, where $\Gamma(p)$ is the Galois group of the maximal p-extension of K. Since the one-to-one correspondences π_s and π_1 from our theorem agree with each other under the reduction mapping $r_{s,1}$ (see Remark in 1.1.4), we conclude that the composition $\Gamma(p) \to G_{\mathbb{F}_p}(s) \to G_{\mathbb{F}_p}(1)$ is an epimorphism. But

$$\mathrm{Ker}(G_{\mathbb{F}_p}(s) \to G_{\mathbb{F}_p}(1)) = [G_{\mathbb{F}_p}(s), G_{\mathbb{F}_p}(s)],$$

so our proposition follows from the following well-known property of p-groups:

let Γ_1, Γ_2 *be profinite p-groups, then the homomorphism* $\pi: \Gamma_1 \to \Gamma_2$ *is an epimorphism if and only if it induces an epimorphism*

$$\Gamma_1/\Gamma_1^p[\Gamma_1, \Gamma_1] \to \Gamma_2/\Gamma_2^p[\Gamma_2, \Gamma_2]$$

(see [Se2, Chapter 1, §4]).

EXAMPLE. Let us apply the above considerations to the explicit construction of extensions of K with noncommutative Galois groups of order p^3 for $p > 2$. Let $\mathcal{L}$ be a Lie algebra over $\mathbb{F}_p$ with $\mathbb{F}_p$-basis l_1, l_2, l_3 and relations $[l_1, l_2] = l_3$, $[l_1, l_3] = [l_2, l_3] = 0$. Assume that $w_1, w_2 \in K$ have linearly independent images in the quotient $K/(F - \mathrm{id})K$; take a 2-diagonal element $e = E \bmod J_3(A_K)$, where $E = \widetilde{\exp}(w_1 l_1 + w_2 l_2)$. The corresponding extension $L = K(T_1, T_2, T_3)$ of K is given by $\mathcal{F}^{(p)} \equiv \mathcal{F}E \bmod J_3(A_{K_{\mathrm{sep}}})$, where $\mathcal{F} = \widetilde{\exp}(T_1 l_1 + T_2 l_2 + T_3 l_3)$.

By the Campbell-Hausdorff formula, we obtain

$$\widetilde{\exp}(T_1^p l_1 + T_2^p l_2 + T_3^p l_3) \equiv \widetilde{\exp}(T_1 l_1 + T_2 l_2 + T_3 l_3) \widetilde{\exp}(w_1 l_1 + w_2 l_2)$$
$$\equiv \widetilde{\exp}((T_1 + w_1)l_1 + (T_2 + w_2)l_2 + (w_2 T_1/2 - w_1 T_2/2 + T_3)l_3).$$

Now we have the explicit equations for this extension:

$$T_1^p = T_1 + w_1, \qquad T_2^p = T_2 + w_2, \qquad T_3^p = T_3 + (w_2 T_1 - w_1 T_2)/2.$$

The action of $\operatorname{Gal}(L/K) \simeq G_{\mathbb{F}_p}(2)$ is given by the relation $\mathcal{F} \mapsto u\mathcal{F}$, $u \in G_{\mathbb{F}_p}(2)$. For example, the generators $u_1 = \widetilde{\exp}(l_1)$ and $u_2 = \widetilde{\exp}(l_2)$ of $G_{\mathbb{F}_p}(2)$ act in the following manner:

$$u_1 \colon (T_1, T_2, T_3) \mapsto (T_1 + 1, T_2, T_3 + T_2/2),$$
$$u_2 \colon (T_1, T_2, T_3) \mapsto (T_1, T_2 + 1, T_3 - T_1/2).$$

§2. Explicit construction of the Galois action (case of an arbitrary field)

2.1. Let K be a field of characteristic $p > 0$, $\Gamma_K = \operatorname{Gal}(K_{\text{sep}}/K)$, let $\mathcal{L}$ be a finite-dimensional Lie algebra over $\mathbb{F}_p$ and A be the enveloping algebra of $\mathcal{L}$ with the structure of an f.a.b., defined in 1.3. We use the notation of 1.3.

Let $s_0 < p$, $e \in G_K(s_0)$ and $\mathcal{M}_{s_0}(e) = \{f \in G_{K_{\text{sep}}}(s_0) \mid f^{(p)} = fe\}$ be the set from the proof of the theorem in 1.1.4. Denote by $d \colon A \to \mathbb{N}$ the grading of A and by $\{c_\alpha\}_{\alpha \in I}$ the special $\mathbb{F}_p$-basis of A that was defined in the proof of Lemma 1.2.1.

Let $f \in M_{s_0}(e)$. We can take its (uniquely defined) representative

$$\widehat{f} = 1 + \sum_{d(c_\alpha) \leqslant s_0} f_\alpha c_\alpha$$

in $A_{K_{\text{sep}}}$ and denote by $\mathcal{M}_{s_0}(e)$ the $\mathbb{F}_p$-submodule in K_{sep} generated by all f_α, where $d(c_\alpha) \leqslant s_0$, and $1 \in K_{\text{sep}}$.

We have:

2.1a. $\mathcal{M}_{s_0}(e)$ does not depend either on the choice of the special basis $\{c_\alpha\}_{\alpha \in I}$, or on the choice of $f \in M_{s_0}(e)$.

2.1b. $\mathcal{M}_{s_0}(e)$ is a Γ_K-invariant submodule of K_{sep}.

2.1c. If the homomorphism $F \colon \Gamma_K \to \operatorname{Aut} \mathcal{M}_{s_0}(e)$ gives an action of Γ_K on $\mathcal{M}_{s_0}(e)$, then $\operatorname{Ker} F = \operatorname{Gal}(K_{\text{sep}}/K_{s_0}(e))$, where $K_{s_0}(e)$ is the minimal extension of K in K_{sep} such that the following implication is true: $f \in M_{s_0}(e) \iff f \in G_{K_{s_0}(e)}(s_0)$, see 1.2.

2.1d. For any $\tau \in \Gamma_K$, $F(\tau)$ is a unipotent automorphism of $\mathcal{M}_{s_0}(e)$ such that $(F(\tau) - \operatorname{id})^{s_0+1} = 0$. Therefore it defines an endomorphism $L(\tau) = \widetilde{\log} F(\tau) \in \operatorname{End} \mathcal{M}_{s_0}(e)$.

Let the representative $\widehat{e}$ of e be of the form

$$\widehat{e} = 1 + \sum_{d(c_\alpha) \leqslant s_0} e_\alpha c_\alpha$$

and suppose that the images of elements of the set $\{e_\alpha \mid \alpha \in I,\ d(c_\alpha) = 1\}$ in the quotient $K/(F - \mathrm{id})K$ are linearly independent. We fix $f \in M_{s_0}(e)$. Then the homomorphism $\pi_{e,f,s_0}: \Gamma_K \to G_{\mathbb{F}_p}(s_0)$ from 1.2.5 is an epimorphism and defines an isomorphism $\mathrm{Gal}(K_{s_0}(e)/K) \simeq G_{\mathbb{F}_p}(s_0)$. By Proposition 1.1.3, for any $\tau \in \Gamma_K$ there exists an $l_\tau \in \mathcal{L}$ such that $\pi_{e,f,s_0}(\tau) = \widetilde{\exp}(l_\tau) \bmod J_{s_0+1}(A)$. Clearly, l_τ is unique modulo $C_{s_0+1}(\mathcal{L})$.

We have:

2.1e. The correspondence $l_\tau \mapsto L(\tau)$ for $\tau \in \Gamma_K$ (see 2.1d) defines a homomorphism of Lie algebras $LF: \mathcal{L} \to \mathrm{End}\, M_{s_0}(e)$, i.e., yields an action of the Lie algebra $\mathcal{L}$ on $M_{s_0}(e)$.

2.2. Let us treat the previous construction in the case of a free Lie algebra and $s_0 = p - 1$.

Let $\mathcal{L}$ be a free Lie algebra over $\mathbb{F}_p$ with free generators $D_1, \ldots, D_N$. Take the system
$$\{D_{i_1} \ldots D_{i_s} \mid 1 \leqslant i_1, \ldots, i_s \leqslant N,\ s \geqslant 1\}$$
as a special basis $\{c_\alpha\}_{\alpha \in I}$. It is easy to see that if $\mathcal{G}$ is a free group with free generators $g_1, \ldots, g_N$, then the correspondence $g_i \mapsto \widetilde{\exp}(D_i) \bmod J_p(A)$, where $i = 1, \ldots, N$, defines an epimorphism $h: \mathcal{G} \to G_{\mathbb{F}_p}(p-1)$ and $\mathrm{Ker}\, h = \mathcal{G}^p C_p(\mathcal{G})$, where $C_p(\mathcal{G})$ is the subgroup of $\mathcal{G}$, generated by all commutators of order p.

Let $w_1, \ldots, w_N \in K$ be such that their images in $K/(F - \mathrm{id})K$ are linearly independent. If we take
$$e = \widetilde{\exp}\left(\sum_{1 \leqslant i \leqslant N} w_i D_i\right) \bmod J_p(A_K) \in G_K(p-1),$$
then
$$\widehat{e} = 1 + \sum_{\substack{1 \leqslant s < p \\ 1 \leqslant i_1, \ldots, i_s \leqslant N}} \frac{1}{s!} w_{i_1} \ldots w_{i_s} D_{i_1} \ldots D_{i_s}.$$

Let $f \in M_{p-1}(e)$ and
$$\widehat{f} = 1 + \sum_{\substack{1 \leqslant s < p \\ 1 \leqslant i_1, \ldots, i_s \leqslant N}} T_{i_1, \ldots, i_s} D_{i_1} \ldots D_{i_s}.$$

By definition, we set $T_{i_1, \ldots, i_s} = 1$, if $s = 0$.

The elements $T_{i_1, \ldots, i_s}$, where $0 \leqslant s < p$ and $1 \leqslant i_1, \ldots, i_s \leqslant N$, generate $\mathcal{M}_{p-1}(e)$ and the relation $f^{(p)} = fe$ yields the following equations for these elements
$$T_{i_1, \ldots, i_s}^p = T_{i_1 \ldots i_s} + T_{i_1 \ldots i_{s-1}} \frac{w_{i_s}}{1!} + \cdots + T_{i_1} \frac{w_{i_2} \ldots w_{i_s}}{(s-1)!} + \frac{w_{i_1} \ldots w_{i_s}}{s!}.$$

The action of the Lie algebra $\mathcal{L}$ on $\mathcal{M}_{p-1}(e)$ is given by the relations
$$LF(D_i)(T_{i_1 \ldots i_s}) = \delta(i, i_1) T_{i_2 \ldots i_s}$$

for any $1 \leqslant i, i_1, \ldots, i_s \leqslant N$ and $1 \leqslant s < p$, where $\delta(i, i_1)$ is the Kronecker symbol. This yields a faithful action of $\widetilde{\mathcal{L}} = \mathcal{L}/C_p(\mathcal{L})$ on $\mathcal{M}_{p-1}(e)$ (where $C_p(\mathcal{L})$ is the ideal in $\mathcal{L}$ generated by all commutators of order p).

We can identify $\mathrm{Gal}(K_{p-1}(e)/K)$ and $G_{\mathbb{F}_p}(p-1)$ by means of $\pi_{e,f,p-1}$. Then there is an explicit description of the Galois action, which is given on generators $\tau_i = \widetilde{\exp}(D_i) \bmod J_p(A)$, $1 \leqslant i \leqslant n$, by the following relation:

$$\tau_i(T_{i_1\ldots i_s}) = T_{i_1\ldots i_s} + \frac{1}{1!}\delta(i, i_1)T_{i_2\ldots i_s} + \frac{1}{2!}\delta(i, i_1, i_2)T_{i_3\ldots i_s} + \ldots$$
$$+ \frac{1}{s!}\delta(i, i_1, \ldots, i_s),$$

where $\delta(i, i_1, \ldots, i_l)$ is equal to 1 if $i = i_1 = \ldots i_l$ and equal to 0 otherwise.

PROPOSITION. *The system* $\{T_{i_1\ldots i_s} \mid 0 \leqslant s < p, 1 \leqslant i_1, \ldots, i_s \leqslant N\}$ *is linearly independent over* K.

PROOF. Let

$$\sum_{\substack{0 \leqslant s < p \\ 1 \leqslant i_1, \ldots, i_s \leqslant N}} \alpha_{i_1\ldots i_s} T_{i_1\ldots i_s} = 0$$

be any nontrivial linear relation. Let us choose the $\alpha_{i'_1\ldots i'_{s'}} \neq 0$ with the largest s'. Then the relation

$$LF(D_{i'_1}) \ldots LF(D_{i'_{s'}})\left(\sum \alpha_{i_1\ldots i_s} T_{i_1\ldots i_s}\right) = \alpha_{i'_1\ldots i'_{s'}}$$

leads to a contradiction.

2.3. We can give the following profinite version of the previous construction. Let $V \subset K$ be a $\mathbb{F}_p$-subspace such that $V + (F - \mathrm{id})K = K$ and $V \cap (F - \mathrm{id})K = 0$, where $F: K \to K$ is the absolute Frobenius map on K. Let us choose an $\mathbb{F}_p$-basis $\{w_i\}_{i \in I}$ of V.

For any finite subset $R \subset I$ we denote by V_R the subspace of V generated by w_i, $i \in R$. Obviously, $V = \varinjlim V_R$. Let $V_R^* = \mathrm{Hom}(V_R, \mathbb{F}_p)$ be the dual vector space for V_R. Then $V^* = \varprojlim V_R^*$ is the topological vector space over $\mathbb{F}_p$ dual to V. Let us denote by $\{D_i\}_{i \in I}$ the topological $\mathbb{F}_p$-basis of V^* dual to the basis $\{w_i\}_{i \in I}$.

For any finite subset $R \subset I$ we denote by $\mathcal{L}_R$ the free Lie algebra with the system of free generators $\{D_i\}_{i \in R}$. Then $\mathcal{L} = \varprojlim \mathcal{L}_R$ is a profinite free Lie algebra over $\mathbb{F}_p$ with module of free topological generators V^*. Let A_R be the enveloping algebra of $\mathcal{L}_R$, then $A = \varprojlim A_R$ is the (topological) enveloping algebra of $\mathcal{L}$. Assume that all A_R and A are equipped with the structure of f.a.b. (see 1.3.1). We shall use the notation of 1.3 for all constructions related to A. The notation for all similar constructions related to A_R will be supplied with the index R.

Let
$$e_R = \widetilde{\exp}\left(\sum_{i \in R} w_i D_i\right) \bmod J_p(A_{K,R}) \in G_{K,R}(p-1)$$

and $e = \varprojlim_R e_R \in G_K(p-1)$. Choose $f \in \varprojlim_R M_{p-1}(e_R)$ and denote by f_R the projections of f to $M_{p-1}(e_R)$ for any finite subset $R \subset I$. Then we have a system of epimorphisms $\pi_{e,f,p-1,R} \colon \Gamma_K \to G_{\mathbb{F}_p,R}(p-1)$ which gives an epimorphism

$$\pi_f := \varprojlim_R \pi_{e,f,p-1,R} \colon \Gamma_K \to G_{\mathbb{F}_p}(p-1).$$

It is clear that π_f can be factored through the quotient $\Gamma_K \to \Gamma_K(p)$, where $\Gamma_K(p)$ is the Galois group of the maximal p-extension $K(p)$ of K. It is well known, see [Se2, Chapter 2, §2], that $\Gamma(p)$ is a free pro-p-group and from [Se2, Chapter 1, §4] we deduce that π defines an epimorphism $\pi_f(p) \colon \Gamma(p) \to G_{\mathbb{F}_p}(p-1)$ such that $\operatorname{Ker} \pi_f(p) = \Gamma^p(p) C_p(\Gamma(p))$.

Let
$$\widehat{f} = 1 + \sum_{\substack{1 \leqslant s < p \\ i_1, \dots, i_s \in I}} T_{i_1 \dots i_s} D_{i_1} \dots D_{i_s}$$

be the representative of f. Then we have:

2.3a. All the $T_{i_1 \dots i_s}$ are in $K(p)^{\Gamma^p(p) C_p(\Gamma(p))}$.

2.3b. The system $\{T_{i_1 \dots i_s} \mid 0 \leqslant s < p, \; i_1, \dots, i_s \in I\}$ is linearly independent over K.

2.3c. The $\mathbb{F}_p$-module $\mathcal{M}$ generated by all $T_{i_1 \dots i_s}$ is invariant under the Galois action.

2.3d. There is an action $LF \colon \mathcal{L} \to \operatorname{End} \mathcal{M}$ of the profinite Lie algebra $\mathcal{L}$ on $\mathcal{M}$, given by the following relation on free generators D_i, $i \in I$:

$$LF(D_i)(T_{i_1 \dots i_s}) = \delta(i, i_1) T_{i_2 \dots i_s}.$$

Let $T(w_{i_1}, \dots, w_{i_s}) = T_{i_1 \dots i_s}$, where $1 \leqslant s < p$, $i_1, \dots, i_s \in I$. Define $T(v_1, \dots, v_s) \in \mathcal{M}$ for $v_1, \dots, v_s \in V$ by multilinearity: if $v_i = \sum_{j \in J} \alpha_{ij} w_j$, for $i = 1, \dots, s$, where $\alpha_{ij} \in \mathbb{F}_p$ and almost all the α_{ij} are equal to 0, then

$$T(v_1, \dots, v_s) = \sum_{j_1, \dots, j_s} \alpha_{1 j_1} \dots \alpha_{s j_s} T(w_{j_1}, \dots, w_{j_s}).$$

In this notation we have:

2.3e. $\mathcal{M} = \oplus_{0 \leqslant s < p} \mathcal{M}_s$, where $\mathcal{M}_s = \{T(v_1, \dots, v_s) \mid v_1, \dots, v_s \in V\}$ for $1 \leqslant s < p$ and $\mathcal{M}_0 = \mathbb{F}_p$.

2.3f. If $D \in V^*$, then

$$LF(D)(T(v_1, \dots, v_s)) = \langle D, v_1 \rangle T(v_2, \dots, v_s)$$

for any $v_1, \dots, v_s \in V$, $1 \leqslant s < p$ (by definition $T(v_2, \dots, v_s) = 1$ if $s = 1$).

2.3g. $T(v_1, \dots, v_s)$ satisfies the following equation:

$$T(v_1, \dots, v_s)^p = T(v_1, \dots, v_s) + \frac{v_1}{1!} T(v_2, \dots, v_s) + \dots$$
$$+ \frac{v_1 \dots v_{s-1}}{(s-1)!} T(v_s) + \frac{v_1 \dots v_s}{s!}.$$

2.3h. Let $\tau \in \Gamma_K$ and $l_\tau \in \mathcal{L}$ be such that

$$\pi_f(\tau) = \widetilde{\exp}(l_\tau) \bmod J_p(A).$$

Then l_τ is uniquely defined modulo $C_p(\mathcal{L})$ and $\tau|_{\mathcal{M}} = \widetilde{\exp}(LF(l_\tau))$.

§3. Explicit construction of the Galois action (case of a local field)

Let K be a local field of characteristic $p > 0$, complete with respect to a discrete valuation and with residue field $k \simeq \bar{\mathbb{F}}_p$. Then K is isomorphic to the fraction field of the power series ring over k. Fix some uniformizing element of this ring in the form t^{-1}, $t \in K$.

We now present a modification of the previous construction, which will be useful later in the study of the ramification filtration.

3.1. The structural element e° and the constants $\eta(r_1, \ldots, r_s)$. Let $\mathbb{Q}^+(p) = \{ r \in \mathbb{Q} \mid r > 0, (r, p) = 1 \}$.

For any finite subset $R \subset \mathbb{Q}^+(p)$ consider a free Lie $\mathbb{F}_p$-algebra $\mathcal{L}_R^\circ$ having the set of free topological generators $\{ D_r^\circ \mid r \in R \}$. Then $\varprojlim_R \mathcal{L}_R^\circ = \mathcal{L}^\circ$ is a pro-free Lie $\mathbb{F}_p$-algebra with the set of free generators $\{ D_r^\circ \mid r \in \mathbb{Q}^+(p) \}$. If A_R is an f.a.b. related to $\mathcal{L}_R^\circ$, then $A^\circ = \varprojlim_R A_R^\circ$ is an f.a.b. related to $\mathcal{L}^\circ$.

We call an element $e^\circ \in A^\circ \bmod J_p(A^\circ)$ *structural* if
(1) $e^\circ \in G_{A^\circ, \mathbb{F}_p}(p-1)$;
(2) $e^\circ \equiv 1 + \sum_{r \in \mathbb{Q}^+(p)} D_r^\circ \bmod J_2(A^\circ)$.

It is clear that $e^\circ = \varprojlim_R e_R^\circ$, where $e_R^\circ \in G_{A_R^\circ, \mathbb{F}_p}(p-1)$ and

$$e_R^\circ \equiv 1 + \sum_{r \in R} D_r^\circ \bmod J_2(A_R^\circ).$$

As before, we can consider a uniquely-defined representative $E^\circ \in A^\circ$ of e° of the form

$$E^\circ = 1 + \sum_{\substack{1 \leqslant s < p \\ r_1, \ldots, r_s \in \mathbb{Q}^+(p)}} \eta(r_1, \ldots, r_s) D_{r_1}^\circ \ldots D_{r_s}^\circ$$

where $\eta(r_1, \ldots, r_s) \in \mathbb{F}_p$ for any $r_1, \ldots, r_s \in \mathbb{Q}^+(p)$.

These coefficients $\eta(r_1, \ldots, r_s)$ will be called the *structural constants* (related to a structural element e°).

EXAMPLES. (1) If we take

$$E^\circ = \widetilde{\exp}\Big(\sum_{r \in \mathbb{Q}^+(p)} D_r^\circ \Big),$$

then $e^\circ = E^\circ \bmod J_p(A^\circ)$ is a structural element and its structural constants satisfy

$$\eta(r_1, \ldots, r_s) = \frac{1}{s!}.$$

(2) If we take

$$E^\circ = \prod_{r \in \mathbb{Q}^+(p)} \widetilde{\exp}(D_r^\circ)$$

with respect to the natural ordering in $\mathbb{Q}^+(p)$, then $e^\circ = E^\circ \bmod J_p(A^\circ)$ is structural and its structural constants are given by the following relations:

$$\eta(r_1, \ldots, r_s) = \frac{1}{s_1!(s_2 - s_1)! \ldots (s_l - s_{l-1})!},$$

if $r_1 = \cdots = r_{s_1} < r_{s_1+1} = \cdots = r_{s_2} < \cdots < r_{s_{l-1}+1} = \cdots = r_{s_l}$, where $1 \leqslant s_1 < s_2 < \cdots < s_l = s$, and $\eta(r_1, \ldots, r_s) = 0$ otherwise, i.e., if $r_1 \leqslant r_2 \leqslant \ldots \leqslant r_s$ is not true.

PROPOSITION. *A collection of constants* $\eta(r_1, \ldots, r_s) \in \mathbb{F}_p$, *where* $r_1, \ldots, r_s \in \mathbb{Q}^+(p)$, $1 \leqslant s < p$, *is a collection of structural constants if and only if*:

(1) *for any* $r_1 \in \mathbb{Q}^+(p)$, $\eta(r_1) = 1$;
(2) *for any* $1 \leqslant s_1 < s < p$ *and* $r_1, \ldots, r_s \in \mathbb{Q}^+(p)$

$$\eta(r_1, \ldots r_{s_1})\eta(r_{s_1+1}, \ldots, r_s) = \sum_{\sigma \in P_{s_1,s}} \eta(r_{\sigma^{-1}(1)}, \ldots, r_{\sigma^{-1}(s)}),$$

where $P_{s_1,s}$ *is the subset of permutations of order* s *such that* $\sigma(i) < \sigma(j)$, *where either* $1 \leqslant i < j \leqslant s_1$ *or* $s_1 + 1 \leqslant i < j \leqslant s$.

PROOF. This follows easily from the fact that in the coalgebra A° we have:

$$\Delta(D_{r_1} \ldots D_{r_{s_2}}) = \sum_{\substack{0 \leqslant s_1 \leqslant s \\ \sigma \in P_{s_1,s}}} D_{r_{\sigma(1)}} \ldots D_{r_{\sigma(s_1)}} \otimes D_{r_{\sigma(s_1+1)}} \ldots D_{r_{\sigma(s)}}.$$

We assume until the end of this paper that the choice of some structural element e° *is fixed and denote its structural constants by* $\eta(r_1, \ldots, r_s)$, *where* $1 \leqslant s < p$, $r_1, \ldots, r_s \in \mathbb{Q}^+(p)$.

3.2.. Let K_{sep} be any fixed separable closure of K, $\Gamma = \Gamma_K = \text{Gal}(K_{\text{sep}}/K)$. For any natural number N we consider the extension $K_N = K(t_N) \subset K_{\text{sep}}$, where $t_N^{p^N-1} = t$. The system of these fields K_N is an inductive system of subfields in K_{sep} and $\varinjlim_N K_N = K_{\text{tr}}$ is the maximal tamely ramified extension of K. Now K_{sep} can be regarded as a maximal p-extension of K_{tr}. Its Galois group $I = \text{Gal}(K_{\text{sep}}/K_{\text{tr}})$ is said to be the *subgroup of higher ramification* of Γ and, as we mentioned earlier, is a free pro-p-group.

In order to apply the construction of 2.3, we can assume that the elements $t_N \in K_{\text{tr}}$, $N \geqslant 1$, satisfy the following condition: for any natural numbers N_1, N_2 such that $N_2 \mid N_1$ we have $t_{N_2} = t_{N_1}^{1+p^{N_2}+\cdots+p^{(l-1)N_2}}$, where $N_1 = lN_2$.

Let $\mathbb{Q}^+(p)$ be the set defined in 3.1. Obviously, every $r \in \mathbb{Q}^+(p)$ can be written in the form $r = m/p^N - 1$ for some natural numbers m, N, where $(m, p) = 1$. We use this fact to define $t^r := t_N^m$ for $r \in \mathbb{Q}^+(p)$. It is easy to see that this definition does not depend on the above choice of m and N.

Now consider the vector space

$$V = \bigoplus_{r \in \mathbb{Q}^+(p)} kt^r \subset K_{\text{tr}}$$

over $\mathbb{F}_p$; then $V + (F - \text{id})K_{\text{tr}} = K_{\text{tr}}$ and $V \cap (F - \text{id})K_{\text{tr}} = 0$, where $F: K \to K$ is the absolute Frobenius endomorphism of K. Let $\{w_i\}_{i \in I}$ be some basis of k over $\mathbb{F}_p$, then

$$\{w_i t^r \mid i \in I, \ r \in \mathbb{Q}^+(p)\}$$

is an $\mathbb{F}_p$-basis of V. As earlier, we consider the dual vector space

$$V^* = \mathrm{Hom}(V, \mathbb{F}_p) = \prod_{r \in \mathbb{Q}^+(p)} \mathrm{Hom}(k, \mathbb{F}_p)_r$$

for V and the profinite free Lie algebra $\mathcal{L}$ with the $\mathbb{F}_p$-module of free generators V^*. Let

$$E^\circ = \sum \eta(r_1, \ldots, r_s) D^\circ_{r_1} \ldots D^\circ_{r_s}$$

be a representative of a fixed structural element e° (see 3.1). We write $E^\circ = E^\circ(\{D^\circ_r\}_{r \in \mathbb{Q}^+(p)})$ if we want to regard E° as a function of the variables D°_r, $r \in \mathbb{Q}^+(p)$.

Consider the element

$$E = E^\circ\left(\left\{\sum_{i \in I} w_i t^r D_{i,r}\right\}_{r \in \mathbb{Q}^+(p)}\right)$$

of $A_{K_{\mathrm{tr}}} = A \otimes K_{\mathrm{tr}}$, where A is the f.a.b. related to $\mathcal{L}$, and $\{D_{i,r} \mid i \in I, r \in \mathbb{Q}^+(p)\}$ is the basis of V^* dual to the basis $\{w_i t^r \mid i \in I, r \in \mathbb{Q}^+(p)\}$ of V.

It is clear that E does not depend on the choice of basis $\{w_i \mid i \in I\}$ of k over $\mathbb{F}_p$,

$$e = E \bmod J_p(A_{K_{\mathrm{tr}}}) \in G_{K_{\mathrm{tr}}}(p-1),$$

$$E \equiv 1 + \sum_{\substack{1 \leqslant s < p \\ i_1, \ldots, i_s \in I \\ r_1, \ldots, r_s \in \mathbb{Q}^+(p)}} \eta(r_1, \ldots, r_s) w_{i_1} \ldots w_{i_s} t^{r_1 + \cdots + r_s} D_{i_1 r_1} \ldots D_{i_s r_s}.$$

3.3. Let $\widetilde{K} = K_{\mathrm{sep}}^{I^p C_p(I)}$. As earlier we have:

3.3a. *The set*

$$\{T_{i_1 r_1 \ldots i_s r_s} \mid i_1, \ldots, i_s \in I, \ r_1, \ldots, r_s \in \mathbb{Q}^+(p), \ 0 \leqslant s < p\}$$

(as before, $T_{i_1 r_1 \ldots i_s r_s} = 1$ if $s = 0$) generates an I-invariant $\mathbb{F}_p$-submodule $\mathcal{M}$ in $\widetilde{K}$. This set is linearly independent over K_{tr} (therefore, this set is $\mathbb{F}_p$-basis of $\mathcal{M}$).

3.3b. *The elements $T_{i_1 r_1 \ldots i_s r_s}$, where $i_1, \ldots, i_s \in I$, $r_1, \ldots, r_s \in \mathbb{Q}^+(p)$, and $1 \leqslant s < p$, satisfy the relations*

$$T^p_{i_1 r_1 \ldots i_s r_s} = T_{i_1 r_1 \ldots i_s r_s} + T_{i_1 r_1 \ldots i_{s-1} r_{s-1}} w_{i_s} t^{r_s} \eta(r_s) + \ldots$$
$$+ T_{i_1 r_1} w_{i_2} \ldots w_{i_s} t^{r_2 + \cdots + r_s} \eta(r_2, \ldots, r_s) + w_{i_1} \ldots w_{i_s} t^{r_1 + \cdots + r_s} \eta(r_1, \ldots, r_s).$$

3.3c. *The Lie algebra $\mathcal{L}$ acts on $\mathcal{M}$ and this action $LF: \mathcal{L} \to \mathrm{End}\,\mathcal{M}$ is given on its generators $D_{i,r}$, $i \in I$, $r \in \mathbb{Q}^+(p)$, by the relation*

$$L(D_{i,r})(T_{i_1 r_1 \ldots i_s r_s}) = \delta(i, i_1)\delta(r, r_1) T_{i_2 r_2 \ldots i_s r_s}.$$

3.3d. *For any $\tau \in I$ there exists an $l_\tau \in \mathcal{L}$, uniquely defined modulo $C_p(\mathcal{L})$, such that $\tau|_\mathcal{M} = \widetilde{\exp} LF(l_\tau)$.*

As earlier, we define

$$T_{i_1 r_1 \ldots i_s r_s} = T(w_{i_1}, r_1, \ldots, w_{i_s}, r_s)$$

for all $i_1, \ldots, i_s \in I$, $r_1, \ldots, r_s \in \mathbb{Q}^+(p)$, $1 \leqslant s < p$ and elements $T(\alpha_1, r_1, \ldots, \alpha_s, r_s) \in \mathcal{M}$ for any $\alpha_1, \ldots, \alpha_s \in k$ by multilinearity.

For any $\alpha_1, \ldots, \alpha_s \in k$, $r_1, \ldots, r_s \in \mathbb{Q}^+(p)$ we have

3.3e.

$$T(\alpha_1, r_1, \ldots, \alpha_s, r_s)^p = T(\alpha_1, r_1, \ldots, \alpha_s, r_s)$$
$$+ T(\alpha_1, r_1, \ldots, \alpha_{s-1}, r_{s-1})\alpha_s t^{r_s}\eta(r_s) + \cdots + \alpha_1 \ldots \alpha_s t^{r_1+\cdots+r_s}\eta(r_1, \ldots, r_s).$$

3.3f. If

$$D = (D(r))_{r \in \mathbb{Q}^+(p)} \in V^* = \prod_{r \in \mathbb{Q}^+(p)} \operatorname{Hom}(k, \mathbb{F}_p)_r$$

then

$$LF(D)(T(\alpha_1, r_1, \ldots, \alpha_s, r_s)) = \langle D(r_1), \alpha_1 \rangle T(\alpha_2, r_2, \ldots, \alpha_s, r_s).$$

REMARK. The element $f \in \operatorname{Hom}(k, \mathbb{F}_p)_r$ in the above decomposition

$$V^* = \prod_{r \in \mathbb{Q}^+(p)} \operatorname{Hom}(k, \mathbb{F}_p)_r$$

will be denoted below by $D_{r,f}$.

REMARK. We obtain a similar description of the maximal p-extension of K (modulo pth commutators) if we replace K_{tr} by K and $\mathbb{Q}^+(p)$ by $\mathbb{Z}^+(p) = \{n \in \mathbb{N} \mid (n, p) = 1\}$ everywhere.

§4. The "ramification" filtration of the Lie algebra $\mathcal{L}$

Let $\mathcal{L}$ be the profinite free Lie algebra over $\mathbb{F}_p$ defined in 3.2 and $\widetilde{\mathcal{L}} = \mathcal{L}/C_p(\mathcal{L})$, where $C_p(\mathcal{L})$ is the ideal of $\mathcal{L}$ generated by all commutators of order p. In this section we define a decreasing filtration $\{\widetilde{\mathcal{L}}^{(v)}\}_{v > 0}$ of $\widetilde{\mathcal{L}}$ by its ideals $\widetilde{\mathcal{L}}^{(v)}$, where $v \in \mathbb{Q}$, $v > 0$. This filtration will be related to the ramification filtration of the $\operatorname{Gal}(K_{\mathrm{sep}}/K)$ in §7 below. We use the notation of §3.

4.1. Let $V = \bigoplus_{r \in \mathbb{Q}^+(p)} kt^r \subset K_{\mathrm{tr}}$ be the vector space over $\mathbb{F}_p$ from 3.2. For any finite subset R in $\mathbb{Q}^+(p)$ and natural number N, introduce the vector space $V_{R,N} = \bigoplus_{r \in R} \mathbb{F}_q t^r$ over $\mathbb{F}_p$, where $q = p^N$. Obviously, each $V_{R,N}$ can be identified with a subspace in V and $V = \varinjlim_{R,N} V_{R,N}$. Let $\mathcal{L}_{R,N}$ be a free Lie algebra over $\mathbb{F}_p$ with an $\mathbb{F}_p$-module of free generators $V_{R,N}^* = \operatorname{Hom}(V_{R,N}, \mathbb{F}_p)$. Then $\{\mathcal{L}_{R,N}\}_{R,N}$ is a projective system and $\varprojlim_{R,N} \mathcal{L}_{R,N} = \mathcal{L}$, where $\mathcal{L}$ is the free profinite Lie algebra over $\mathbb{F}_p$ from 3.2. We set also $\widetilde{\mathcal{L}}_{R,N} = \mathcal{L}_{R,N}/C_p(\mathcal{L}_{R,N})$. It is clear that $\varprojlim \widetilde{\mathcal{L}}_{R,N} = \widetilde{\mathcal{L}}$.

4.2. The elements $D_\lambda(r_1, 0, r_2, m_2, \ldots, r_s, m_s)$ **of** $\mathcal{L}_{R,N}$. Let $N \geqslant 1$, $q = p^N$. Then

$$\text{Hom}(\mathbb{F}_q, \mathbb{F}_p) \subset \text{Hom}(\mathbb{F}_q \otimes \mathbb{F}_q, \mathbb{F}_q) = \bigoplus_{n \bmod N} \text{Hom}_n(\mathbb{F}_q, \mathbb{F}_q)$$

where $\text{Hom}_n(\mathbb{F}_q, \mathbb{F}_q)$ consists of all additive morphisms $\varphi \colon \mathbb{F}_q \to \mathbb{F}_q$ such that $\varphi(\alpha) = \alpha^{p^n} \varphi(1)$ for any $\alpha \in \mathbb{F}_q$. Now any $f \in \text{Hom}(\mathbb{F}_q, \mathbb{F}_p)$ can be identified with the sum $\sum_{n \bmod N} f_n$, where all $f_n \in \text{Hom}_n(\mathbb{F}_q, \mathbb{F}_q)$ and the conjugacy condition $f_{n+1} = f_n^p$ holds for all $n \bmod N$. We note that for any $f \in \text{Hom}(\mathbb{F}_q, \mathbb{F}_p)$ there exists a unique $\beta_f \in \mathbb{F}_q$ such that $f(\alpha) = Tr_{\mathbb{F}_q/\mathbb{F}_p}(\alpha \beta_f)$ for any $\alpha \in \mathbb{F}_q$. It is easy to see that in the above decomposition $f = \sum_n f_n$, one has $f_n(1) = \beta_f^{p^n}$.

In the same way we can consider tensors $F \in \text{Hom}(\mathbb{F}_q^{\otimes s}, \mathbb{F}_p)$, where s is any natural number. Such an F may be identified with the sum

$$\sum_{\text{all } n_i \bmod N} F_{n_1 \ldots n_s}, \quad \text{where } F_{n_1 \ldots n_s} \in \text{Hom}_{n_1, \ldots, n_s}(\mathbb{F}_q^{\otimes s}, \mathbb{F}_q)$$

and $\text{Hom}_{n_1 \ldots n_s}(\mathbb{F}_q^{\otimes s}, \mathbb{F}_q)$ is the group of all multilinear mappings such that

$$F_{n_1 \ldots n_s}(\alpha_1, \ldots, \alpha_s) = \alpha_1^{p^{n_1}} \ldots \alpha_s^{p^{n_s}} F_{n_1 \ldots n_s}(1, \ldots, 1)$$

for all $\alpha_i \in \mathbb{F}_q$ and the conjugation conditions $F_{n_1 \ldots n_s}^p = F_{n_1+1, \ldots, n_s+1}$ hold. If $F = f_1 \otimes \cdots \otimes f_s$ for $f_i \in \text{Hom}(\mathbb{F}_q, \mathbb{F}_p)$, $1 \leqslant i \leqslant s$, then

$$F_{n_1 \ldots n_s}(1, \ldots, 1) = \beta_{f_1}^{p^{n_1}} \ldots \beta_{f_s}^{p^{n_s}}.$$

Let $\lambda \in \mathbb{F}_q$ and $m_2^\circ, \ldots, m_s^\circ$ be any integers such that $0 \leqslant m_2^\circ, \ldots, m_s^\circ < N$. We shall use the same notation for their residues $\bmod N$. Introduce the tensor

$$F_\lambda(m_2^\circ, \ldots, m_s^\circ) \in \text{Hom}(\mathbb{F}_q^{\otimes s}, \mathbb{F}_p)$$

defined by the following conditions:

$$F_\lambda(m_2^\circ, \ldots, m_s^\circ)_{0, m_2^\circ, \ldots, m_s^\circ}(1, \ldots, 1) = \lambda, \quad F_\lambda(m_2^\circ, \ldots, m_s^\circ)_{0, m_2, \ldots, m_s} = 0$$

for any residues $m_2, \ldots, m_s \bmod N$ such that $(m_2, \ldots, m_s) \neq (m_2^\circ, \ldots, m_s^\circ)$.

The above tensor can be expressed as a sum of elementary tensors

$$F_\lambda(m_2^\circ, \ldots, m_s^\circ) = \sum_i f_{1i} \otimes \cdots \otimes f_{si}$$

where $f_{li} \in \text{Hom}(\mathbb{F}_q, \mathbb{F}_p)$. If R is some finite subset of $\mathbb{Q}^+(p)$, $r_1, \ldots, r_s \in R$, we use the above expression to define the element $D_\lambda(r_1, 0, r_2, m_2^\circ, \ldots, r_s, m_s^\circ)$ of $\mathcal{L}_{R,N}$ in the following way:

$$D_\lambda(r_1, 0, r_2, m_2^\circ, \ldots, r_s, m_s^\circ) = \sum_i [\ldots [D_{r_1, f_{1i}}, D_{r_2, f_{2i}}], \ldots, D_{r_s, f_{si}}].$$

It is clear that this element does not depend on the expression of $F_\lambda(m_2^\circ, \ldots, m_s^\circ)$ as a sum of elementary tensors chosen above.

4.3. The constants $\widehat{\eta}(r_1, n_1, \ldots, r_s, n_s)$. Recall that for a natural number $s < p$ and $r_1, \ldots, r_s \in \mathbb{Q}^+(p)$, we have the structural constants $\eta(r_1, \ldots, r_s) \in \mathbb{F}_p$ (see 3.1). We set

$$\widehat{\eta}(r_1, \ldots, r_s) := \eta(r_s, \ldots, r_1).$$

Consider the collection $(r_1, m_1, r_2, m_2, \ldots, r_s, m_s)$, where $s < p$, $r_1, \ldots, r_s \in \mathbb{Q}^+(p)$ and $m_1, \ldots, m_s$ are nonnegative integers and set

$$\widehat{\eta}(r_1, m_1, r_2, m_2, \ldots, r_s, m_s)$$
$$= \widehat{\eta}(r_1, \ldots, r_{s_1})\widehat{\eta}(r_{s_1+1}, \ldots, r_{s_2})\ldots\widehat{\eta}(r_{s_{l-1}}+1, \ldots, r_{s_l})$$

if $m_1 = \cdots = m_{s_1} < m_{s_1+1} = \cdots = m_{s_2} < \cdots < m_{s_{l-1}} = \cdots = m_{s_l}$ for $1 \leqslant s_1 < \cdots < s_l = s$, and

$$\widehat{\eta}(r_1, m_1, r_2, m_2, \ldots, r_s, m_s) = 0$$

otherwise, i.e., if $m_1 \leqslant m_2 \leqslant \ldots \leqslant m_s$ is not true.

4.4. Definition of the filtration $\{\widetilde{\mathcal{L}}^{(v)}\}_{v>0}$. Let $R \subset \mathbb{Q}^+(p)$ be a finite subset, $N \geqslant 1$, $q = p^N$. For any $\gamma_0 \in \mathbb{Q}$, $\gamma_0 > 0$, $\lambda \in \mathbb{F}_q$ define an element $\mathcal{F}_{R,N}(\gamma_0, \lambda) \in \mathcal{L}_{R,N}$ by putting:

$$\mathcal{F}_{R,N}(\gamma_0, \lambda) = \sum (-1)^{s+1} r_1 \widehat{\eta}(r_1, 0, r_2, m_2, \ldots, r_s, m_s)$$
$$\times D_\lambda(r_1, 0, r_2, m_2, \ldots, r_s, m_s),$$

where the sum is taken over

$$1 \leqslant s < p, \quad r_1, \ldots, r_s \in R, \quad 0 \leqslant m_2, \ldots, m_s < N,$$
$$r_1 + \frac{r_2}{p^{m_2}} + \cdots + \frac{r_s}{p^{m_s}} = \gamma_0.$$

It is clear from the definition of the constants $\widehat{\eta}(r_1, 0, r_2, m_2, \ldots, r_s, m_s)$ that among all possible presentations of γ_0 in the form

$$\gamma_0 = r_1 + \frac{r_2}{p^{m_2}} + \cdots + \frac{r_s}{p^{m_s}}$$

only the ordered ones are important.

Let $v_0 \in \mathbb{Q}$, $v_0 > 0$. Define the ideals $\widetilde{\mathcal{L}}_{R,N}^{(v_0)}$ of the Lie algebra $\widetilde{\mathcal{L}}_{R,N}$ as the ideals generated by all $\mathcal{F}_{R,N}(\gamma_0, \lambda) \bmod C_p(\mathcal{L}_{R,N})$, where $\gamma_0 \geqslant v_0$ and $\lambda \in \mathbb{F}_q$, $q = p^N$.

It is clear that these ideals determine a decreasing filtration in $\widetilde{\mathcal{L}}_{R,N}$. We want to use them to define the "ramification" filtration of the Lie algebra $\widetilde{\mathcal{L}} = \varprojlim \widetilde{\mathcal{L}}_{R,N}$. But a priori it is not clear that for any fixed $v_0 \in \mathbb{Q}$ a system of ideals $\{\widetilde{\mathcal{L}}_{R,N}^{(v_0)}\}$ can be included in a projective system $\{\widetilde{\mathcal{L}}_{R,N}\}$. The following proposition ensures this property.

PROPOSITION. *For any finite subset $R \subset \mathbb{Q}^+(p)$ and $v_0 \in \mathbb{Q}$, $v_0 > 0$, there exists a natural number $N_0(R, v_0)$ such that the connecting morphisms $\widetilde{\mathcal{L}}_{R,N_1} \to \widetilde{\mathcal{L}}_{R,N_2}$, $N_2 \mid N_1$ of the projective system $\{\widetilde{\mathcal{L}}_{R,N}\}$ induce the following epimorphisms for $N_2 \geqslant N_0(R, v_0)$:*

$$\widetilde{\mathcal{L}}_{R,N_1}^{(v_0)} \to \widetilde{\mathcal{L}}_{R,N_2}^{(v_0)}.$$

The proof of this proposition will be given in §5 below.

We use this proposition in order to set, for any $v_0 \in \mathbb{Q}$, $v_0 > 0$,

$$\widetilde{\mathcal{L}}^{(v_0)} = \varprojlim_{R,N} \widetilde{\mathcal{L}}_{R,N}^{(v_0)}.$$

§5. Proof of Proposition 4.4

Let R be some finite set in $\mathbb{Q}^+(p)$, $N \geqslant 1$, $k = \overline{\mathbb{F}_p}$. It is clear that it is sufficient to prove the proposition for ideals $\widetilde{\mathcal{L}}_{R,N}^{(v_0)} \otimes k$ in a projective system $\{\widetilde{\mathcal{L}}_{R,N} \otimes k\}_{R,N}$ of Lie algebras over k.

5.1. Let $q = p^N$.

LEMMA. *There exist two $\mathbb{F}_p$-bases $\{\alpha_i\}_{1 \leqslant i \leqslant N}$ and $\{\beta_i\}_{1 \leqslant i \leqslant N}$ of $\mathbb{F}_q$ such that for any natural number n we have*

$$\sum_{1 \leqslant i \leqslant N} \beta_i^{p^n} \alpha_i = \delta(n, 0),$$

where $\delta(n, 0) = 1$ if $n \equiv 0 \mod N$, and $\delta(n, 0) = 0$ otherwise.

PROOF. Choose $\alpha_0 \in \mathbb{F}_q$ so that the elements of $\{\alpha_0^{p^i}\}_{0 \leqslant i < N}$ give a (normal) basis of $\mathbb{F}_q$ over $\mathbb{F}_p$. It is easy to see that the basis $\{\alpha_i\}_{1 \leqslant i \leqslant N}$, where $\alpha_i = \alpha_0^{p^i}$, $1 \leqslant i \leqslant N$, and its dual basis $\{\beta_i\}_{1 \leqslant i \leqslant N}$ satisfy the conclusion of our lemma.

Let $\{\alpha_i\}_{1 \leqslant i \leqslant N}$ and $\{\beta_i\}_{1 \leqslant i \leqslant N}$ be some bases from the above lemma. Construct a basis $\{f_i\}_{1 \leqslant i \leqslant N}$ of $\mathrm{Hom}(\mathbb{F}_q, \mathbb{F}_p)$ by taking $f_i \in \mathrm{Hom}(\mathbb{F}_q, \mathbb{F}_p)$ so that $f_i(\alpha) = \mathrm{Tr}_{\mathbb{F}_q/\mathbb{F}_p}(\alpha \beta_i)$ for all $\alpha \in \mathbb{F}_q$. Then for any $r \in R$ and $0 \leqslant n < N$ we define the elements

$$D_{r,n} = \sum \alpha_i^{p^n} D_{r,f_i} \in \mathcal{L}_{R,N} \otimes k.$$

It is clear that the family $\{D_{r,n}\}_{r \in R, \, 0 \leqslant n < N}$ can be taken as a system of free generators of the Lie algebra $\mathcal{L}_{R,N} \otimes k$ over k. One can check that these elements do not depend on the above choice of bases $\{\alpha_i\}_{1 \leqslant i \leqslant N}$ and $\{\beta_i\}_{1 \leqslant i \leqslant N}$.

Now the tensors $F_\lambda(m_2^\circ, \ldots, m_s^\circ)$ from 4.2 can be written in the following form

$$F_\lambda(m_2^\circ, \ldots, m_s^\circ) = \sum_{\substack{1 \leqslant i_1, \ldots, i_s \leqslant N \\ 0 \leqslant n < N}} \lambda^{p^n} (\alpha_{i_1} \alpha_{i_2}^{p^{m_2^\circ}} \ldots \alpha_{i_s}^{p^{m_s^\circ}})^{p^n} f_{i_1} \otimes \cdots \otimes f_{i_s}.$$

Therefore,

$$D_\lambda(r_1, 0, r_2, m_2^\circ, \ldots, r_s, m_s^\circ) = \sum_{0 \leqslant n < N} \lambda^{p^n} [\ldots [D_{r_1, n}, D_{r_2, \widetilde{n+m_2^\circ}}], \ldots, D_{r_s, \widetilde{n+m_s^\circ}}],$$

where $\widetilde{n+m_j^\circ}$ are residues of $n+m_j^\circ$ from $[0, N)$. For $\gamma_0 \in \mathbb{Q}$, $\gamma_0 > 0$ and $0 \leq n < N$ introduce the elements of $\mathcal{L}_{R,N} \otimes k$:

$$\mathcal{F}_{R,N}(\gamma_0, n) = \sum (-1)^{s+1}$$
$$\times \{r_1 \widehat{\eta}(r_1, 0, r_2, m_2, \ldots, r_s, m_s)[\ldots[D_{r_1,n}, D_{r_2,\widetilde{n+m_2}}], \ldots, D_{r_s,\widetilde{n+m_s}}]\},$$

where the sum is taken over

$$1 \leq s < p, \quad r_1, \ldots, r_s \in R, \quad 0 \leq m_2, \ldots, m_s < N,$$
$$r_1 + \frac{r_2}{p^{m_2}} + \cdots + \frac{r_s}{p^{m_s}} = \gamma_0.$$

It follows from

$$\mathcal{F}_{R,N}(\gamma_0, \lambda) = \sum_{0 \leq n < N} \lambda^{p^n} \mathcal{F}_{R,N}(\gamma_0, n), \quad \lambda \in \mathbb{F}_q,$$

that the ideal $\widetilde{\mathcal{L}}_{R,N}^{(v_0)} \otimes k$ is generated by $\mathcal{F}_{R,N}(\gamma_0, n) \bmod C_p(\mathcal{L}_{R,N} \otimes k)$ for all $\gamma_0 \geq v_0$ and $0 \leq n < N$.

5.2. In order to write the generators $\mathcal{F}_{R,N}(\gamma_0, n)$ in a more symmetric form, we shall change some notation.

For every integer n such that $0 \leq n < N$ we now use the same symbol when it is regarded as its residue modulo N. For any collection $n_1, \ldots, n_s$ of integers from $[0, N)$, we define integers n_{ij}, where $1 \leq i, j \leq s$, by the conditions: $n_{ij} \equiv n_i - n_j \bmod N$, $n_{ij} \in [0, N)$.

From 4.3 we shall also use another notation for the constants $\widehat{\eta}(r_1, m_1, r_2, m_2, \ldots, r_s, m_s)$. For every collection $(r_1, n_1, \ldots, r_s, n_s)$, where $r_1, \ldots, r_s \in R$ and all the n_i are residues modulo N, we set

$$\widetilde{\eta}(r_1, n_1, \ldots, r_s, n_s) = \widehat{\eta}(r_1, n_{11}, r_2, n_{12}, \ldots, r_s, n_{1s}).$$

REMARK. These constants $\widetilde{\eta}(r_1, n_1, \ldots, r_s, n_s)$ reflect the idea of "circular" ordering of residues $n_i \bmod N$ regarded as lying on a unit circle via the map:

$$n \bmod N \mapsto \exp(2\pi i n/N) \in \{z \in \mathbb{C} \mid |z| = 1\}.$$

Now the generators $\mathcal{F}_{R,N}(\gamma_0, n_1)$ can be written in the following form:

$$\mathcal{F}_{R,N}(\gamma_0, n_1)$$
$$= \sum (-1)^{s+1} r_1 \widetilde{\eta}(r_1, n_1, r_2, n_2, \ldots, r_s, n_s)[\ldots[D_{r_1,n_1}, D_{r_2,n_2}], \ldots, D_{r_s,n_s}],$$

where the sum is taken over

$$1 \leq s < p, \quad r_1, \ldots, r_s \in R, \quad 0 \leq n_2, \ldots, n_s < N,$$
$$r_1 + \frac{r_2}{p^{n_{12}}} + \cdots + \frac{r_s}{p^{n_{1s}}} = \gamma_0.$$

5.3. We want to investigate the representation of any given $\gamma \in \mathbb{Q}$ in the form
$$\gamma = r_1 + \frac{r_2}{p^{m_2}} + \cdots + \frac{r_s}{p^{m_s}},$$
where $r_1, \ldots, r_s \in \mathbb{Q}^+(p)$, $m_2, \ldots, m_s \in \mathbb{N} \cup \{0\}$.

5.3.1. As usual, R is a finite subset of $\mathbb{Q}^+(p)$. For any rational $\gamma > 0$ and integer $s \geq 0$, consider the set

$$M_{\gamma,s}(R) = \Big\{ (r_1, \ldots, r_s; m_2, \ldots, m_s) \in R^s \times \mathbb{Z}^{s-1}$$
$$| 0 \leq m_2 \leq \ldots \leq m_s, \ \gamma = r_1 + \frac{r_2}{p^{m_2}} + \cdots + \frac{r_s}{p^{m_s}} \Big\}.$$

The elements of $M_{\gamma,s}(R)$ will be called *decompositions* of γ.

LEMMA. $M_{\gamma,s}(R)$ *is finite*.

PROOF. We use induction on s. For $s = 1$ this is evident. Let $s > 1$ and let the subset $M_{\gamma,s}(r_1, m_2) \subset M_{\gamma,s} = M_{\gamma,s}(R)$ consist of decompositions $(r_1, \ldots, r_s; m_2, \ldots, m_s)$ with fixed values of r_1 and m_2. The mapping

$$(r_1, \ldots, r_s; m_2, \ldots, m_s) \mapsto (r_2, \ldots, r_s; m_3 - m_2, \ldots, m_s - m_2)$$

defines a one-to-one correspondence

$$M_{\gamma,s}(r_1, m_2) \to M_{(\gamma-r_1)p^{m_2}, s-1}.$$

Since R is finite, there exists a natural number N_0 such that for every $m_2 > N_0$, $M_{(\gamma-r_1)p^{m_2}, s-1} = \varnothing$. Therefore,

$$M_{\gamma,s} = \bigcup_{\substack{r_1 \in R \\ m_2 \leq N_0}} M_{\gamma,s}(r_1, m_2)$$

is a finite union of finite sets. The lemma is proved.

It follows now that the set $M_\gamma(R) = \cup_{1 \leq s < p} M_{\gamma,s}(r)$ of all representations of γ in the form $r_1 + r_2/p^{m_2} + \cdots + r_s/p^{m_s}$, where $s < p$, $r_1, \ldots, r_s \in R$ and $0 \leq m_2 \leq \ldots \leq m_s$ is finite.

5.3.2. Fix a rational number $v_0 > 0$ and a finite set $R \subset \mathbb{Q}^+(p)$. Let $\gamma \in \mathbb{Q}$, $\gamma > 0$.

DEFINITION. $N(R, \gamma) = \max\{ m_s \mid (r_1, \ldots, r_s; m_2, \ldots, m_s) \in M_\gamma(R) \}$.

DEFINITION. A decomposition $(r_1, \ldots, r_s; m_2, \ldots, m_s) \in M_\gamma(R)$ is called (v_0, R)-*bad* if $\gamma \geq v_0$ and for all $1 \leq t \leq s$ and all numbers

$$\gamma'_t = r_1 + \frac{r_2}{p^{m_2}} + \cdots + \frac{r_{s-t}}{p^{m_{s-t}}}$$

the following implication is true:

if $\gamma'_t \geqslant v_0$, then $N(R, \gamma'_t) \geqslant m_{s-t+1}$, i.e., there exists a decomposition

$$(r'_1, \ldots, r'_{s'}; n_2, \ldots, n_{s'}) \in M_{\gamma'_t}(r)$$

such that $n'_{s'} \geqslant m_{s-t+1}$ (by definition $m_1 = 0$).

The following properties are immediate consequences of this definition:
(a) any decomposition $\gamma = r_1$, where $r_1 \in R$, $r_1 \geqslant v_0$, is (v_0, R)-bad;
(b) if $(r_1, \ldots, r_s; m_2, \ldots, m_s) \in M_\gamma(R)$ and $\gamma - r_s/p^{n_s} < v_0$, then this decomposition of γ is (v_0, R)-bad;
(c) if $(r_1, \ldots, r_s; m_2, \ldots, m_s) \in M_\gamma(R)$ is (v_0, R)-bad and $\gamma'_1 = \gamma - r_s/p^{m_s} \geqslant v_0$, then $(r_1, \ldots, r_{s-1}; m_2, \ldots, m_{s-1}) \in M_{\gamma'_1}(R)$ is also (v_0, R)-bad;
(d) from (b) and (c) it follows that if $\gamma \geqslant v_0$ and $(r_1, \ldots, r_s; m_2, \ldots, m_s) \in M_\gamma(R)$ is not (v_0, R)-bad, then there exists a unique index $s_1 < s$ such that the decompositions

$$(r_1, \ldots, r_{s-1}; m_2, \ldots, m_{s-1}) \in M_{\gamma'_1}(R), \quad \ldots,$$
$$(r_1, \ldots, r_{s_1+1}; m_2, \ldots, m_{s_1+1}) \in M_{\gamma'_{s-s_1+1}}(R)$$

are not (v_0, R)-bad and $(r_1, \ldots, r_{s_1}; m_2, \ldots, m_{s_1}) \in M_{\gamma'_{s-s_1}}(R)$ is (v_0, R)-bad. So, $\gamma'_{s-s_1} \geqslant v_0$ and $N(R, \gamma'_{s-s_1}) < m_{s_1+1} \leqslant \ldots \leqslant m_s$.

DEFINITION. A rational number γ is called (v_0, R)-*bad* if there exists a (v_0, R)-bad decomposition $(r_1, \ldots, r_s; m_2, \ldots, m_s) \in M_\gamma(R)$.

DEFINITION. For any natural number N, we set

$$M_\gamma(R, N) = \{(r_1, \ldots, r_s; m_2, \ldots, m_s) \in M_\gamma(R) \mid m_s < N\}.$$

From (d) we can easily obtain:
(e) For any given rational number γ_0 and natural number N there exists a finite set J, (v_0, R)-bad numbers $\gamma^{(\alpha)}$, and collections

$$\bar{r}^{(\alpha)} = (r_1^{(\alpha)}, \ldots, r_{t_\alpha}^{(\alpha)}; m_1^{(\alpha)}, \ldots, m_{t_\alpha}^{(\alpha)}),$$

where $\alpha \in J$, $t_\alpha < p$, $r_1^{(\alpha)}, \ldots, r_{t_\alpha}^{(\alpha)} \in R$, and $0 \leqslant m_1^{(\alpha)} \leqslant \ldots \leqslant m_{t_\alpha}^{(\alpha)}$ are integers, which satisfy the following properties:
(e$_1$) $N(R, \gamma^{(\alpha)}) < m_1^{(\alpha)}$ for any $\alpha \in J$;
(e$_2$) if for $\alpha \in J$, M_α is the set of all decompositions of the form

$$(r_1, \ldots, r_{s_1}, r_1^{(\alpha)}, \ldots, r_{t_\alpha}^{(\alpha)}; m_2, \ldots, m_{s_1}, m_1^{(\alpha)}, \ldots, m_{t_\alpha}^{(\alpha)}),$$

where $(r_1, \ldots, r_{s_1}; m_2, \ldots, m_{s_1}) \in M_{\gamma^{(\alpha)}}(R)$ and $s_1 + t_\alpha < p$, then $M_\alpha \subset M_{\gamma_0}(R, N)$;
(e$_3$) for any $\alpha_1, \alpha_2 \in J$, $\alpha_1 \neq \alpha_2$, we have $M_{\alpha_1} \cap M_{\alpha_2} = \varnothing$;
(e$_4$) $\bigcup_{\alpha \in J} M_\alpha = M_{\gamma_0}(R, N)$.

5.3.3. PROPOSITION. *For any finite subset $R \subset \mathbb{Q}^+(p)$ and any rational number $v_0 > 0$ the set of all (v_0, R)-bad numbers is finite.*

PROOF. By 5.3.1 it is sufficient to prove the finiteness of the set of all (v_0, R)-bad decompositions. For any decomposition

$$\pi = (r_1, \ldots, r_s; m_2, \ldots, m_s) \in M_\gamma(R),$$

we define

$$m_0(\pi) = \max\{ t \mid \gamma_t' \geqslant v_0 \}$$

if this set is not empty and $m_0(\pi) = 0$ otherwise, where the numbers γ_t' are taken from the definition of a (v_0, R)-bad decomposition.

Now we take any (v_0, R)-bad decomposition

$$\pi = (r_1, \ldots, r_s; m_2, \ldots, m_s) \in M_\gamma(R)$$

and use induction on $m_0(\pi)$.

If $m_0(\pi) = 0$, then

$$\gamma_1' = \gamma - \frac{r_s}{p^{m_s}} < v_0.$$

LEMMA. *There exists a $\delta = \delta(R, v_0) > 0$ such that*

$$\delta = \min\Big\{ X = v_0 - \Big(\frac{r_1}{p^{m_1}} + \cdots + \frac{r_l}{p^{m_l}}\Big) \;\Big|\; l < p,\; r_1, \ldots, r_l \in R,\; m_1, \ldots, m_s \geqslant 0,\; X > 0 \Big\}.$$

PROOF. This is obvious.

We have $r_s/p^{m_s} \geqslant \delta$ from this lemma, so m_s can only run through a finite set of values. So there exists only a finite number of (v_0, R)-bad decompositions π with $m_0(\pi) = 0$.

Now let $\pi = (r_1, \ldots, r_s; m_2, \ldots, m_s)$ be a (v_0, R)-bad decomposition and suppose that our proposition is proved for all (v_0, R)-bad decompositions π' with $m_0(\pi') < m_0^*$, where $m_0^* = m_0(\pi) \geqslant 1$. By property 5.3.2c), $\pi_1 = (r_1, \ldots, r_{s-1}; m_2, \ldots, m_{s-1})$ is (v_0, R)-bad. By the inductive assumption, such decompositions create only a finite set and we can take

$$N^* = \max\{ N(\gamma', R) \mid \text{there exists } (v_0, R)\text{-bad } \pi_1 \in M_{\gamma'}(R)$$

$$\text{such that } m_0(\pi_1) < m_0^* \}.$$

We have $m_s \leqslant N^*$, because π is (v_0, R)-bad. Again, there is only a finite number of decompositions $(r_1, \ldots, r_s; m_2, \ldots, m_s)$ such that $s < p$ and $m_s \leqslant N^*$. The proposition is proved.

5.3.4. Let

$$N_0(R, v_0) = \max\{ N(R, \gamma) \mid \gamma \text{ is } (v_0, R)\text{-bad} \} + 1.$$

LEMMA. *Let* $N \geqslant N_0(R, v_0)$. *Then the ideal* $\widetilde{\mathcal{L}}_{R,N}^{(v_0)} \otimes k$ *is generated by elements* $\mathcal{F}_{R,N}(\gamma, n)$ mod $C_p(\mathcal{L}_{R,N} \otimes k)$, *where* $0 \leqslant n < N$, *and* γ *is* (v_0, R)-*bad*.

PROOF. As was shown in 5.1, $\widetilde{\mathcal{L}}_{R,N}^{(v_0)} \otimes k$ is generated by elements

$$\mathcal{F}_{R,N}(\gamma_0, n) = \sum_{\pi \in M_{\gamma_0}(R,N)} (-1)^{s+1} r_1 \widehat{\eta}(r_1, 0, r_2, m_2, \ldots, r_s, m_s)$$
$$[\ldots [D_{r_1,n}, D_{r_2, \widetilde{n+m_2}}], \ldots, D_{r_s, \widetilde{n+m_s}}]$$

where $\gamma_0 \geqslant v_0$, $0 \leqslant n < N$, $\pi = (r_1, \ldots, r_s; m_2, \ldots, m_s)$ and, for $2 \leqslant i \leqslant s$, $\widetilde{n+m_i}$ are the representatives of $(n+m_i)$ mod N in $[0, N)$.

Now we apply property 5.3.2e). From the definition of the constants $\widehat{\eta}$ (see 4.3) and $\widetilde{\eta}$ (see 5.2) for any decomposition

$$(r_1, \ldots, r_{s_1}, r_1^{(\alpha)}, \ldots, r_{t_\alpha}^{(\alpha)}; m_2, \ldots, m_{s_1}, m_1^{(\alpha)}, \ldots, m_{t_\alpha}^{(\alpha)}) \in M_\alpha$$

we have

$$\widehat{\eta}(r_1, 0, r_2, m_2, \ldots, r_{s_1}, m_{s_1}, r_1^{(\alpha)}, m_1^{(\alpha)}, \ldots, r_{t_\alpha}^{(\alpha)}, m_{t_\alpha}^{(\alpha)})$$
$$= \widehat{\eta}(r_1, 0, r_2, m_2, \ldots, r_{s_1}, m_{s_1}) \widetilde{\eta}(r_1^{(\alpha)}, m_1^{(\alpha)}, \ldots, r_{t_\alpha}^{(\alpha)}, m_{t_\alpha}^{(\alpha)}).$$

Then the decomposition $\bigcup_{\alpha \in J} M_\alpha = M_{\gamma_0}(R, N)$ gives the following relation:

$$\mathcal{F}_{R,N}(\gamma_0, n) = \sum_{\substack{\alpha \in J \\ \pi \in M_\alpha}} (-1)^{s_1+t_\alpha+1} r_1 \widehat{\eta}(r_1, 0, r_2, m_2, \ldots, r_{t_\alpha}^{(\alpha)}, m_{t_\alpha}^{(\alpha)})$$
$$\times [\ldots [D_{r_1, n}, D_{r_2, \widetilde{n+m_2}}], \ldots, D_{r_{t_\alpha}^{(\alpha)}, m_{t_\alpha}^{(\alpha)}}]$$
$$\equiv \sum_{\alpha \in J} (-1)^{t_\alpha} \widetilde{\eta}(r_1^{(\alpha)}, m_1^{(\alpha)} \ldots, r_{t_\alpha}^{(\alpha)}, m_{t_\alpha}^{(\alpha)})$$
$$\times [\ldots [\mathcal{F}_{R,N}(\gamma^{(\alpha)}, n), D_{r_1^{(\alpha)}, \widetilde{n+m_1^{(\alpha)}}}], \ldots, D_{r_{t_\alpha}^{(\alpha)}, \widetilde{n+m_{t_\alpha}^{(\alpha)}}}] \mod C_p(\mathcal{L}_{R,N} \otimes k).$$

This relation proves our lemma.

In order to finish the proof of the proposition from §4, we need only state the following:

LEMMA. *Let* $N_2 \geqslant N_0(R, v_0)$, $N_2 | N_1$ *and* $\theta_{N_1, N_2}: \widetilde{\mathcal{L}}_{R,N_1} \otimes k \to \widetilde{\mathcal{L}}_{R,N_2} \otimes k$ *be connecting morphisms of the projective system* $\{\widetilde{\mathcal{L}}_{R,N} \otimes k\}$. *If* γ_0 *is* (v_0, R)-*bad and* $0 \leqslant n < N_1$, *then*

$$\theta_{N_1, N_2}(\mathcal{F}_{R,N_1}(\gamma_0, n)) \mod C_p(\mathcal{L}_{R,N_1} \otimes k) = \mathcal{F}_{R,N_2}(\gamma_0, \widetilde{n}) \mod C_p(\mathcal{L}_{R,N_2} \otimes k),$$

where $\widetilde{n} \equiv n \mod N_2$ *and* $0 \leqslant \widetilde{n} < N_2$.

PROOF. This follows from the relation $M_{\gamma_0}(R) = M_{\gamma_0}(R, N)$, for any (v_0, R)-bad number γ_0 and $N \geqslant N_0(R, v_0)$.

The Proposition from 4.4 is proved.

§6. Some standard facts about ramification filtrations

Let K be a local complete discrete valuation field with perfect residue field k of characteristic $p > 0$. For simplicity we suppose k is algebraically closed. We denote the separable closure of K by K_{sep} and $\Gamma = \text{Gal}(K_{\text{sep}}/K)$ will be the absolute Galois group of K.

6.1. Definition of a ramification filtration, [Se3, De]. Let L be a finite Galois extension of K, $\Gamma_{L/K} = \text{Gal}(L/K)$, v_L be a valuation of L such that $v_L(\pi) = 1$, where π is some uniformizer of L. For any real number $x \geqslant 0$, we set

$$\Gamma_{L/K, x} = \{\tau \in \Gamma_{L/K} \mid v_L(\tau\pi - \pi) \geqslant x + 1\}.$$

Then all $\Gamma_{L/K, x}$ are normal subgroups of $\Gamma_{L/K}$. Because k is algebraically closed, $\Gamma_{L/K, 0} = \Gamma_{L/K}$. So we have a *ramification filtration* of $\Gamma_{L/K}$ *in the lower numbering*. Let

$$\varphi_{L/K}(x) = \int_0^x [\Gamma_{L/K} : \Gamma_{L/K, x}]^{-1} dx$$

be the Herbrandt function. The relation $\Gamma_{L/K, x} = \Gamma_{L/K}^{(v)}$, where $v = \varphi_{L/K}(x)$ for $x \geqslant 0$, gives the *ramification filtration* $\{\Gamma_{L/K}^{(v)}\}_{v \geqslant 0}$ of $\Gamma_{L/K}$ *in the upper numbering*. Now for every tower of Galois extensions $L_1 \supset L_2 \supset K$ the natural epimorphism $\Gamma_{L_1/K} \to \Gamma_{L_2/K}$ gives an epimorphism $\Gamma_{L_1/K}^{(v)} \to \Gamma_{L_2/K}^{(v)}$ for any $v \geqslant 0$. Hence it is possible to define a ramification filtration $\{\Gamma_K^{(v)}\}_{v > 0}$ of the absolute Galois group Γ_K by setting:

$$\Gamma_K^{(v)} = \varprojlim_L \Gamma_{L/K}^{(v)} \quad \text{for any } v > 0.$$

The ramification filtration of any separable extension of K may be defined in the same way. So we have:

(1) a decreasing filtration $\{\Gamma_K^{(v)}\}_{v \geqslant 0}$ of normal subgroups in Γ_K, such that $\Gamma_K^{(0)} = \Gamma_K$, $\bigcap_{v > 0} \Gamma_K^{(v)} = \{e\}$;

(2) for every separable extension L/K with Galois group $\Gamma_{L/K}$, the natural morphism $\Gamma_K \to \Gamma_{L/K}$ determines an epimorphism $\Gamma_K^{(v)} \to \Gamma_{L/K}^{(v)}$ for any $v \geqslant 0$;

(3) $I = \bigcup_{v > 0} \Gamma_K^{(v)}$ is a pro-p-group and $K_{\text{sep}}^I = K_{\text{tr}}$ is the maximal tamely ramified extension of K.

6.2. Let L/K be arbitrary finite separable extension. The number $v(L/K)$ is called the *largest upper ramification number* of L/K if the following implication is true:

$$\Gamma_K^{(v)} \text{ acts trivially on } L \iff v > v(L/K).$$

The existence of $v(L/K)$ follows from the left-continuity of the ramification filtration.

The above definition of the Herbrandt function was given in the case when L/K is a Galois extension. Deligne [De] extended this definition to the case of arbitrary finite separable extensions. We have the following properties:

(1) $\varphi_{L/K}(x)$ is a piecewise-linear convex function;

(2) if $(a, \varphi_{L/K}(a))$ is the last edge point of the graph of $\varphi_{L/K}$, then $v(L/K) = \varphi_{L/K}(a)$;

(3) if $K \subset L \subset L_1$ is a tower of finite separable extensions, then

$$\varphi_{L_1/K} = \varphi_{L_1/L}\varphi_{L/K}$$

(for the Galois extensions see [Se3], for general case see [De]).

6.3. We say that L/K has the *unique ramification number* $y_0 > 0$, if $(y_0, \varphi_{L/K}(y_0))$ is the unique edge point of the graph of $\varphi_{L/K}(x)$. In this case:

$$\varphi_{L/K}(x) = \begin{cases} x & \text{for } 0 \leqslant x \leqslant y_0, \\ \dfrac{x - y_0}{[L:K]} + y_0 & \text{for } x \geqslant y_0. \end{cases}$$

It is clear that here $y_0 = v(L/K)$.

LEMMA. *Let* $\operatorname{char} K = p > 0$, $N \in \mathbb{N}$, $q = p^N$ *and* $r^* \in \mathbb{Q}^+(p)$ *be such that* $r^*(q-1) \in \mathbb{N}$. *Then there exists an extension* K' *of* K *such that*

(1) $[K':K] = q$;

(2) K'/K *has the unique ramification number* r^*.

PROOF. Let $r^* = m/(q-1)$, where $m \in \mathbb{N}$, $(m, p) = 1$. Choose some $t \in K$ such that t^{-1} is an uniformizer of K. Consider extensions $K \subset K_N \subset K'_N$, where $K_N = K(t_N)$, $t_N^{q-1} = t$ and $K'_N = K_N(T_N)$, where $T^q - T = t_N^m$.

If $\Gamma_N = \operatorname{Gal}(K_N/K)$ and $\Gamma' = \operatorname{Gal}(K'_N/K)$, then the natural epimorphism $\Gamma' \to \Gamma$ has a section $s: \Gamma \to \Gamma'$. It is easy to see that the field $K' = K_N^{\prime s(\Gamma_N)}$ satisfies the conclusion of the lemma.

REMARK. One can choose T in a such a way that $K' = K(T^{q-1})$.

From 6.2 we obtain the following properties.

(1) Let $K \subset K_0 \subset K_1 \subset \cdots \subset K_n = L$ be a tower of finite separable extensions such that K_0/K is tamely ramified (we write $e_0 = [K_0 : K]$) and for any $1 \leqslant t \leqslant n$, K_{t+1}/K_t is the Galois extension with unique ramification number $x_t > 0$. If $x_1 \leqslant x_2 \leqslant \ldots \leqslant x_n$, then

$$v(K_n/K) = \frac{1}{e_0}\left(x_1 + \frac{x_2 - x_1}{[K_1 : K_0]} + \cdots + \frac{x_n - x_{n-1}}{[K_{n-1} : K_0]}\right).$$

(2) Let $K \subset L_1 \subset L_2$ be a tower of finite separable extensions, L_1/K has the unique ramification number y_0 and $v(L_2/L_1) = v_1$. Then

$$v(L_2/K) = \max\left\{y_0, \frac{v_1 - y_0}{[L_1 : K]} + y_0\right\}.$$

6.4. The following example will be useful in §7 below.

EXAMPLE. Let char $K = p$ and $t \in K$ be such that t^{-1} is a uniformizer of K.

(1) Let $r \in \mathbb{Q}^+(p)$, $N \in \mathbb{N}$, $q = p^N$, $\alpha \in k \setminus \{0\}$ and $L = K_{\text{tr}}(T)$, where $T^p - T = \alpha t^r$. Then $v(L/K) = r$.

(2) Let $A = \sum_{r \in \mathbb{Q}^+(p)} \alpha_r t^r \in K$, where $\alpha_r \in k$ and almost all are equal to 0. If $L_A = K(T)$, where $T^p - T = A$, then

$$v(L_A/K) = \max\{r \mid \alpha_r \neq 0\}.$$

(3) We have also a slight generalisation of (2): let

$$N \geqslant 1, \quad q = p^N, \quad B = \sum_{\substack{r \in \mathbb{Q}^+(p) \\ 0 \leqslant n < N}} \alpha_{r,n} t^{rp^n},$$

where $\alpha_{r,n} \in k$ and almost all are equal to 0. Then for $L_B = K(T)$, where $T^q - T = B$, we have

$$v(L_B/K) = \max\{r \mid \alpha_{r,n} \neq 0 \text{ for some } 0 \leqslant n < N\}.$$

§7. The main theorem

Let K be a complete local discrete valuation field of characteristic $p > 0$, with residue field $k \simeq \overline{\mathbb{F}_p}$. As before, let $\Gamma = \text{Gal}(K_{\text{sep}}/K)$, $\{\Gamma^{(v)}\}_{v>0}$ be the ramification filtration of Γ. If I is the subgroup of higher ramification, we set $\widetilde{\Gamma} = \Gamma/I^p C_p(I)$ and denote by $\{\widetilde{\Gamma}^{(v)}\}_{v>0}$ the image of the ramification filtration of Γ in $\widetilde{\Gamma}$. We also fix $t \in K$ so that t^{-1} is uniformizer of K.

Let e_0 be structural element from 3.1 and let $\eta(r_1, \ldots, r_s)$, where $1 \leqslant s < p$, $r_1, \ldots, r_s \in \mathbb{Q}^+(p)$, be its structural constants.

Let $\mathcal{L}$ be a profree Lie $\mathbb{F}_p$-algebra from §3, $\widetilde{\mathcal{L}} = \mathcal{L}/C_p(\mathcal{L})$ and let A be an f.a.b. related to $\mathcal{L}$. Then the $(p-1)$-diagonal element $e \in G_{\mathcal{L}, K_{\text{tr}}}(p-1)$, which has a representative element of the form

$$E = 1 + \sum_{\substack{1 \leqslant s < p \\ r_1, \ldots, r_s \in \mathbb{Q}^+(p) \\ i_1, \ldots, i_s \in I}} \eta(r_1, \ldots, r_s) w_{i_1} \ldots w_{i_s} t^{r_1 + \cdots + r_s} D_{i_1, r_1} \ldots D_{i_s, r_s}$$

(see 3.2), determines a conjugacy class of isomorphisms of the groups $\widetilde{I} = I/I^p C_p(I)$ and $G_{\mathbb{F}_p, \mathcal{L}}(p-1)$. Fix one of them by choosing $f \in G_{K_{\text{sep}}, \mathcal{L}}(p-1)$ so that $f^{(p)} = fe$ (see §1). We use this isomorphism below for the identification of the groups $\widetilde{I}$ and $G_{\mathcal{L}, \mathbb{F}_p}(p-1)$.

Under this assumption, we have the one-to-one mapping

$$\widetilde{\exp}: \widetilde{\mathcal{L}} \to \widetilde{I} = \bigcup_{v>0} \widetilde{\Gamma}^{(v)}.$$

For any positive rational number $v > 0$, we set $\widetilde{\mathcal{L}}(v) = \widetilde{\exp}^{-1}(\widetilde{\Gamma}^{(v)})$. Then $\widetilde{\mathcal{L}}(v)$ is an ideal of the Lie algebra $\widetilde{\mathcal{L}}$. So, we have a decreasing filtration of the ideals $\widetilde{\mathcal{L}}(v)$ in $\widetilde{\mathcal{L}}$.

THEOREM. *The filtration $\{\widetilde{\mathcal{L}}^{(v)}\}_{v>0}$ of $\widetilde{\mathcal{L}}$, defined in §4, coincides with the above filtration $\{\widetilde{\mathcal{L}}(v)\}_{v>0}$.*

PROOF. The proof of the theorem occupies the rest of this section.

7.1. Characteristic properties. Let $J \subset \widetilde{\mathcal{L}}$ be any ideal and let A_J be the f.a.b. over $\mathbb{F}_p$ related to the Lie algebra $\widetilde{\mathcal{L}}/J$. It is clear that the quotient morphism $\mathcal{L} \to \widetilde{\mathcal{L}}/J$ gives a morphism of f.a.b. objects $A \to A_J$. For any field L of characteristic p, we also have the surjective homomorphism of groups

$$G_{\mathcal{L},L}(p-1) \to G_{\widetilde{\mathcal{L}}/J,L}(p-1).$$

Let e_J be the image of e under the homomorphism

$$G_{\mathcal{L},K_{\mathrm{tr}}}(p-1) \to G_{\widetilde{\mathcal{L}}/J,K_{\mathrm{tr}}}(p-1)$$

and let f_J be the image of f under the homomorphism

$$G_{\mathcal{L},K_{\mathrm{sep}}}(p-1) \to G_{\widetilde{\mathcal{L}}/J,K_{\mathrm{sep}}}(p-1).$$

We have $f_J^{(p)} = f_J e_J$; the homomorphism f_J determines an identification of the groups $\widetilde{I}/\widetilde{\exp}(J)$ and $G_{\widetilde{\mathcal{L}}/J,\mathbb{F}_p}(p-1)$ and this identification agrees in the obvious sense with the above identification of $\widetilde{I}$ and $G_{\mathcal{L},\mathbb{F}_p}(p-1)$ defined by f.

Let $K_{p-1}(e_J)$ be the field of definition of f_J (see §1). The following proposition follows immediately from the above construction.

7.1.1. PROPOSITION. *For any $v_0 \in \mathbb{Q}$, $v_0 > 0$, the ideal $\widetilde{\mathcal{L}}(v_0)$ is the minimal element in the set of ideals $J \subset \widetilde{\mathcal{L}}$, such that the largest upper ramification number $v(K_{p-1}(e_J)/K)$ of the extension $K_{p-1}(e_J)/K$ is less than v_0.*

Let R be any finite subset in $\mathbb{Q}^+(p)$, $N \in \mathbb{N}$ and $\mathcal{L}_{R,N}$ be the Lie $\mathbb{F}_p$-algebra from 4.1. Then $\mathcal{L} = \varprojlim \mathcal{L}_{R,N}$ and for any $v_0 \in \mathbb{Q}$, $v > 0$, $\mathcal{L}(v_0) = \varprojlim \mathcal{L}_{R,N}(v_0)$, where $\mathcal{L}(v_0)$ is the inverse image of $\widetilde{\mathcal{L}}(v_0)$ under the quotient $\mathcal{L} \to \widetilde{\mathcal{L}}$ and $\mathcal{L}_{R,N}(v_0)$ is the image of $\mathcal{L}(v_0)$ under the projection $\mathcal{L} \to \mathcal{L}_{R,N}$.

Similarly, define elements

$$e_{R,N} \in G_{\mathcal{L}_{R,N},K_{\mathrm{tr}}}(p-1) \quad \text{and} \quad f_{R,N} \in G_{\mathcal{L}_{R,N},K_{\mathrm{sep}}}(p-1),$$

such that $e = \varprojlim e_{R,N}$ and $f = \varprojlim f_{R,N}$. The field of definition of f is equal to $\widetilde{K} = K_{\mathrm{sep}}^{I^p C_p(I)}$.

Let $K_{R,N}$ be the field of definition of $f_{R,N}$ (in the notation of 1.2.3 we have $\widetilde{K} = K_e(p-1)$ and $K_{R,N} = K_{e_{R,N}}(p-1)$), then $\widetilde{K} = \varinjlim K_{R,N}$.

For the corresponding ideals $\mathcal{L}_{R,N}(v_0)$ of $\mathcal{L}_{R,N}$, we have the same minimal property as in Proposition 7.1.

For any $1 \leqslant s < p$ denote by $C_{s+1}(\mathcal{L}_{R,N})$ the ideal of $\mathcal{L}_{R,N}$ generated by all commutators of order $\geqslant s+1$ and set

$$\mathcal{L}_{R,N,s}(v_0) = \mathcal{L}_{R,N}(v_0) + C_{s+1}(\mathcal{L}_{R,N}).$$

By $K_{R,N,s}$ denote the field of definition of
$$f_{R,N} \in G_{\mathcal{L}_{R,N},K_{\text{sep}}}(s) \subset (A_{R,N} \otimes K_{\text{sep}}) \bmod J_{s+1},$$
where $A_{R,N}$ is an f.a.b. related to $\mathcal{L}_{R,N}$ and $J_{s+1} = J_{s+1}(A_{R,N}) \otimes K_{\text{sep}}$. Obviously, $K_{R,N,s} \subset K_{R,N}$ and $K_{R,N,s}$ is the maximal Galois extension of K inside $K_{R,N}$ having higher ramification subgroup of nilpotency class s.

For any ideal I such that $C_{s+1}(\mathcal{L}_{R,N}) \subset I \subset \mathcal{L}_{R,N}$, denote by $K_{R,N,s}(I)$ the field of definition of $f_{R,N} \bmod (IA_{R,N} \otimes K_{\text{sep}} + J_{s+1})$. As earlier, we have the following proposition:

7.1.2. PROPOSITION. $\mathcal{L}_{R,N,s}(v_0)$ *is the minimal element in the set of ideals* I, *such that* $C_{s+1}(\mathcal{L}_{R,N}) \subset I \subset \mathcal{L}_{R,N}$ *and*
$$v(K_{R,N,s}(I)/K) < v_0.$$

7.2. Restatement of the main theorem. Let R be a fixed finite subset in $\mathbb{Q}^+(p)$. Let $\delta = \delta(R, v_0) > 0$ be the minimum of all positive values of the expression
$$v_0 - \left(\frac{r_1}{p^{m_1}} + \cdots + \frac{r_l}{p^{m_l}}\right),$$
where $1 \leqslant l < p$, $r_1, \ldots, r_l$ run over R and $m_1, \ldots, m_l$ run over $\mathbb{N} \cup \{0\}$ (see 5.3.3).

Choose $N(R, v_0) \in \mathbb{N}$ such that for any $N \geqslant N(R, v_0)$ there exists $r^* = r^*(N) \in \mathbb{Q}^+(p)$ satisfying the following conditions:
(1) $r^*(q-1) \in \mathbb{N}$, where $q = p^N$;
(2) $r^* < v_0$;
(3) $r^* > \dfrac{q}{q - 1 - (p-1)p^{N_0}}(v_0 - \delta)$, where $N_0 = N_0(R, v_0)$ is the natural number from Proposition 4.4.

Now Proposition 7.1.2 shows that the following proposition implies our theorem.

PROPOSITION. *For any* $N \geqslant N(R, v_0)$, $1 \leqslant s < p$, *and ideal* I *such that*
$$C_{s+1}(\mathcal{L}_{R,N}) \subset I \subset \mathcal{L}_{R,N},$$
we have:
$$v(K_{R,N,s}(I)/K) < v_0 \iff \widetilde{\mathcal{L}}_{R,N}^{(v_0)} \bmod C_{s+1}(\mathcal{L}_{R,N}) \subset I \bmod C_{s+1}(\mathcal{L}_{R,N}).$$

REMARK. Until the end of §7 we use the following simplified notation:
- C_s for the ideal $C_s(\mathcal{L}_{R,N})$ of commutators of order $\geqslant s$ in $\mathcal{L}_{R,N}$, $1 \leqslant s \leqslant p$;
- A for the f.a.b. $A_{\mathcal{L}_{R,N}}$ over $\mathbb{F}_p$ related to $\mathcal{L}_{R,N}$;
- A_L for the f.a.b. $A_{\mathcal{L}_{R,N}} \otimes L$, where L is a field, char $L = p$;
- A_{sep} for $A_{\mathcal{L}_{R,N}, K_{\text{sep}}}$;
- J_s for $J_s(A_{\mathcal{L}_{R,N}, K_{\text{sep}}})$, $1 \leqslant s \leqslant p$;
- $J_s(O_{\text{sep}})$ for the O_{sep}-submodule $J_s(A_{\mathcal{L}_{R,N}}) \otimes O_{\text{sep}}$ in $A_{\mathcal{L}_{R,N}, K_{\text{sep}}}$, where $1 \leqslant s \leqslant p$ and O_{sep} is the valuation ring of K_{sep};
- K_s for the field $K_{R,N,s}$ of definition of $f_{R,N} \bmod J_{s+1}$, $1 \leqslant s < p$, see 7.1;
- $\mathcal{L}_s(v_0)$ for the ideal $\mathcal{L}_{R,N,s}(v_0)$ from 7.1;
- $K_s(v_0)$ for the field $K_{R,N,s}(\mathcal{L}_{R,N,s}(v_0))$.

7.3. Some identities.

7.3.1. Let $\{D_{r,n} \mid r \in R,\ 0 \leqslant n < N\}$ be the system of generators of the Lie k-algebra $\mathcal{L}_{R,N} \otimes k$ that was introduced in 5.1. It is clear that the representative $E \in A_{K_{\mathrm{tr}}}$ of $e_{R,N} \in G_{\mathcal{L}_{R,N}, K_{\mathrm{tr}}}(p-1)$ can be written in the form

$$E = 1 + \sum_{\substack{1 \leqslant s < p \\ r_1, \ldots, r_s \in R}} \eta(r_1, \ldots, r_s) t^{r_1 + \cdots + r_s} D_{r_1, 0} \ldots D_{r_s, 0}.$$

Let $\mathcal{F}$ be a representative of $f_{R,N}$; then we have $\mathcal{F}^{(p)} \equiv \mathcal{F}E \mod J_p$.

7.3.2. Let $E_N = E E^{(p)} \ldots E^{(p^{N-1})}$. Then

$$E_N \equiv 1 + \sum_{\substack{1 \leqslant s < p \\ r_1, \ldots, r_s \in R \\ 0 \leqslant n_1, \ldots, n_s < N}} \eta(r_1, n_1, \ldots, r_s, n_s) t^{r_1 p^{n_1} + \cdots + r_s p^{n_s}} D_{r_1, n_1} \ldots D_{r_s, n_s} \mod J_p,$$

where the constants $\eta(r_1, n_1, \ldots, r_s, n_s)$ are defined as follows:

$$\eta(r_1, n_1, \ldots, r_s, n_s) = \eta(r_1, \ldots, r_{s_1}) \eta(r_{s_1+1}, \ldots, r_{s_2}) \ldots \eta(r_{s_{l-1}+1}, \ldots, r_{s_l}),$$

if $n_1 = \cdots = n_{s_1} < n_{s_1+1} = \cdots = n_{s_2} < \cdots < n_{s_{l-1}+1} = \cdots = n_{s_l}$, where $1 \leqslant s_1 < s_2 < \cdots < s_l = s$ and

$$\eta(r_1, n_1, \ldots, r_s, n_s) = 0,$$

otherwise.

REMARK. The constants $\eta(r_1, n_1, \ldots, r_s, n_s)$ are obtained from the constants $\eta(r_1, \ldots, r_s)$ in the same way as the constants $\widehat{\eta}(r_1, n_1, \ldots, r_s, n_s)$ were obtained from the constants $\widehat{\eta}(r_1, \ldots, r_s) = \eta(r_s, \ldots, r_1)$ in 4.3.

For $q = p^N$ and the above element E_N, we have the following equivalence:

$$\mathcal{F}^{(q)} \equiv \mathcal{F} E_N \mod J_p.$$

7.3.3. Let $\mathcal{F}^* = \mathcal{F}^{(p)}$, then $\mathcal{F}^{(q)} \equiv \mathcal{F}^* E^{(p)} \ldots E^{(p^{N-1})} \mod J_p$, i.e.,

$$\mathcal{F}^{(q)} \equiv \mathcal{F}^* \left(1 + \sum_{\substack{1 \leqslant s < p \\ r_1 \ldots r_s \in R \\ 0 < n_1 \ldots n_s < N}} \eta(r_1, n_1, \ldots, r_s, n_s) t^{r_1 p^{n_1} + \cdots + r_s p^{n_s}} D_{r_1, n_1} \ldots D_{r_s, n_s} \right) \mod J_p.$$

7.3.4. Let

$$E_0 = E - 1 = \sum_{\substack{1 \leqslant s < p \\ r_1, \ldots, r_s \in R}} \eta(r_1, \ldots, r_s) t^{r_1 + \cdots + r_s} D_{r_1, 0} \ldots D_{r_s, 0}.$$

From the equivalence $\mathcal{F}^{(p)} - \mathcal{F} \equiv \mathcal{F} E_0 \mod J_p$, we obtain

$$\mathcal{F}^{(q)} - \mathcal{F} \equiv \sum_{0 \leqslant n < N} (\mathcal{F} E_0)^{(p^n)} \mod J_p.$$

7.4. The field K'.

Let $N \geq N(R, v_0)$, $q = p^N$ and $r^* \in \mathbb{Q}^+(p)$ be some number, related to N in the definition of $N(R, v_0)$, see 7.2.

Denote by K' the extension of K, whose Herbrandt function is of the form:

$$\varphi_{K'/K} = \begin{cases} x & \text{for } 0 \leq x \leq r^*, \\ r^* + \dfrac{x - r^*}{q} & \text{for } x \geq r^* \end{cases}$$

(see 6.3).

7.4.1. LEMMA. *There exists a $t_1 \in K'$ such that*
(1) t_1^{-1} *is a uniformizer of* K';
(2) $t = t_1^q e_1$, *where* $e_1 = \widetilde{\exp}(-\frac{1}{r^*} t_1^{-r^*(q-1)})$.

PROOF. It follows from the explicit construction of the field K', see 6.3, that $K'K_{\mathrm{tr}} = K_{\mathrm{tr}}(T)$, where $T^q - T = t^{r^*}$ and $T = t_1'^{r^*}$, for some uniformizing element $t_1'^{-1}$ of K'. Therefore,

$$t = t_1'^q (1 - t_1'^{-r^*(q-1)})^{1/r^*}$$

and the existence of $t_1 \in K'$ that satisfy the requirements of our lemma may be established by using Hensel's lemma.

7.4.2. It is clear that there exists (a unique) isomorphism f of the fields K and K' which is the identity on their residue fields and sends t to t_1. The following property of the extension K'/K will be useful later.

LEMMA. *Let L/K and L'/K' be finite extensions such that there exists an isomorphism of fields $g: L \to L'$ which prolongs f, i.e., $g|_K = f$. Then either $v(L/K)$ and $v(L'/K)$ are both less than v_0 or $v(L'/K) < v(L/K)$.*

PROOF. Let $v(L'/K) \geq v(L/K)$. Applying property (2) of 6.3 to the tower of fields
$$K \subset K' \subset L',$$
we obtain
$$v(L'/K) = \max\left\{r^*, \frac{v_1 - r^*}{q} + r^*\right\},$$
where $v_1 = v(L'/K') = v(L/K)$. This means that
$$v_1 \leq \max\left\{r^*, \frac{v_1 - r^*}{q} + r^*\right\}.$$

Obviously, this is equivalent to the inequality $v_1 \leq r^*$, which implies $v(L/K) \leq v(L'/K) = r^* < v_0$.

7.4.3. The following property is related to a special choice of r^* and will be useful below. Let

$$M_{p-1}(R) = \left\{\gamma \in \mathbb{Q} \,\Big|\, \gamma = \frac{r_1}{p^{m_1}} + \cdots + \frac{r_s}{p^{m_s}}, \right.$$
$$\left. 1 \leq s < p, \ r_1, \ldots, r_s \in R, \ m_1, \ldots, m_s \in \mathbb{N} \cup \{0\}\right\}$$

and let $O'_{\mathrm{tr}} = O_{K'_{\mathrm{tr}}}$ be the valuation ring of the field $K'_{\mathrm{tr}} = K_{\mathrm{tr}} K'$.

LEMMA. *If $\gamma \in M_{p-1}(R)$, $\gamma < v_0$, then*

$$t_1^{q\gamma - r^*(q-1)} \in t_1^{-r^*(p-1)p^{N_0}} O'_{\mathrm{tr}} \subset O'_{\mathrm{tr}},$$

where $N_0 = N_0(R, v_0)$ is the natural number from Proposition 4.4.

PROOF. This follows immediately from condition (3) of 7.2.

7.5. Some identities.

7.5.1. Let $1 \leqslant s < p$, $r_1, \ldots, r_s \in \mathbb{Q}^+(p)$, $0 \leqslant n_1, \ldots, n_s < N$. We use the constants $\eta(r_1, n_1, \ldots, r_s, n_s)$, $\widehat{\eta}(r_1, n_1, \ldots, r_s, n_s)$, and $\widetilde{\eta}(r_1, n_1, \ldots, r_s, n_s)$, defined in 7.3.2, 4.3, and 5.2, respectively.

Let us use the agreement about indices from 5.2, i.e., for any natural numbers $n_1, \ldots, n_s$, let us denote by n_{ij}, where $1 \leqslant i, j \leqslant s$, the reduced residue of $n_i - n_j$ modulo N, i.e., n_{ij} is uniquely defined by the conditions:

$$n_{ij} \equiv n_i - n_j \mod N, \quad 0 \leqslant n_{ij} < N.$$

We then have:

$$\widetilde{\eta}(r_1, n_1, \ldots, r_s, n_s) = \eta(r_s, \ldots, r_1), \quad \text{if } n_1 = \cdots = n_s,$$
$$\widetilde{\eta}(r_1, n_1, \ldots, r_s, n_s) = \widehat{\eta}(r_1, n_{11}, \ldots, r_s, n_{1s}).$$

Introduce the new constants $\eta^*(r_1, n_1^*, \ldots, r_s, n_s^*)$, where $1 \leqslant s < p$, $r_1, \ldots, r_s \in \mathbb{Q}^+(p)$, $n_1^*, \ldots, n_s^* \in (0, N]$.

DEFINITION.

$$\eta^*(r_1, n_1^*, \ldots, r_s, n_s^*) = \widehat{\eta}(r_1, \ldots, r_{s_1})\widehat{\eta}(r_{s_1+1}, \ldots, r_{s_2}) \ldots \widehat{\eta}(r_{s_{l-1}+1}, \ldots, r_{s_l})$$

if $n_1^* = \cdots = n_{s_1}^* > n_{s_1+1}^* = \cdots = n_{s_2}^* > \cdots > n_{s_{l-1}+1}^* = \cdots = n_{s_l}^*$, where $1 \leqslant s_1 < s_2 < \cdots < s_l = s$ (recall that $\widehat{\eta}(r_1, \ldots, r_{s_1}) = \eta(r_{s_1}, \ldots, r_1)$, see 4.3), and

$$\eta^*(r_1, n_1^*, \ldots, r_s, n_s^*) = 0$$

otherwise, i.e., if $n_1^* \geqslant \ldots \geqslant n_s^*$ is not true.

We have:

$$\widetilde{\eta}(r_1, n_1, \ldots, r_s, n_s) = \eta^*(r_1, n_{11}^*, \ldots, r_s, n_{s1}^*)$$
$$= \widehat{\eta}(r_1, N - n_{11}^*, r_2, N - n_{21}^*, \ldots, r_s, N - n_{s1}^*),$$

where n_{ij}^* are the residues modulo N of $n_i - n_j$ from $(0, N]$ (it is sufficient to remark that for any i, j, we have $n_{ij}^* = N - n_{ji}$).

7.5.2. For the constants $\eta(r_1, n_1, \ldots, r_s, n_s)$, the following analog of Proposition 3.1 holds:

LEMMA. *If s_1, s are natural numbers such that $s_1 < s < p$, then*

$$\eta(r_1, n_1, \ldots r_{s_1}, n_{s_1})\eta(r_{s_1+1}, n_{s_1+1}, \ldots, r_s, n_s)$$
$$= \sum_{\sigma \in P_{s_1,s}^*} \eta(r_{\sigma^{-1}(1)}, n_{\sigma^{-1}(1)}, \ldots, r_{\sigma^{-1}(s)}, n_{\sigma^{-1}(s)}),$$

where $P_{s_1,s}^*$ is the subset of permutations of order s such that $\sigma(i) < \sigma(j)$, where either $1 \leqslant i < j \leqslant s_1$ or $s_1 + 1 \leqslant i < j \leqslant s$.

PROOF. This follows from the fact that

$$E = 1 + \sum_{\substack{1 \leqslant s < p \\ r_1, \ldots, r_s \in R \\ 0 \leqslant n_1, \ldots, n_s < N}} \eta(r_1, n_1, \ldots, r_s, n_s) D_{r_1, n_1} \ldots D_{r_s, n_s}$$

is the representative of a $(p-1)$-diagonal element (see 3.1).

REMARK. The meaning of the right-hand side of the above formula is very simple: the collections of variables are numbered by all inclusions of the first set of indices $\{1, \ldots, s_1\}$ into the second set $\{s_1+1, \ldots, s\}$ that preserve the natural orderings of these sets.

REMARK. By the same reasoning, a similar statement is true for the constants $\widehat{\eta}(r_1, n_1, \ldots, r_s, n_s)$ and $\eta^*(r_1, n_1^*, \ldots, r_s, n_s^*)$.

7.5.3. Let s be any natural number.

DEFINITION. A subset Φ_s of *connected* permutations of order s consists of all one-to-one mappings $\sigma: \{1, \ldots, s\} \to \{1, \ldots, s\}$ such that for any $1 \leqslant s_1 \leqslant s$ the set $\{\sigma(1), \ldots, \sigma(s_1)\}$ consists of s_1 successive integers.

LEMMA. *For any indeterminates $D_1, \ldots, D_s$ we have:*

$$[\ldots [D_1, D_2], \ldots, D_s] = \sum_{\sigma \in \Phi_s} (-1)^{\sigma(1)+1} D_{\sigma^{-1}(1)} D_{\sigma^{-1}(2)} \cdots D_{\sigma^{-1}(s)}.$$

PROOF. This may be proved by straightforward combinatorial arguments.

7.5.4. Let $1 \leqslant s < p$, $r_1, \ldots, r_s \in \mathbb{Q}^+(p)$, $0 \leqslant n_1, \ldots, n_s < N$.

DEFINITION. For $1 \leqslant t \leqslant s$, we set

$$B_t(r_1, n_1, \ldots, r_s, n_s) = \sum_{\substack{\sigma \in \Phi_s \\ \sigma(1)=t}} \widetilde{\eta}(r_{\sigma(1)}, n_{\sigma(1)}, \ldots, r_{\sigma(s)}, n_{\sigma(s)}).$$

EXAMPLE.

$$B_1(r_1, n_1) = \widetilde{\eta}(r_1, n_1) = 1,$$
$$B_2(r_1, n_1, r_2, n_2, r_3, n_3) = \widetilde{\eta}(r_2, n_2, r_1, n_1, r_3, n_3) + \widetilde{\eta}(r_2, n_2, r_3, n_3, r_1, n_1),$$
$$B_1(r_1, n_1, \ldots, r_s, n_s) = \widetilde{\eta}(r_1, n_1, \ldots, r_s, n_s),$$
$$B_s(r_1, n_1, \ldots, r_s, n_s) = \widetilde{\eta}(r_s, n_s, \ldots, r_1, n_1).$$

LEMMA. *For any $\gamma_0 \in \mathbb{Q}$, $\gamma_0 > 0$ and natural number $n^* \leqslant N$, we have:*

$$\sum r_1 \widetilde{\eta}(r_1, n_1, \ldots, r_s, n_s)[\ldots[D_{r_1,n_1}, D_{r_2,n_2}], \ldots, D_{r_s,n_s}]$$
$$= \sum_{1 \leqslant t \leqslant s} \sum (-1)^{t+1} r_t B_t(r_1, n_1, \ldots, r_s, n_s) D_{r_1,n_1} \ldots D_{r_s,n_s},$$

where the first sum is taken over

$$r_1, \ldots, r_s \in R, \quad 0 < n_2, \ldots, n_s \leqslant n^*, \quad n_1 = n^*, \quad r_1 + \frac{r_2}{p^{n_{12}}} + \cdots + \frac{r_s}{p^{n_{1s}}} = \gamma$$

and the last one over

$$r_1, \ldots, r_s \in R, \quad 0 < n_1, \ldots, n_{t-1}, n_{t+1}, \ldots, n_s \leqslant n^*, \quad n_t = n^*,$$
$$\frac{r_1}{p^{n_{t1}}} + \cdots + \frac{r_s}{p^{n_{ts}}} = \gamma.$$

PROOF. This follows from Lemma 7.5.3.

7.5.5. LEMMA.

$$B_t(r_1, n_1, \ldots, r_s, n_s) + \delta(n_t, n_{t+1}) B_{t+1}(r_1, n_1, \ldots, r_s, n_s)$$
$$= \eta^*(r_t, n_{tt}^*, \ldots, r_1, n_{1t}^*) \eta^*(r_{t+1}, n_{t+1,t}^*, \ldots, r_s, n_{s,t}^*),$$

where

$$\delta(n_t, n_{t+1}) = \begin{cases} 1, & \text{if } n_t = n_{t+1}, \\ 0, & \text{otherwise.} \end{cases}$$

PROOF. Let $n_t \neq n_{t+1}$, then

$$B_t(r_1, n_1, \ldots, r_s, n_s)$$
$$= \sum_{\substack{\sigma \in \Phi_s \\ \sigma(1)=t}} \widetilde{\eta}(r_t, n_t, \ldots, r_{\sigma(i)}, n_{\sigma(i)}, \ldots, r_{\sigma(s)}, n_{\sigma(s)})$$
$$= \sum_{\substack{\sigma \in \Phi_s \\ \sigma(1)=t}} \eta^*(r_t, n_{tt}^*, \ldots, r_{\sigma(i)}, n_{\sigma(i),t}^*, \ldots)$$
$$+ \sum_{\substack{\sigma \in \Phi_s \\ \sigma(1)=t+1}} \eta^*(r_{t+1}, n_{t+1,t}^*, \ldots, r_{\sigma(i)}, n_{\sigma(i),t}^*, \ldots)$$
$$= \eta^*(r_t, n_{tt}^*, \ldots, r_1, n_{1t}^*) \eta^*(r_{t+1}, n_{t+1,t}^*, \ldots, r_s, n_{s,t}^*)$$

(all summands of the last sum are equal to 0, because $n_{t+1,t}^* < n_{t,t}^* = 0$) by the lemma and the remarks 7.5.2.

The same arguments yield the proof in the case $n_t = n_{t+1}$.

REMARK. In these propositions we use the following property:

$$\{\sigma \in \Phi_s \mid \sigma(1) = t\} \cup \{\sigma \in \Phi_s \mid \sigma(1) = t+1\}$$
$$= \{\text{all ordered insertions of } \{t, \ldots, 1\} \text{ to } \{t+1, \ldots, s\}\}.$$

7.5.6. Let $1 \leqslant t \leqslant s < p$, $r_1, \ldots, r_s \in R$, $0 \leqslant n_1, \ldots, n_s < N$.

DEFINITION.

$$B_t^*(r_1, n_1, \ldots, r_s, n_s) = \begin{cases} B_t(r_1, n_1, \ldots, r_s, n_s) & \text{for } n_t \geq \ldots \geq n_s, \\ 0, & \text{otherwise.} \end{cases}$$

EXAMPLE.

$$B_1^*(r_1, n_1, \ldots, r_s, n_s) = \eta(r_s, n_s, \ldots, r_1, n_1),$$
$$B_s^*(r_1, n_1, \ldots, r_s, n_s) = B_s(r_1, n_1, \ldots, r_s, n_s),$$
$$B_t^*(r_1, n_1, \ldots, r_{s-1}, n_{s-1}, r_s, 0) = B_t(r_1, n_1, \ldots, r_{s-1}, n_{s-1}, r_s, 0).$$

REMARK. If $n_t \geq n_s$, then

$$B_t^*(r_1, n_1, \ldots, r_s, n_s) = B_t(r_1, n_1, \ldots, r_s, n_s).$$

LEMMA. *If $n_{t+1} \leq n_t$, $s \geq t+1$, then*

$$B_t^*(r_1, n_1, \ldots, r_s, n_s) + \delta(n_t, n_{t+1}) B_{t+1}^*(r_1, n_1, \ldots, r_s, n_s)$$
$$= \widetilde{\eta}(r_t, n_t, \ldots, r_1, n_1) B_1^*(r_{t+1}, n_{t+1}, \ldots, r_s, n_s).$$

PROOF. If $n_{t+1} \geq \ldots \geq n_s$ is not true, then both sides are equal to 0. If $n_{t+1} \geq \ldots \geq n_s$, then for any $t+1 \leq u \leq s$ we have $n_{t,u} = n_{t,t+1} + n_{t+1,u}$ or (equivalently) $n_{u,t}^* = (n_{t+1,t}^* - N) + n_{u,t+1}^*$.
Therefore,

$$B_1^*(r_{t+1}, n_{t+1}, \ldots, r_s, n_s) = \eta^*(n_{t+1,t+1}^*, n_{t+2,t+1}^*, \ldots, n_{s,t+1}^*)$$
$$= \eta^*(n_{t+1,t}^*, n_{t+2,t}^*, \ldots, n_{s,t}^*).$$

Now our lemma follows from Lemma 7.5.5, because

$$\widetilde{\eta}(r_t, n_t, \ldots, r_1, n_1) = \eta^*(r_t, n_{tt}^*, \ldots, r_1, n_{1t}^*).$$

7.5.7. PROPOSITION. *Let*

$$1 \leq t \leq s < p, \quad r_1, \ldots, r_s \in R, \quad 0 \leq n_1, \ldots, n_s < N.$$

Then we have the following identity:

$$\sum_{t \leq m < s} (-1)^{m+t+1} B_t^*(r_1, n_{1s}, \ldots, r_m, n_{ms}) \eta(r_{m+1}, n_{m+1,s}, \ldots, r_s, n_{ss})$$
$$+ \widetilde{\eta}(r_s, n_s, \ldots, r_1, n_1)|_{n_t = \cdots = n_s} = (-1)^{s+t} B_t(r_1, n_1, \ldots, r_s, n_s),$$

where by definition $\widetilde{\eta}(r_s, n_s, \ldots, r_1, n_1)|_{n_t = \cdots = n_s}$ is equal to $\widetilde{\eta}(r_s, n_s, \ldots, r_1, n_1)$ if $n_t = \cdots = n_s$, and is equal to 0 otherwise.

PROOF. We have:

$$B_t(r_1, n_1, \ldots, r_s, n_s) = B_t(r_1, n_{1s}, \ldots, r_s, n_{ss}) \quad \text{(by the definition from 7.5.4)}$$
$$= B_t^*(r_1, n_{1s}, \ldots, r_s, n_{ss}) \quad \text{(see the example in 7.5.6)}.$$

So, we can write our identity in the form:

$$(*) \quad \sum_{t \leqslant m \leqslant s} (-1)^{m+s} B_t^*(r_1, n_{1s}, \ldots, r_m, n_{ms}) \eta(r_{m+1}, n_{m+1,s}, \ldots, r_s, n_{ss})$$
$$= (-1)^{s+t} \widetilde{\eta}(r_s, n_s, \ldots, r_1, n_1)|_{n_1 = \cdots = n_s}.$$

Step 1. Let $t = 1$. Then

$$B_1^*(r_1, n_{1s}, \ldots, r_m, n_{ms}) = \eta(r_m, n_{ms}, \ldots, r_1, n_{1s})$$

for $1 \leqslant m \leqslant s$ (see the example in 7.5.6). In self-evident notation, we must prove:

$$\eta(s, \ldots, 1) - \eta(s-1, \ldots, 1)\eta(s) + \cdots$$
$$+ (-1)^l \eta(s-l, \ldots, 1)\eta(s-l+1, \ldots, s) + \cdots$$
$$+ (-1)^{s+1} \eta(1)\eta(2, \ldots, s) = (-1)^{s+1} \widetilde{\eta}(r_s, n_s, \ldots, r_1, n_1)|_{n_1 = \cdots = n_s}.$$

It follows from the lemma and the remark in 7.5.2 that the right-hand side of the above equality is equal to

$$(-1)^{s+1} \eta(1, \ldots, s) = (-1)^{s+1} \eta(r_1, n_{1s}, \ldots, r_s, n_{ss}).$$

It follows from the definition of the constants η (see 7.3.2), that

$$\eta(r_1, n_{1s}, \ldots, r_s, n_{ss}) \neq 0 \iff n_1 = \cdots = n_s$$

and, if $n_1 = \cdots = n_s$, then

$$\eta(r_1, n_{1s}, \ldots, r_s, n_{ss}) = \widetilde{\eta}(r_s, n_s, \ldots, r_1, n_1).$$

Step 2. Let $t = s$. Then the left-hand side of $(*)$ is equal to

$$B_s^*(r_1, n_{1s}, \ldots, r_s, n_{ss}) = \widetilde{\eta}(r_s, n_{ss}, \ldots, r_1, n_{1s}) = \widetilde{\eta}(r_s, n_s, \ldots, r_1, n_1)$$

and $(*)$ is obviously true.

Step 3. Let $1 < t < s$ and $n_{ts} \neq n_{t+1,s}$.
If $n_{ts} < n_{t+1,s}$, there is nothing to prove, because both sides of $(*)$ vanish.
If $n_{ts} > n_{t+1,s}$, then we can apply Lemma 7.5.6:

$$B_t^*(1, \ldots, m) = \widetilde{\eta}(t, \ldots, 1) B_1^*(t+1, \ldots, m).$$

Therefore the left-hand side of $(*)$ is equal to

$$\widetilde{\eta}(t, \ldots, 1)[B_1^*(t+1, \ldots, s) - B_1^*(t+1, \ldots, s-1)\eta(s) + \cdots$$
$$+ (-1)^{s+t+1} B_1^*(t+1)\eta(t+2, \ldots, s) + (-1)^{s+t} \eta(t+1, \ldots, s)] = 0,$$

by the first step.

Obviously, the right-hand side of $(*)$ also vanishes.

Step 4. Let $n_t = n_{t+1}$. Then we can assume that $n_t = n_{t+1} = \cdots = n_{t+l} \neq n_{t+l+1}$. By the third step, we have the assertion of our lemma with t replaced by $t + l$. Now we can apply the lemma from 7.5.6 and obtain the assertion of our lemma by inverse induction on l.

7.6. Consider the extension K' of K from 7.4. We can assume that K_{sep} is chosen in such a way that $K \subset K' \subset K_{\text{sep}}$. Then $A_K \subset A_{K'} \subset A_{K_{\text{sep}}}$ and we use these inclusions for the identification of $A_{K'_{\text{sep}}}$ and $A_{K_{\text{sep}}}$.

On the other hand, consider the isomorphism f of the fields K and K' from 7.4.2. The map f can be extended to isomorphisms K_{sep} and K'_{sep}, A_K and $A_{K'}$, $A_{K_{\text{sep}}}$ and $A_{K'_{\text{sep}}}$, respectively. The composition of the isomorphism $A_{K_{\text{sep}}} \to A_{K'_{\text{sep}}}$ with the above identification of $A_{K'_{\text{sep}}}$ and $A_{K_{\text{sep}}}$ will be denoted by the same symbol f.

The following facts are obvious consequences of this definition.

(1) Let

$$E = 1 + \sum_{\substack{1 \leqslant s < p \\ r_1, \ldots, r_s \in R}} \eta(r_1, \ldots, r_s) t^{r_1 + \cdots + r_s} D_{r_1, 0} \ldots D_{r_s, 0} \in A_{K_{\text{tr}}}$$

(see 7.3). Then

$$E' = f(E) = 1 + \sum_{\substack{1 \leqslant s < p \\ r_1, \ldots, r_s \in R}} \eta(r_1, \ldots, r_s) t_1^{r_1 + \cdots + r_s} D_{r_1, 0} \ldots D_{r_s, 0} \in A_{K'_{\text{tr}}},$$

where $K'_{\text{tr}} = K' K_{\text{tr}}$.

(2) Consider $\mathcal{F} \in A_{K_{\text{sep}}}$ from 7.3 and set $f(\mathcal{F}) = \mathcal{F}'$, $E'_0 = f(E_0) = E' - 1$, $\mathcal{F}'^* = \mathcal{F}'^{(p)}$. Then

$$\mathcal{F}'^{(p)} \equiv \mathcal{F}' E' \mod J_p,$$

$$\mathcal{F}'^{(q)} \equiv \mathcal{F}'^* \left(1 + \sum_{\substack{1 \leqslant s < p \\ r_1 \ldots r_s \in R \\ 0 < n_1 \ldots n_s < N}} \eta(r_1, n_1, \ldots, r_s, n_s) t_1^{r_1 p^{n_1} + \cdots + r_s p^{n_s}} D_{r_1, n_1} \ldots D_{r_s, n_s} \right)$$

$$\mod J_p,$$

$$\mathcal{F}'^{(q)} - \mathcal{F}' \equiv \sum_{0 \leqslant n < N} (\mathcal{F}' E'_0)^{(p^n)} \mod J_p.$$

(3) For $1 \leqslant s < p$, the field of definition of $\mathcal{F}'$ mod J_{s+1} is $K'_{R, N, s} = f(K_{R, N, s})$. The field of definition of $\mathcal{F}'$ mod $(\mathcal{L}_s(v_0) A_{\text{sep}} + J_{s+1})$ is $K'_s(v_0) = f(K_s(v_0))$, i.e., the maximal Galois extension of K' inside $K'_{R, N}$ that has the higher ramification subgroup of class of nilpotency $\leqslant s$ and upper ramification numbers $< v_0$.

7.7. Inductive assumption. We use an induction on s^* in order to prove the subsequent assertions for $1 \leqslant s^* < p$. Obviously, our theorem is a consequence of the following statement.

PROPOSITION. *Let* $1 \leqslant s^* < p$. *Then*
(a) $\mathcal{L}_{s^*}(v_0) = \mathcal{L}_{R, N}^{(v_0)} + C_{s^* + 1}$;
(b) $K_{s^*}(v_0) K' = K'_{s^*}(v_0)$;

(c)
$$\mathcal{F} \equiv \mathcal{F}'^{(q)} + X(s^*) \mod \left(\mathcal{L}_{s^*}(v_0) A_{\text{sep}} + \sum_{1 \leq s \leq s^*} t^{-r^*(p-s)} J_s(O_{\text{sep}}) + J_{s^*+1} \right),$$

where

$$X(s^*) \equiv \sum_{\substack{1 \leq t \leq s \leq s^* \\ r_1, \ldots, r_s \in R \\ 0 \leq n_1, \ldots, n_s < N}} \mathcal{F}'^{*(p^{n_t})} (-1)^{s+t+1} B_t^*(r_1, n_1, \ldots, r_s, n_s)$$

$$\times \left[t_1^{q\left(\frac{r_1}{p^{n_t 1}} + \cdots + \frac{r_s}{p^{n_t s}}\right)} (e_1^{r_t} - 1) \right]^{p^{n_t}} e_1^{r_{t+1} p^{n_t+1}} \cdots e_1^{r_s p^{n_s}} D_{r_1, n_1} \cdots D_{r_s, n_s} \pmod{J_{s^*+1}}$$

(we use the agreement about indices from 7.5.1 and the element $e_1 \in K'_{\text{tr}}$ from Lemma 7.4.1);

(d) we note that $B_t(r_1, n_1, \ldots, r_s, n_s)$ and $D_{r_1, n_1}, \ldots, D_{r_s, n_s}$ depend only on the residues of $n_1, \ldots, n_s$ modulo N. Then

$$A(s^*)_0 = \sum (-1)^{s+t} B_t(r_1, n_1^*, \ldots, r_s, n_s^*) \left[t_1^{q\left(\frac{r_1}{p^{N-n_1^*}} + \cdots + \frac{r_s}{p^{N-n_s^*}}\right)} (e_1^{r_t} - 1) \right]^{p^N}$$

$$\times e_1^{r_{t+1} p^{n_{t+1}^*}} \cdots e_1^{r_s p^{n_s^*}} D_{r_1, n_1^*} \cdots D_{r_s, n_s^*},$$

where the sum is taken over

$$1 \leq t \leq s \leq s^*, \quad r_1, \ldots, r_s \in R, \quad 0 < n_1^*, \ldots, n_s^* \leq N, \quad n_t^* = N$$

is an element of

$$\mathcal{L}_{s^*}(v_0) A_{\text{sep}} + \sum_{1 \leq s \leq s^*} t^{-r^*(p-s)} J_s(O_{\text{sep}}) + J_{s^*+1}.$$

7.8. The case $s^* = 1$. This case is very simple. We take an element $\mathcal{F}$ in the form
$$\mathcal{F} \equiv 1 + \sum_{\substack{r \in R \\ 0 \leq n < N}} T_{r,n} D_{r,n} \mod J_2.$$

Then the equivalence (see 7.3.4)

$$\mathcal{F}^{(q)} - \mathcal{F} \equiv \sum_{0 \leq n < N} (\mathcal{F} E_0)^{(p^n)} \equiv \sum_{\substack{0 \leq n < N \\ r \in R}} t^{rp^n} D_{r,n} \mod J_2$$

yields the equations $T_{r,n}^q - T_{r,n} = t^{rp^n}$, where $r \in R$, $0 \leq n < N$ and we conclude from 6.4 that $\mathcal{L}_1(v_0) \otimes k \mod J_2$ is generated by

$$\{ D_{r,n} \mid r \geq v_0, \ 0 \leq n < N \}.$$

This set is the set of generators of $(\widetilde{\mathcal{L}}_{R,N}^{(v_0)} \otimes k) \bmod C_2 \otimes k$, see 5.1. So, we have $\mathcal{L}_1(v_0) \bmod C_2 = \widetilde{\mathcal{L}}_{R,N}^{(v_0)} \bmod C_2$. We also have

$$E_0 \equiv E_0'^{(q)} + \sum_{r \in R} t_1^{qr}(e_1^r - 1) D_{r,0} \bmod J_2,$$

where E_0' was defined in 7.6. Let $\mathcal{F}' \in A_{K_{\text{sep}}}$ be the element from 7.6. Then the identity from 7.3.4 gives

$$\mathcal{F}^{(q)} - \mathcal{F} \equiv \sum_{0 \leqslant n < N} E_0^{(p^n)} \equiv \sum_{0 \leqslant n < N} \left[(E_0')^{(p^n)}\right]^{(q)} - X(1)$$

$$\equiv \left[(\mathcal{F}')^{(q)} - \mathcal{F}'\right]^{(q)} - X(1) \pmod{J_2},$$

where

$$X(1) = - \sum_{\substack{0 \leqslant n < N \\ r \in R}} \left[t_1^{qr}(e_1^r - 1)\right]^{p^n} D_{r,n}.$$

We have $\mathcal{F} \equiv \mathcal{F}'^{(q)} + X(1) + Y \bmod J_2$, where

$$Y^{(q)} - Y = -X(1)^{(q)} = \sum_{0 \leqslant n < N} A(1)_0^{(p^n)}, \qquad A(1)_0 = \sum_{r \in R} \left[t_1^{qr}(e_1^r - 1)\right]^q D_{r,0}.$$

One may check that $A(1)_0 \in \mathcal{L}_1(v_0) A_{\text{sep}} + t^{-r^*(p-1)} J_1(O_{\text{sep}})$. Indeed, if $r \geqslant v_0$, then $D_{r,0} \in \mathcal{L}_1(v_0) \otimes k$, as shown earlier. If $r < v_0$, then $t_1^{qr-r^*(q-1)} \in t_1^{-r^*(p-1)} O_{\text{tr}}'$ (see 7.4.3), so that

$$\left[t_1^{qr}(e_1^r - 1)\right]^q D_{r,0} \in t^{-r^*(p-1)} J_1(O_{\text{sep}}).$$

Now it is clear that $Y \in \mathcal{L}_1(v_0) A_{\text{sep}} + t^{-r^*(p-1)} J_1(O_{\text{sep}})$. Therefore,

$$\mathcal{F} \equiv \mathcal{F}'^{(q)} + X(1) \bmod (\mathcal{L}_1(v_0) A_{\text{sep}} + t^{-r^*(p-1)} J_1(O_{\text{sep}}) + J_2).$$

The fact that $X(1)$ is defined over K' implies that $K_1(v_0) K' = K_1'(v_0)$.

7.9. Some calculations. Let s_0 satisfy $1 < s_0 < p$ and assume that our inductive assumption (see the proposition in 7.7) is valid for all $1 \leqslant s^* < s_0$.

7.9.1. PROPOSITION. *Let E_0 be the element from 7.3.4. Then*

$$E_0 \in \mathcal{L}_{s_0-1}(v_0) A_{\text{sep}} + \sum_{1 \leqslant s < s_0} t^{r^* s} J_s(O_{\text{sep}}) + J_{s_0}.$$

PROOF. We use the following lemma.

LEMMA. *Let* $r \in R$, $s_1 \in \mathbb{N}$, $0 \leqslant n < N$ *and* $r \geqslant s_1 r^*$. *Then*
$$D_{r,n} \in (\mathcal{L}_{s_0-1}(v_0) + C_s) \otimes k,$$
where $s = \min\{s_1 + 1, s_0\}$.

PROOF. By 7.7 (a), $\mathcal{L}_{s_0-1}(v_0) = \mathcal{L}_{R,N}^{(v_0)} + C_{s_0}$, so $\mathcal{L}_{s_0-1}(v_0) \otimes k \bmod C_{s_0} \otimes k$ is generated (as an ideal) by the elements $\mathcal{F}_{R,N}(\gamma, n)$, where $\gamma \geqslant v_0$, $0 \leqslant n < N$ (see §5). Now we can apply induction on s_1 to show that, if $r \geqslant s_1 r^*$, then $\mathcal{F}_{R,N}(r, n) \equiv D_{r,n} \bmod (\mathcal{L}_{s_0-1} + C_s) \otimes k$, where $s = \min\{s_1 + 1, s_0\}$. The lemma is proved.

Now the above proposition can be proved as follows. The expression for $E_0 \bmod J_{s_0}$ is a linear combination over $\mathbb{F}_p$ of the terms
$$t^{r_1 + \cdots + r_l} D_{r_1, 0} \ldots D_{r_l, 0},$$
where $1 \leqslant l < s_0$, $r_1, \ldots, r_l \in R$. We use induction on l to show that these terms are in $\mathcal{L}_{s_0-1}(v_0) A_{\text{sep}} + \sum_{1 \leqslant s < s_0} t^{r^* s} J_s(O_{\text{sep}})$.

If $l = 1$ and $(s_1 + 1)r^* > r_1 \geqslant s_1 r^*$, the above lemma gives
$$D_{r_1, 0} \in \mathcal{L}_{s_0-1}(v_0) \otimes k + J_{s_1+1}(O_{\text{sep}}) + J_{s_0},$$
therefore, $t^{r_1} D_{r_1, 0} \in \mathcal{L}_{s_0-1}(v_0) A_{\text{sep}} + t^{(s_1+1)r^*} J_{s_1+1}(O_{\text{sep}}) + J_{s_0}$.

Let $l > 1$ and $(s+1)r^* > r_1 + \cdots + r_l \geqslant sr^*$. By the inductive assumption we have
$$(s_1 + 1)r^* > r_1 + \cdots + r_{l-1} \geqslant s_1 r^*$$
$$\implies t^{r_1 + \cdots + r_{l-1}} D_{r_1, 0} \ldots D_{r_{l-1}, 0} \in \mathcal{L}_{s_0-1}(v_0) A_{\text{sep}} + t^{(s_1+1)r^*} J_{s_1+1}(O_{\text{sep}}) + J_{s_0}.$$

The above inequalities imply $r_l \geqslant (s - s_1 - 1)r^*$. Therefore,
$$t^{r_l} D_{r_l, 0} \in \mathcal{L}_{s_0-1}(v_0) A_{\text{sep}} + t^{(s-s_1)r^*} J_{s-s_1}(O_{\text{sep}}) + J_{s_0},$$
and we obtain $t^{r_1 + \cdots + r_l} D_{r_1, 0} \ldots D_{r_l, 0} \in \mathcal{L}_{s_0-1}(v_0) A_{\text{sep}} + t^{(s+1)r^*} J_{s+1}(O_{\text{sep}}) + J_{s_0}$.
The proposition is proved.

COROLLARY.
$$\mathcal{F} E_0 \equiv (\mathcal{F}'^{(q)} + X(s_0 - 1)) E_0$$
$$\bmod \left(\mathcal{L}_{s_0-1}(v_0) J_1 + \sum_{1 \leqslant s \leqslant s_0} t^{-r^*(p-s)} J_s(O_{\text{sep}}) + J_{s_0+1} \right).$$

7.9.2. *Computation of* $X(s_0 - 1) E_0$. Write $X(s_0 - 1)$ in the following form:
$$X(s_0 - 1) \equiv \sum_{\substack{1 \leqslant t \leqslant m < s_0 \\ 0 \leqslant n_1, \ldots, n_m < N \\ r_1, \ldots, r_m \in R}} \mathcal{F}'^{*(p^{n_t})} (-1)^{m+t+1} B_t^*(r_1, n_1, \ldots, r_m, n_m)$$
$$\times \left[t_1^{q\left(\frac{r_1}{p^{n_{t_1}}} + \cdots + \frac{r_m}{p^{n_{t_m}}}\right)} (e_1^{r_t} - 1) \right]^{p^{n_t}}$$
$$\times e_1^{r_{t+1} p^{n_{t+1}}} \ldots e_1^{r_m p^{n_m}} D_{r_1, n_1} \ldots D_{r_m, n_m} \pmod{J_{s_0}}.$$

For a fixed m we have

$$E_0 = \sum_{\substack{m<s<m+s_0 \\ r_{m+1},\ldots,r_s \in R}} \eta(r_{m+1},\ldots,r_s) t_1^{q(r_{m+1}+\cdots+r_s)} e_1^{r_{m+1}} \ldots e_1^{r_s} D_{r_{m+1},0} \ldots D_{r_s,0} \pmod{J_{s_0}}.$$

This relation can be written in the following form:

$$E_0 = \sum_{\substack{m<s<m+s_0 \\ r_{m+1},\ldots,r_s \in R \\ 0 \leqslant n_{m+1,s},\ldots,n_{s-1,s}<N}} \eta(r_{m+1}, n_{m+1,s}, \ldots, r_s, n_{ss}) t_1^{q(r_{m+1}+\cdots+r_s)}$$
$$\times e_1^{r_{m+1}p^{n_{m+1,s}}} \ldots e_1^{r_s p^{n_{ss}}} D_{r_{m+1},n_{m+1,s}} \ldots D_{r_s,n_{ss}} \pmod{J_{s_0}},$$

because

$$\eta(r_{m+1}, n_{m+1,s}, \ldots, r_s, n_{ss}) = \begin{cases} \eta(r_{m+1},\ldots,r_s) & \text{for } n_{m+1} = \cdots = n_s, \\ 0, & \text{otherwise.} \end{cases}$$

Therefore,

$$X(s_0-1)E_0 \equiv \sum \mathcal{F}'^{*(p^{n_{ts}})}(-1)^{m+t+1} B_t^*(r_1, n_{1s}, \ldots, r_s, n_{ms})$$
$$\times \eta(r_{m+1}, n_{m+1,s}, \ldots, r_s, n_{ss})$$
$$\times \left[t_1^{q\left(\frac{r_1}{p^{n_{t1}}}+\cdots+\frac{r_s}{p^{n_{ts}}}\right)}(e_1^{r_t}-1) \right]^{p^{n_{ts}}} e_1^{r_{t+1}p^{n_{t+1,s}}} \ldots e_1^{r_s p^{n_{ss}}}$$
$$\times D_{r_1,n_{1,s}} \ldots D_{r_s,n_{s,s}} \pmod{J_{s_0+1}},$$

where the sum is taken over

$$1 \leqslant t \leqslant m < s < s_0, \quad 0 \leqslant n_{1s}, \ldots, n_{s-1,s} < N,$$
$$n_{m+1,s} = \cdots = n_{s-1,s} = 0, \quad r_1, \ldots, r_s \in R$$

(multiplying $X(s_0 - 1)$ by the component of E_0 with index s, we use the indices $n_{1s}, \ldots, n_{ms}$ in the expression of $X(s_0 - 1)$).

7.9.3. *Computation of $\mathcal{F}'^{(q)}E_0$*. We use the convention that the empty sum is equal to 1. Then

$$\mathcal{F}'^{(q)} = \mathcal{F}'^* \sum_{\substack{1 \leqslant m < s_0 \\ 0 < n_1, \ldots, n_{m-1} < N \\ r_1, \ldots, r_{m-1} \in R}} t_1^{r_1 p^{n_1}+\cdots+r_{m-1}p^{n_{m-1}}} \eta(r_1, n_1, \ldots, r_{m-1}, n_{m-1})$$
$$\times D_{r_1,n_1} \ldots D_{r_{m-1},n_{m-1}} \pmod{J_{s_0}}.$$

For fixed m we have

$$E_0 = E_0'^{(q)} + \sum_{\substack{m \leqslant t \leqslant s<m+s_0 \\ r_m,\ldots,r_s}} t_1^{q(r_m+\cdots+r_s)} \eta(r_m,\ldots,r_s)(e_1^{r_t}-1) e_1^{r_{t+1}} \ldots e_1^{r_s} D_{r_m,0} \ldots D_{r_s,0}.$$

and, as before, this can be written in the form:

$$E_0 = E_0^{\prime(q)} + \sum_{m \leqslant t \leqslant s < m+s_0} t_1^{q(r_m+\cdots+r_s)} \eta(r_m, n_{m,s}, \ldots, r_s, n_{ss})$$
$$\times (e_1^{r_t} - 1)^{p^{n_{t,s}}} e_1^{r_{t+1} p^{n_{t+1,s}}} \ldots e_1^{r_s p^{n_{ss}}} D_{r_m, n_{ms}} \ldots D_{r_s, n_{ss}}.$$

Therefore,

$$\mathcal{F}^{\prime(q)} E_0 = (\mathcal{F}' E_0')^{(q)} + \sum_{\substack{1 \leqslant m \leqslant t \leqslant s \leqslant s_0 \\ n_{1s}, \ldots, n_{m-1,s} \neq 0 \\ r_1, \ldots, r_s \in R}} \mathcal{F}'^* \eta(r_1, n_{1s}, \ldots, r_{m-1}, n_{m-1,s})$$
$$\times \eta(r_{m+1}, n_{m+1,s}, \ldots, r_s, n_{s,s}) \left[t_1^{q\left(\frac{r_1}{p^{n_{t1}}} + \cdots + \frac{r_s}{p^{n_{ts}}}\right)} \right]^{p^{n_{ts}}}$$
$$\times (e_1^{r_t} - 1)^{p^{n_{ts}}} e_1^{r_{t+1} p^{n_{t+1,s}}} \ldots e_1^{r_s p^{n_{ss}}} D_{r_1, n_{1s}} \ldots D_{r_s, n_{ss}} \pmod{J_{s_0+1}}.$$

Note that

$$\sum_{1 \leqslant m \leqslant t} \eta(r_1, n_{1s}, \ldots, r_{m-1}, n_{m-1,s})|_{n_{1s}, \ldots, n_{m-1,s} \neq 0} \eta(r_m, n_{m,s}, \ldots, r_s, n_{ss})$$
$$= \widetilde{\eta}(r_s, n_s, \ldots, r_1, n_1)|_{n_t = \cdots = n_s},$$

where $\widetilde{\eta}(r_s, n_s, \ldots, r_1, n_1)|_{n_t=\cdots=n_s}$ is equal to $\widetilde{\eta}(r_s, n_s, \ldots, r_1, n_1)$, if $n_t = \cdots = n_s$, and is equal to 0, otherwise.

7.9.4. One can apply the identity of Proposition 7.5.7 to calculate the sum of $\mathcal{F}^{\prime(q)} E_0$ and $X E_0$. We obtain:

$$\mathcal{F} E_0 \equiv (\mathcal{F}' E_0')^{(q)} + \sum_{\substack{1 \leqslant t \leqslant s \leqslant s_0 \\ 0 \leqslant n_1, \ldots, n_s < N \\ r_1, \ldots, r_s \in R}} \mathcal{F}'^{*(p^{n_{ts}})} (-1)^{s+t} B_t(r_1, n_1, \ldots, r_s, n_s)$$
$$\times \left[t_1^{q\left(\frac{r_1}{p^{n_{t1}}} + \cdots + \frac{r_s}{p^{n_{ts}}}\right)} (e_1^{r_t} - 1) \right]^{p^{n_{ts}}} e_1^{r_{t+1} p^{n_{t+1,s}}} \ldots e_1^{r_s p^{n_{ss}}}$$
$$\times D_{r_1, n_{1,s}} \ldots D_{r_s, n_{s,s}} \pmod{J_{s_0+1}}.$$

Therefore,

$$\mathcal{F}^{(q)} - \mathcal{F} = \left[\mathcal{F}^{\prime(q)} - \mathcal{F}' \right]^{(q)} + A_1$$
$$\mod \left(\mathcal{L}_{s_0-1}(v_0) J_1 + \sum_{1 \leqslant s \leqslant s_0} t^{-r^*(p-s)} J_s(O_{\text{sep}}) + J_{s_0+1} \right),$$

where

$$A_1 \equiv \sum_{\substack{1 \leqslant t \leqslant s \leqslant s_0 \\ 0 \leqslant n_1, \ldots, n_s < N \\ r_1, \ldots, r_s \in R}} \mathcal{F}'^{*(p^{n_{ts}+n_s})} (-1)^{s+t} B_t(r_1, n_1, \ldots, r_s, n_s)$$
$$\times \left[t_1^{q\left(\frac{r_1}{p^{n_{t1}}} + \cdots + \frac{r_s}{p^{n_{ts}}}\right)} (e_1^{r_t} - 1) \right]^{p^{n_{ts}+n_s}} e_1^{r_{t+1} p^{n_{t+1,s}+n_s}} \ldots e_1^{r_s p^{n_{ss}+n_s}}$$
$$\times D_{r_1, n_1} \ldots D_{r_s, n_s} \pmod{J_{s_0+1}}.$$

7.10. Let $X_1 \in A_{K_{\text{sep}}}$ satisfy $X_1^{(q)} - X_1 = A_1$. Then the above calculation gives

$$\mathcal{F} = \mathcal{F}'^{(q)} + X_1 \mod \left(\mathcal{L}_{s_0-1}(v_0)J_1 + \sum_{1 \leqslant s \leqslant s_0} t^{-r^*(p-s)} J_s(O_{\text{sep}}) + J_{s_0+1} \right).$$

Let I be any ideal of the Lie algebra $\mathcal{L}$ such that $I \supset C_{s_0+1}(\mathcal{L})$. It is clear from the proposition in 7.2, that $\mathcal{L}_{s_0}(v_0)$ is the minimal element in the subset of such ideals having the following property:

the field of definition of $\mathcal{F}$ mod $(IA_{\text{sep}} + J_{s_0+1})$

has the largest upper ramification number less than v_0.

By induction we can assume that $I \supset (\mathcal{L}_{s_0-1}(v_0)J_1) \cap \mathcal{L}$ or (equivalently) $IA_{\text{sep}} \supset \mathcal{L}_{s_0-1}(v_0)J_1 + J_{s_0+1}$.

PROPOSITION. $\mathcal{L}_{s_0}(v_0)$ *is the minimal element in the set of all ideals of the Lie algebra $\mathcal{L}$ such that*
(a) $IA_{\text{sep}} \supset \mathcal{L}_{s_0-1}(v_0)J_1 + J_{s_0+1}$;
(b) *the field of definition of X_1 mod $(IA_{\text{sep}} + J_{s_0+1})$ has the largest upper ramification number less than v_0.*

PROOF. It is clear that $\mathcal{L}_{s_0}(v_0)$ satisfies condition (a) of the proposition. Let I be an arbitrary ideal of $\mathcal{L}$ satisfying (a). Let $\mathcal{F} = \mathcal{F}'^{(q)} + Y_1$. Then

$$X_1 \equiv Y_1 \mod \left(\mathcal{L}_{s_0-1}(v_0)J_1 + \sum_{1 \leqslant s \leqslant s_0} t^{-r^*(p-s)} J_s(O_{\text{sep}}) + J_{s_0+1} \right).$$

The field of definition of X_1 mod $(IA_{\text{sep}} + J_{s_0+1})$ has the largest upper ramification number less than v_0 if and only if the field of definition of Y_1 mod $(IA_{\text{sep}} + J_{s_0+1})$ has the largest upper ramification number $< v_0$. Let $\mathcal{L}(I)$ be the field of definition of Y_1 mod $(IA_{\text{sep}} + J_{s_0+1})$, $K(I)$ be the field of definition of $\mathcal{F}$ mod $(IA_{\text{sep}} + J_{s_0+1})$, then $K'(I) := f(K(I))$ will be the field of definition of $\mathcal{F}'$ mod $(IA_{\text{sep}} + J_{s_0+1})$ (isomorphism $f : K_{\text{sep}} \to K'_{\text{sep}}$ was defined in 7.6). The equality $\mathcal{F} = \mathcal{F}'^{(q)} + Y_1$ gives $K(I) \subset K'(I)L(I)$ and $L(I) \subset K(I)K'(I)$. So, our proposition follows from Lemma 7.4.2.

7.11. Some calculations.

7.11.1. Let (see 7.7(c))

$$X(s_0) = \sum_{\substack{1 \leqslant t \leqslant s \leqslant s_0 \\ 0 \leqslant n_1, \ldots, n_s < N \\ r_1, \ldots, r_s \in R}} \mathcal{F}'^{*(p^{n_t})}(-1)^{s+t+1} B_t^*(r_1, n_1, \ldots, r_s, n_s)$$

$$\times \left[t_1^{q\left(\frac{r_1}{p^{n_1}} + \cdots + \frac{r_s}{p^{n_s}}\right)} (e_1^{r_1} - 1) \right]^{p^{n_t}} e_1^{r_{t+1}p^{n_{t+1}}} \ldots e_1^{r_s p^{n_s}} D_{r_1, n_1} \ldots D_{r_s, n_s}.$$

This sum consists of all terms of the above expression for $-A_1$ (see 7.9) that satisfy the additional condition $n_t \geqslant n_s$.

Let $X_1' = X_1 - X(s_0)$, then $X_1'^{(q)} - X_1' = A_1 - (X(s_0)^{(q)} - X(s_0)) = A_1'$, where

$$A_1' \equiv \sum_{\substack{1 \leq t \leq s \leq s_0 \\ 0 \leq n_1, \ldots, n_s < N \\ r_1, \ldots, r_s \in R}} \mathcal{F}'^{*(p^{n_t + N})}(-1)^{s+t} B_t(r_1, n_1, \ldots, r_s, n_s)$$

$$\times \left[t_1^{q\left(\frac{r_1}{p^{n_t 1}} + \cdots + \frac{r_s}{p^{n_{ts}}}\right)} (e_1^{r_t} - 1) \right]^{p^{n_t + N}}$$

$$\times e_1^{r_{t+1} p^{n_t + N - n_{t,t+1}}} \cdots e_1^{r_s p^{n_t + N - n_{t,s}}} D_{r_1, n_1} \ldots D_{r_s, n_s} \pmod{J_{s_0 + 1}}.$$

It is easy to see that

$$A_1' \equiv \sum_{0 \leq m < N} \left[\mathcal{F}'^{*(q)} A(s_0)_0 \right]^{(p^m)},$$

where $A(s_0)_0$ is given by the formula in 7.7(d) with $s^* = s_0$.

LEMMA. *We have*

$$\mathcal{F}'^* - 1 \in \mathcal{L}_{s_0 - 1}(v_0) A_{\text{sep}} + \sum_{1 \leq s < s_0} t_1^{r^* s} J_s(O_{\text{sep}}) + J_{s_0}.$$

PROOF. This follows from the equality $\mathcal{F}'^* = \mathcal{F}'^{(p)}$, from the equivalence $\mathcal{F}'^{(p)} \equiv \mathcal{F}' E' \mod J_p$ (see 7.6), and Proposition 7.9.1.

From this lemma and the inductive assumption 7.7(d) it follows that

$$\sum_{0 \leq m < N} \left[\mathcal{F}'^{*(q)} - 1 \right]^{p^m} A(s_0)_0^{(p^m)} \in \mathcal{L}_{s_0 - 1}(v_0) J_1 + \sum_{1 \leq s \leq s_0} t^{-r^*(p-s)} J_s(O_{\text{sep}}) + J_{s_0 + 1}$$

(we use the fact that $A(s_0)_0 \equiv A(s_0 - 1)_0 \mod J_{s_0}$), therefore,

$$A_1' \equiv \sum_{0 \leq m < N} A(s_0)_0^{(p^m)} \mod \left(\mathcal{L}_{s_0 - 1}(v_0) J_1 + \sum_{1 \leq s \leq s_0} t^{-r^*(p-s)} J_s(O_{\text{sep}}) + J_{s_0 + 1} \right).$$

Let $X_1'' \in A_{\text{sep}}$ satisfy

$$X_1''^{(q)} - X_1'' = \sum_{0 \leq m < N} A(s_0)_0^{(p^m)}.$$

Obviously,

$$X_1'' \equiv X_1' \mod \left(\mathcal{L}_{s_0 - 1}(v_0) J_1 + \sum_{1 \leq s \leq s_0} t^{-r^*(p-s)} J_s(O_{\text{sep}}) + J_{s_0 + 1} \right)$$

and we have the following reduction.

PROPOSITION. $\mathcal{L}_{s_0}(v_0)$ *is the minimal element in the set of all ideals of the Lie algebra $\mathcal{L}$ such that*
(a) $IA_{\text{sep}} \supset \mathcal{L}_{s_0-1}(v_0)J_1 + J_{s_0+1}$;
(b) *the field of definition of X_1'' mod $(IA_{\text{sep}} + J_{s_0+1})$ has the largest upper ramification number less than v_0.*

7.11.2. One can rewrite the expression for $A(s_0)_0$ in the following form:

$$A(s_0)_0 = \sum_{\gamma \in \mathbb{Q}} A(\gamma)_0 \left[t_1^{q\gamma}\right]^q,$$

$$A(\gamma)_0 = \sum \sum_{\substack{1 \leqslant t \leqslant s \\ n_t = N}} (-1)^{s+t} B_t(r_1, n_1, \ldots, r_s, n_s)$$

$$\times (e_1^{r_t} - 1)^{p^{n_t}} e_1^{r_{t+1}p^{n_t+1}} \ldots e_1^{r_s p^{n_s}} D_{r_1,n_1} \ldots D_{r_s,n_s},$$

where the first sum is taken over

$$1 \leqslant s \leqslant s_0, \quad 0 < n_1, \ldots, n_s \leqslant N,$$
$$r_1, \ldots, r_s \in R, \quad \frac{r_1}{p^{N-n_1}} + \cdots + \frac{r_s}{p^{N-n_s}} = \gamma.$$

7.11.3. For a positive rational number γ and a natural number n^* such that $0 < n^* \leqslant N$, we introduce the elements A_{γ,n^*} of $A_{K_{\text{tr}}}$ given by the following expression:

$$A_{\gamma,n^*} = \sum \sum (-1)^{s+t} B_t(r_1, n_1, \ldots, r_s, n_s)$$

$$\times \left[(e_1^{r_t} - 1)e_1^{r_{t+1}+\cdots+r_{t_2}}\right]^{p^{n^*}} D_{r_1,n_1} \ldots D_{r_s,n_s},$$

where the first sum is taken over

$$1 \leqslant s \leqslant s_0, \quad 0 < n_1, \ldots, n_s \leqslant n^*, \quad r_1, \ldots, r_s \in R, \quad \frac{r_1}{p^{n_{*1}}} + \cdots + \frac{r_s}{p^{n_{*s}}} = \gamma$$

and the second over

$$t_1 \leqslant t \leqslant t_2, \quad n_{t_1} = \cdots = n_{t_2} = n^*, \quad n_{t_1-1}, n_{t_2+1} \neq n^*$$

(we use the abbreviation $n_{*l} = n_* - n_l$ for $1 \leqslant l \leqslant s$).

PROPOSITION. *We have*

$$A(\gamma)_0 \equiv A_{\gamma,N} + \sum A_{\gamma_0,N} A_{\gamma_1,m_1}$$
$$+ \sum A_{\gamma_0,N} A_{\gamma_1,m_1} A_{\gamma_2,m_2} + \ldots \pmod{J_{s_0+1}},$$

where the first sum is taken over

$$N > m_1 > 0, \quad \gamma_0, \gamma_1 \in \mathbb{Q}, \quad \gamma_0 + \frac{\gamma_1}{p^{N-m_1}} = \gamma$$

and the second over

$$N > m_1 > m_2 > 0, \quad \gamma_0, \gamma_1, \gamma_2 \in \mathbb{Q}, \quad \gamma_0 + \frac{\gamma_1}{p^{N-m_1}} + \frac{\gamma_2}{p^{N-m_2}} = \gamma.$$

PROOF. For any collection $(r_1, n_1, \ldots, r_s, n_s)$, where $r_1, \ldots, r_s \in R$, $0 < n_1, \ldots, n_s \leqslant N$, and index t, such that $t_1 \leqslant t \leqslant t_2$, where $n_{t_1} = n_{t_2} = N$ and $n_{t_1-1}, n_{t_2+1} \neq N$, we set

$$A^{(1)}(r_1, n_1, \ldots, r_s, n_s; t) = (-1)^{s+t} B_t(r_1, n_1, \ldots, r_{t_2}, n_{t_2})$$
$$\times \eta^*(r_{t_2+1}, n_{t_2+1}, \ldots, r_s, n_s)\left[(e_1^{r_t} - 1)e_1^{r_{t+1}+\cdots+r_{t_2}}\right]^{p^N}.$$

REMARK. If t_1, t_2 are not uniquely defined, all $A^{(1)}(r_1, n_1, \ldots, r_s, n_s; t)$ are automatically set equal to 0.

From the definition of the constants $B_t(r_1, n_1, \ldots, r_s, n_s)$ (see 7.5.4) and from remarks to Lemma 7.5.2, it follows that, for $t_1 \leqslant t \leqslant t_2$,

$$B_t(r_1, n_1, \ldots, r_s, n_s) = B_t(r_1, n_1, \ldots, r_{t_2}, n_{t_2})\eta^*(r_{t_2+1}, n_{t_2+1}, \ldots, r_s, n_s)$$

and, therefore,

$$A_{\gamma_0, N} = \sum\sum A^{(1)}(r_1, n_1, \ldots, r_s, n_s; t) D_{r_1, n_1} \ldots D_{r_s, n_s},$$

where the first sum is taken over

$$1 \leqslant s \leqslant s_0, \quad 0 < n_1, \ldots, n_s \leqslant N,$$
$$r_1, \ldots, r_s \in R, \quad \frac{r_1}{p^{N-n_1}} + \cdots + \frac{r_s}{p^{N-n_s}} = \gamma_0$$

and the second over

$$t_1 \leqslant t \leqslant t_2, \quad n_{t_1} = n_{t_2} = N, \quad n_{t_1-1}, n_{t_2+1} \neq N.$$

Let $r_1, \ldots, r_s \in R$, $n_1, \ldots, n_s \in \mathbb{N}$, $n^* = \max\{n_1, \ldots, n_s\}$, $n_i = n^*$ for $t_1 \leqslant i \leqslant t_2$ and $n_{t_1-1}, n_{t_2+1} \neq n^*$. Proceeding as in the proof of Lemma 7.5.5, we obtain the following relations:

$$B_{t_1}(r_1, n_1, \ldots, r_s, n_s) = \eta^*(r_{t_1-1}, n_{t_1-1}, \ldots, r_1, n_1)\eta^*(r_{t_1}, n_{t_1}, \ldots, r_s, n_s),$$
$$B_{t_1}(r_1, n_1, \ldots, r_s, n_s) + B_{t_1+1}(r_1, n_1, \ldots, r_s, n_s)$$
$$= \eta^*(r_{t_1}, n_{t_1}, \ldots, r_1, n_1)\eta^*(r_{t_1+1}, n_{t_1+1}, \ldots, r_s, n_s), \ldots$$
$$B_{t_2-1}(r_1, n_1, \ldots, r_s, n_s) + B_{t_2}(r_1, n_1, \ldots, r_s, n_s)$$
$$= \eta^*(r_{t_2-1}, n_{t_2-1}, \ldots, r_1, n_1)\eta^*(r_{t_2}, n_{t_2}, \ldots, r_s, n_s).$$

These relations and the trivial remark that $\eta^*(r_t, n_t, \ldots, r_s, n_s) = 0$ if $t < t_1$ give the following identity:

$$\sum_{t_1 \leqslant t \leqslant t_2} (-1)^t B_t(r_1, n_1, \ldots, r_s, n_s)\left[e_1^{r_t+\cdots+r_{t_2}} - e_1^{r_{t+1}+\cdots+r_{t_2}}\right]$$
$$= \sum_{1 \leqslant t \leqslant t_2} (-1)^t \eta^*(r_{t-1}, n_{t-1}, \ldots, r_1, n_1)\eta^*(r_t, n_t, \ldots, r_s, n_s)\left(e_1^{r_t+\cdots+r_{t_2}} - 1\right).$$

For a fixed index s, any collection $(r_{s+1}, n_{s+1}, \ldots, r_{\widehat{s}}, n_{\widehat{s}})$ and any index u such that $s+1 \leqslant u \leqslant u_2$, where $n_{u_2} = n^* := \max\{n_{s+1}, \ldots, n_{\widehat{s}}\}$, $n_{u_2+1} \neq n^*$, we set

$$A^{(2)}(r_{s+1}, n_{s+1}, \ldots, r_{\widehat{s}}, n_{\widehat{s}}; u) = (-1)^{\widehat{s}+u}\eta^*(r_{u-1}, n_{u-1}, \ldots, r_{s+1}, n_{s+1})$$
$$\times \eta^*(r_u, n_u, \ldots, r_{\widehat{s}}, n_{\widehat{s}})[e_1^{r_u+\cdots+r_{u_2}} - 1]^{p^{n^*}}.$$

Now the above identity gives

$$A_{\gamma_1, n^*} = \sum\sum (-1)^{\widehat{s}+u} A^{(2)}(r_{s+1}, n_{s+1}, \ldots, r_{\widehat{s}}, n_{\widehat{s}}; u) D_{r_{s+1}, n_{s+1}} \cdots D_{r_{\widehat{s}}, n_{\widehat{s}}},$$

where the first sum is taken over

$$s+1 \leqslant \widehat{s} \leqslant s+s_0, \quad 0 < n_{s+1}, \ldots, n_{\widehat{s}} \leqslant n^*,$$
$$r_{s+1}, \ldots, r_{\widehat{s}} \in R, \quad \frac{r_{s+1}}{p^{n_{*s+1}}} + \cdots + \frac{r_{\widehat{s}}}{p^{n_{*\widehat{s}}}} = \gamma_1$$

and the second over

$$s+1 \leqslant u \leqslant u_2, \quad n_{u_2} = n^*, \quad n_{u_2+1} \neq n^*.$$

The coefficient of $D_{r_1, n_1} \cdots D_{r_{\widehat{s}}, n_{\widehat{s}}}$ in the expression of $\sum_{\gamma_0, \gamma_1} A_{\gamma_0, N} A_{\gamma_1, n^*}$ is equal to the sum $\sum_{t_1 \leqslant t < u \leqslant u_2} C_{t, u}$, where

$$C_{t, u} = A^{(1)}(r_1, n_1, \ldots, r_{t_2}, n_{t_2}; t) A^{(2)}(r_{t_2+1}, n_{t_2+1}, \ldots, r_{\widehat{s}}, n_{\widehat{s}}; u)$$
$$+ A^{(1)}(r_1, n_1, \ldots, r_{t_2+1}, n_{t_2+1}; t) A^{(2)}(r_{t_2+2}, n_{t_2+2}, \ldots, r_{\widehat{s}}, n_{\widehat{s}}; u)$$
$$+ \cdots + A^{(1)}(r_1, n_1, \ldots, r_{u-1}, n_{u-1}; t) A^{(2)}(r_u, n_u, \ldots, r_{\widehat{s}}, n_{\widehat{s}}; u).$$

For $u \neq t_2$ we have the following identity:

$$\sum_{t_2 \leqslant s < u} (-1)^s \eta^*(r_{t_2+1}, n_{t_2+1}, \ldots, r_s, n_s) \eta^*(r_{u-1}, n_{u-1}, \ldots, r_{s+1}, n_{s+1}) = 0,$$

(see the arguments used in the first step of the proof of Lemma 7.5.5). This means that $C_{t, u} = 0$, if $t_2 + 1 \neq u$.
Therefore,

$$\sum A_{\gamma_0, N} A_{\gamma_1, n^*}$$
$$\equiv \sum (-1)^{s+t} \sum B_t(r_1, n_1, \ldots, r_{t_2}, n_{t_2}) [(e_1^{r_t} - 1) e_1^{r_{t+1}+\cdots+r_{t_2}}]^{p^N}$$
$$\times \eta^*(r_{t_2+1}, n_{t_2+1}, \ldots, r_s, n_s) [e_1^{r_{t_2+1}+\cdots+r_{t_3}} - 1]^{p^{n^*}}$$
$$\times D_{r_1, n_1} \cdots D_{r_s, n_s} \mod J_{s_0+1},$$

where the first sum is taken over

$$\gamma_0, \gamma_1: \gamma_0 + \frac{\gamma_1}{p^{N-n^*}} = \gamma,$$

the second over

$$1 \leqslant s \leqslant s_0, \quad 0 < n_1, \ldots, n_s \leqslant N,$$
$$r_1, \ldots, r_s \in R, \quad \frac{r_1}{p^{N-n_1}} + \cdots + \frac{r_s}{p^{N-n_s}} = \gamma_0,$$

the third over

$$t_1 \leqslant t \leqslant t_2, \quad n_{t_1} = \cdots = n_{t_2} = N, \quad n_{t_2+1} = \cdots = n_{t_3} = n^*,$$
$$n_{t_1-1} < N, \quad n_{t_3+1} < n^*.$$

Now we obtain

$$A_{\gamma,N} + \sum A_{\gamma_0,N} A_{\gamma_1,m_1}$$
$$\equiv \sum \sum (-1)^{s+t} B_t(r_1, n_1, \ldots, r_s, n_s) \left[(e_1^{r_t} - 1) e_1^{r_{t+1}+\cdots+r_{t_2}} \right]^{p^N}$$
$$\times \left[e_1^{r_{t_2+1}} \ldots e_1^{r_{t_3}} \right]^{p^{m_1}} D_{r_1,n_1} \ldots D_{r_s,n_s} \mod J_{s_0+1},$$

where the first sum is taken over

$$N > m_1 > 0, \quad \gamma_0, \gamma_1 : \gamma_0 + \frac{\gamma_1}{p^{N-m_1}} = \gamma,$$

the second over

$$1 \leqslant s \leqslant s_0, \quad 0 < n_1, \ldots, n_s \leqslant N,$$
$$r_1, \ldots, r_s \in R, \quad \frac{r_1}{p^{N-n_1}} + \cdots + \frac{r_s}{p^{N-n_s}} = \gamma_0,$$

the third over

$$t_1 \leqslant t \leqslant t_2, \quad n_{t_1} = \cdots = n_{t_2} = N, \quad n_{t_2+1} = \cdots = n_{t_3} = m_1,$$
$$n_{t_1-1} \neq N, \quad n_{t_3+1} \neq m_1.$$

Proceeding in the same manner, we get our proposition.

7.11.4. For any $\gamma \in \mathbb{Q}$, $\gamma > 0$ and $0 < n^* \leqslant N$ we define B_{γ,n^*} by

$$B_{\gamma,n^*} = \sum (-1)^{s+t} B_t(r_1, n_1, \ldots, r_s, n_s)$$
$$\times \left[(e_1^{r_t} - 1) e_1^{r_{t+1}+\cdots+r_s} \right]^{p^{n^*}} D_{r_1,n_1} \ldots D_{r_s,n_s},$$

where the sum is taken over

$$1 \leqslant t \leqslant s \leqslant s_0, \quad n_1 = \ldots n_s = n^*,$$
$$r_1, \ldots, r_s \in R, \quad r_1 + \cdots + r_s = \gamma,$$

$C^{(1)}_{\gamma,n^*}$ by
$$C^{(1)}_{\gamma,n^*} = \sum \eta^*(r_s, n_s, \ldots, r_1, n_1) D_{r_1, n_1} \ldots D_{r_s, n_s},$$
where the sum is taken over
$$1 \leqslant s < s_0, \quad 0 < n_1, \ldots, n_s < n^*,$$
$$r_1, \ldots, r_s \in R, \quad \frac{r_1}{p^{n_{*1}}} + \cdots + \frac{r_s}{p^{n_{*s}}} = \gamma,$$

and finally $C^{(2)}_{\gamma,n^*}$ by
$$C^{(2)}_{\gamma,n^*} = \sum (-1)^s \eta^*(r_1, n_1, \ldots, r_s, n_s) D_{r_1, n_1} \ldots D_{r_s, n_s},$$
where the sum is taken over
$$1 \leqslant s < s_0, \quad 0 < n_1, \ldots, n_s < n^*,$$
$$r_1, \ldots, r_s \in R, \quad \frac{r_1}{p^{n_{*1}}} + \cdots + \frac{r_s}{p^{n_{*s}}} = \gamma.$$

By definition, we set $C^{(1)}_{0,n^*} = C^{(2)}_{0,n^*} = 1$, $B_{0,n^*} = 0$.

REMARKS. (1) If $\gamma \in \mathbb{Q}$, $\gamma > 0$ is not p-entier (i.e., $v_p(\gamma) < 0$, where v_p is p-adic valuation of $\mathbb{Q}$), then $B_{\gamma,n^*} = 0$.

(2) If $\gamma > 0$, then $C^{(1)}_{\gamma,n^*}$, $C^{(2)}_{\gamma,n^*} \in J_1$.

PROPOSITION. *We have*
$$A_{\gamma,n^*} = \sum_{\substack{\gamma_1, \gamma_0, \gamma_2 \in \mathbb{Q} \\ \gamma_1 + \gamma_0 + \gamma_2 = \gamma}} C^{(1)}_{\gamma_1,n^*} B_{\gamma_0,n^*} C^{(2)}_{\gamma_2,n^*}.$$

PROOF. It is sufficient to remark that if $r_1, \ldots, r_s \in R$, $n_1, \ldots, n_s \in \mathbb{N}$, $n_i = n^* = \max\{n_1, \ldots, n_s\}$ for $t_1 \leqslant i \leqslant t_2$ and $n_{t_1-1}, n_{t_2+1} \neq n^*$, then for $t_1 \leqslant t \leqslant t_2$ we have:
$$B_t(r_1, n_1, \ldots, r_s, n_s) = \eta^*(r_{t_1-1}, n_{t_1-1}, \ldots, r_1, n_1)$$
$$\times B_{t-t_1+1}(r_{t_1}, n_{t_1}, \ldots, r_{t_2}, n_{t_2}) \eta^*(r_{t_2+1}, n_{t_2+1}, \ldots, r_s, n_s).$$

7.11.5. Consider the expression for A_{γ,n^*} from 7.11.3. Since
$$e_1 = \widetilde{\exp}\left(-\frac{1}{r^*} t_1^{-r^*(q-1)}\right) \in \mathbb{F}_p\left[t_1^{-r^*(q-1)}\right],$$
we can represent A_{γ,n^*} as a power series of the variable $t_1^{-r^*(q-1)}$:
$$A_{\gamma,n^*} = \sum_{m \geqslant 1} A_{\gamma,n^*}(m) \left[t_1^{-r^*(q-1)}\right]^{mp^{n^*}}.$$

The coefficients $A_{\gamma,n^*}(m)$, where $1 \leqslant m < p$, depend only on the residue

$$A_{\gamma,n^*} \bmod \left(\left[t_1^{-pr^*(q-1)} \right]^{p^{n^*}} J_1(O_{\text{sep}}) \right).$$

Therefore, they can be computed by means of the following equivalences:

$$e_1^{r_{t_1}+\cdots+r_{t_2}} \equiv \widetilde{\exp}\left(-\frac{r_t + \cdots + r_{t_2}}{r^*} t_1^{-r^*(q-1)} \right) \quad \bmod t_1^{-pr^*(q-1)} O'_{\text{tr}}.$$

The same remark applies to the coefficients $B_{\gamma,n^*}(m)$, $1 \leqslant m < p$, of the expression

$$B_{\gamma,n^*} = \sum_{m \geqslant 1} B_{\gamma,n^*}(m) \left[t_1^{-r^*(q-1)} \right]^{mp^{n^*}}.$$

PROPOSITION. *Let* $1 \leqslant m \leqslant p-2$. *Then*

$$-\sum_{\substack{\gamma_1,\gamma_2 \in \mathbb{Q} \\ \gamma_1+\gamma_2=\gamma}} A_{\gamma_1,n^*}(1) A_{\gamma_2,n^*}(m) \equiv (m+1) A_{\gamma,n^*}(m+1)$$

$$+ \frac{1}{r^*} \sum_{\substack{\gamma_1,\gamma_0,\gamma_2 \in \mathbb{Q} \\ \gamma_1+\gamma_0+\gamma_2=\gamma}} C^{(1)}_{\gamma_1,n^*} \gamma_0 B_{\gamma_0,n^*}(m) C^{(2)}_{\gamma_2,n^*} \bmod J_{s_0+1}.$$

PROOF. We have

$$A_{\gamma,n^*}(1) = \frac{(-1)}{1! r^*} \sum \sum (-1)^{s+t} B_t(r_1, n_1, \ldots, r_s, n_s) r_t D_{r_1, n_1} \cdots D_{r_s, n_s},$$

where the first sum is taken over

$$1 \leqslant s \leqslant s_0, \quad 0 < n_1, \ldots, n_s \leqslant n^*,$$
$$r_1, \ldots, r_s \in R, \quad \frac{r_1}{p^{n_{*1}}} + \cdots + \frac{r_s}{p^{n_{*s}}} = \gamma$$

and the second over

$$t_1 \leqslant t \leqslant t_2, \quad n_{t_1} = \cdots = n_{t_2} = n^*, \quad n_{t_1-1}, n_{t_2+1} \neq n^*.$$

In this notation, the arguments of Lemma 7.5.5 give

$$B_t(r_1, n_1, \ldots, r_s, n_s) + B_{t+1}(r_1, n_1, \ldots, r_s, n_s)$$
$$= \eta^*(r_t, n_t, \ldots, r_1, n_1) \eta^*(r_{t+1}, n_{t+1}, \ldots, r_s, n_s)$$

for $t_1 \leqslant t < t_2$, and

$$B_{t_2}(r_1, n_1, \ldots, r_s, n_s) = \eta^*(r_{t_2}, n_{t_2}, \ldots, r_1, n_1) \eta^*(r_{t_2+1}, n_{t_2+1}, \ldots, r_s, n_s).$$

These relations imply the following identity:

$$\sum_{t_1 \leqslant t \leqslant t_2} (-1)^t B_t(r_1, n_1, \ldots, r_s, n_s) r_t$$
$$= \sum_{t_1 \leqslant t \leqslant t_2} (-1)^t \eta^*(r_t, n_t, \ldots, r_1, n_1) \eta^*(r_{t+1}, n_{t+1}, \ldots, r_s, n_s)(r_{t_1} + \cdots + r_t).$$

For any collection $(r_1, n_1, \ldots, r_s, n_s)$ and index t such that $t \geqslant t_1$, where $n_{t_1} = n^* := \max\{n_1, \ldots, n_s\}$, $n_{t_1-1} \neq n^*$, we set

$$E^{(1)}(r_1, n_1, \ldots, r_s, n_s; t)$$
$$= (-1)^{s+t} \eta^*(r_t, n_t, \ldots, r_1, n_1) \eta^*(r_{t+1}, n_{t+1}, \ldots, r_s, n_s)(r_{t_1} + \cdots + r_t).$$

REMARKS. (1) If the index t_1 is not uniquely defined, then the above expression for $E^{(1)}(r_1, n_1, \ldots, r_s, n_s; t)$ is automatically set equal to 0.
(2) If $n_t \neq n^*$, then $E^{(1)}(r_1, n_1, \ldots, r_s, n_s; t) = 0$.
We have:

$$A_{\gamma, n^*}(1)$$
$$= \frac{(-1)}{1! r^*} \sum \sum (-1)^{s+t} E^{(1)}(r_1, n_1, \ldots, r_s, n_s; t) D_{r_1, n_1} \ldots D_{r_s, n_s},$$

where the first sum is taken over

$$1 \leqslant s \leqslant s_0, \quad 0 < n_1, \ldots, n_s \leqslant n^*,$$
$$r_1, \ldots, r_s \in R, \quad \frac{r_1}{p^{n_{*1}}} + \cdots + \frac{r_s}{p^{n_{*s}}} = \gamma_1$$

and the second over

$$t_1 \leqslant t \leqslant s, \quad n_{t_1} = n^*, \quad n_{t_1-1} \neq n^*.$$

For a fixed index s, consider the expression for $A_{\gamma_2, n^*}(m)$ in the following form:

$$A_{\gamma_2, n^*}(m) = \frac{(-1)^m}{m! r^{*m}} \sum \sum (-1)^{\widehat{s}+u} B_{u-s}(r_{s+1}, n_{s+1}, \ldots, r_{\widehat{s}}, n_{\widehat{s}})$$
$$\times \left[(r_u + \cdots + r_{u_2})^m - (r_{u+1} + \cdots + r_{u_2})^m\right] D_{r_{s+1}, n_{s+1}} \ldots D_{r_{\widehat{s}}, n_{\widehat{s}}},$$

where the first sum is taken over

$$s+1 \leqslant \widehat{s} \leqslant s+s_0, \quad 0 < n_{s+1}, \ldots, n_{\widehat{s}} \leqslant n^*,$$
$$r_{s+1}, \ldots, r_{\widehat{s}} \in R, \quad \frac{r_{s+1}}{p^{n_{*s+1}}} + \cdots + \frac{r_{\widehat{s}}}{p^{n_{*\widehat{s}}}} = \gamma_2$$

and the second over

$$u_1 \leqslant u \leqslant u_2, \quad n_{u_1} = \cdots = n_{u_2} = n^*, \quad n_{u_1-1}, n_{u_2+1} \neq n^*.$$

As earlier, we have an identity

$$\sum_{u_1 \leqslant u \leqslant u_2}(-1)^u B_{u-s}(r_{s+1}, n_{s+1}, \ldots, r_{\widehat{s}}, n_{\widehat{s}})\left[(r_u + \cdots + r_{u_2})^m - (r_{u+1} + \cdots + r_{u_2})^m\right]$$
$$= \sum_{s+1 \leqslant u \leqslant u_2}(-1)^u \eta^*(r_{u-1}, n_{u-1}, \ldots, r_{s+1}, n_{s+1})$$
$$\times \eta^*(r_u, n_u, \ldots, r_{\widehat{s}}, n_{\widehat{s}})(r_u + \cdots + r_{u_2})^m.$$

For some fixed index s, any collection $(r_{s+1}, n_{s+1}, \ldots, r_{\widehat{s}}, n_{\widehat{s}})$, and any index u such that $s+1 \leqslant u \leqslant u_2$, where $n_{u_2} = n^* = \max\{n_{s+1}, \ldots, n_{\widehat{s}}\}$, $n_{u_2+1} \neq n^*$, we set:

$$E^{(2)}(r_{s+1}, n_{s+1}, \ldots, r_{\widehat{s}}, n_{\widehat{s}}; u) = (-1)^{u+\widehat{s}} \eta^*(r_{u-1}, n_{u-1}, \ldots, r_{s+1}, n_{s+1})$$
$$\times \eta^*(r_u, n_u, \ldots, r_{\widehat{s}}, n_{\widehat{s}})(r_u + \cdots + r_{u_2})^m.$$

Then,

$$A_{\gamma_2, n^*}(m) = \frac{(-1)^m}{m! r^{*m}} \sum \sum E^{(2)}(r_{s+1}, n_{s+1}, \ldots, r_{\widehat{s}}, n_{\widehat{s}}; u) D_{r_{s+1}, n_{s+1}} \ldots D_{r_{\widehat{s}}, n_{\widehat{s}}},$$

where the first sum is taken over

$$s+1 \leqslant \widehat{s} \leqslant s+s_0, \quad 0 < n_{s+1}, \ldots, n_{\widehat{s}} \leqslant n^*,$$
$$r_{s+1}, \ldots, r_{\widehat{s}} \in R, \quad \frac{r_{s+1}}{p^{n_{*s+1}}} + \cdots + \frac{r_{\widehat{s}}}{p^{n_{*\widehat{s}}}} = \gamma_2$$

and the second over

$$s+1 \leqslant u \leqslant u_2, \quad n_{u_2} = n^*, \quad n_{u_2+1} \neq n^*.$$

Now the coefficient for $D_{r_1, n_1} \ldots D_{r_{\widehat{s}}, n_{\widehat{s}}}$ in the expression of the sum

$$\sum_{\gamma_1, \gamma_2} A_{\gamma_1, n^*}(1) A_{\gamma_2, n^*}(m)$$

is equal to $\sum_{t_1 \leqslant t < u \leqslant u_2} F_{t,u}$, where

$$F_{t,u} = E^{(1)}(r_1, n_1, \ldots, r_t, n_t; t) E^{(2)}(r_{t+1}, n_{t+1}, \ldots, r_{\widehat{s}}, n_{\widehat{s}}; u)$$
$$+ E^{(1)}(r_1, n_1, \ldots, r_{t+1}, n_{t+1}; t) E^{(2)}(r_{t+2}, n_{t+2}, \ldots, r_{\widehat{s}}, n_{\widehat{s}}; u)$$
$$+ \cdots + E^{(1)}(r_1, n_1, \ldots, r_{u-1}, n_{u-1}; t) E^{(2)}(r_u, n_u, \ldots, r_{\widehat{s}}, n_{\widehat{s}}; u).$$

As earlier, we see that $F_{t,u} = 0$, if $u \neq t+1$. Therefore,

$$\sum_{\substack{\gamma_1, \gamma_2 \\ \gamma_1 + \gamma_2 = \gamma}} A_{\gamma_1, n^*}(1) A_{\gamma_2, n^*}(m) \frac{(-1)^{m+1}}{m! r^{*m+1}} = \sum \sum (-1)^{s+t} \eta^*(r_t, n_t, \ldots, r_1, n_1)$$
$$\times \eta^*(r_{t+1}, n_{t+1}, \ldots, r_s, n_s)(r_{t_1} + \cdots + r_t) \times (r_{t+1} + \cdots + r_{t_2})^m D_{r_1 n_1} \ldots D_{r_s n_s},$$

where the the first sum at the right-hand side is taken over

$$1 \leqslant s \leqslant s_0, \quad 0 < n_1, \ldots, n_s \leqslant n^*,$$
$$r_1, \ldots, r_s \in R, \quad \frac{r_1}{p^{n_{*1}}} + \cdots + \frac{r_s}{p^{n_{*s}}} = \gamma$$

and the second over

$$t_1 \leqslant t \leqslant t_2, \quad n_{t_1} = \cdots = n_{t_2} = n^*, \quad n_{t_1-1}, n_{t_2+1} \neq n^*.$$

Now our proposition can be deduced from the following formulas:

$$A_{\gamma, n^*}(m+1) = \frac{(-1)^{m+1}}{(m+1)!r^{*m+1}} \sum\sum (-1)^{s+t}$$
$$\times \eta^*(r_t, n_t, \ldots, r_1, n_1)\eta^*(r_{t+1}, n_{t+1}, \ldots, r_s, n_s)$$
$$\times (r_{t+1} + \cdots + r_{t_2})^{m+1} D_{r_1 n_1} \ldots D_{r_s n_s},$$

$$-\frac{1}{r^*} \sum_{\substack{\gamma_1, \gamma_0, \gamma_2 \in \mathbb{Q} \\ \gamma_1 + \gamma_0 + \gamma_2 = \gamma}} C^{(1)}_{\gamma_1, n^*} \gamma_0 B_{\gamma_0, n^*}(m) C^{(2)}_{\gamma_2, n^*}$$

$$= \frac{(-1)^{m+1}}{m!r^{*m+1}} \sum\sum (-1)^{s+t} \eta^*(r_t, n_t, \ldots, r_1, n_1)$$
$$\times \eta^*(r_{t+1}, n_{t+1}, \ldots, r_s, n_s)(r_{t_1} + \cdots + r_{t_2})$$
$$\times (r_{t+1} + \cdots + r_{t_2})^m D_{r_1 n_1} \ldots D_{r_s n_s},$$

where the first and fourth sum are taken over

$$1 \leqslant s \leqslant s_0, \quad 0 < n_1, \ldots, n_s \leqslant n^*,$$
$$r_1, \ldots, r_s \in R, \quad \frac{r_1}{p^{n_{*1}}} + \cdots + \frac{r_s}{p^{n_{*s}}} = \gamma$$

while the second and fifth are over

$$t_1 \leqslant t \leqslant t_2, \quad n_{t_1} = \cdots = n_{t_2} = n^*, \quad n_{t_1-1}, n_{t_2+1} \neq n^*,$$

and the evident relation

$$(r_{t_1} + \cdots + r_t)(r_{t+1} + \cdots + r_{t_2})^m + (r_{t+1} + \cdots + r_{t_2})^{m+1}$$
$$= (r_{t_1} + \cdots + r_{t_2})(r_{t+1} + \cdots + r_{t_2})^m$$

is used.

7.12. Let $N_0 = N(R, v_0)$ be the natural number from Proposition 4.4.

PROPOSITION. *If $n^* < N_0$, then for any $\gamma \in M_{p-1}(R)$ we have*
$$A_{\gamma, n^*}\left[t_1^{q\gamma}\right]^{p^{n^*}} \in \mathcal{L}_{s_0-1}(v_0)A_{\text{sep}} + \sum_{1 \leqslant s < s_0}\left[t^{sr^*}\right]^{p^{N_0}} J_s(O_{\text{sep}}) + J_{s_0}.$$

PROOF. The arguments of the lemma in 7.9.1 yield the following lemma.

LEMMA. *If $r_1 + \cdots + r_s \geqslant s_1 r^*$, then*
$$D_{r_1, n_1} \ldots D_{r_s, n_s} \in \mathcal{L}_{s_0-1}(v_0)A_{\text{sep}} + J_s(O_{\text{sep}}),$$
where $s = \min\{s_1 + 1, s_0\}$.

Now we prove our proposition as follows. Let s_1 be a natural number such that
$$(s_1 + 1)r^* > \gamma \geqslant s_1 r^*.$$
Applying the above lemma to each term of the expression for A_{γ, n^*}, we obtain
$$A_{\gamma, n^*}[t_1^{q\gamma}]^{p^{n^*}} \in \mathcal{L}_{s_0-1}(v_0)A_{\text{sep}} + [t_1^{q\gamma}]^{p^{n^*}} J_s(O_{\text{sep}}) + J_{s_0},$$
where $s = \min\{s_1 + 1, s_0\}$.

If $s_1 + 1 \geqslant s_0$, our proposition is proved for evident reasons. If $s_1 < s_0$, then the following inclusions
$$[t_1^{q\gamma}]^{p^{n^*}} J_{s_1+1}(O_{\text{sep}}) \subset [t^{(s_1+1)r^*}]^{p^{n^*}} J_{s_1+1}(O_{\text{sep}}) \subset \sum_{1 \leqslant s < s_0} [t^{sr^*}]^{p^{N_0}} J_s(O_{\text{sep}})$$
imply our proposition.

7.13. PROPOSITION. *If $\gamma \geqslant s_0 r^*$, then*
$$A_{\gamma, n^*}\left[t_1^{q\gamma}\right]^{p^{n^*}} \equiv A_{\gamma, n^*}(1)\left[t_1^{q\gamma - r^*(q-1)}\right]^{p^{n^*}} \quad \mod (\mathcal{L}_{s_0-1}(v_0)J_1 + J_{s_0+1}).$$

PROOF. We have the following analog of the lemma in 7.12:

LEMMA. *If $s \geqslant 2$ and $r_1 + \cdots + r_s \geqslant s_0 r^*$, then*
$$D_{r_1, n_1} \ldots D_{r_s, n_s} \in \mathcal{L}_{s_0-1}(v_0)J_1(O_{\text{sep}}) + J_{s_0+1}(O_{\text{sep}}).$$

PROOF. Let $(s_1 + 1)r^* > r_1 + \cdots + r_{s-1} \geqslant s_1 r^*$. If $s_1 + 1 \geqslant s_0$, then
$$D_{r_1, n_1} \ldots D_{r_{s-1}, n_{s-1}} \in \mathcal{L}_{s_0-1}(v_0) \otimes k + J_{s_0},$$
by the lemma in 7.12 and, therefore,
$$D_{r_1, n_1} \ldots D_{r_s, n_s} \in \mathcal{L}_{s_0-1}(v_0)J_1(O_{\text{sep}}) + J_{s_0+1}(O_{\text{sep}}).$$
If $s_1 + 1 < s_0$, then we have $r_s \geqslant (s_0 - (s_1 + 1))r^*$, therefore, by the lemma in 7.12
$$D_{r_1, n_1} \ldots D_{r_{s-1}, n_{s-1}} \in \mathcal{L}_{s_0-1}(v_0) \otimes k + J_{s_1+1}(O_{\text{sep}}),$$
$$D_{r_s, n_s} \in \mathcal{L}_{s_0-1}(v_0) \otimes k + J_{s_0-s_1}(O_{\text{sep}}).$$
These two formulas obviously imply the assertion of our lemma.

From this lemma it follows that
$$A_{\gamma, n^*} \equiv -\frac{\gamma}{r^*}D_{\gamma, n^*}\left[t_1^{-r^*(q-1)}\right]^{p^{n^*}} \quad \mod (\mathcal{L}_{s_0-1}(v_0)J_1(O_{\text{sep}}) + J_{s_0+1}(O_{\text{sep}}))$$
($D_{\gamma, n^*} = 0$, if $\gamma \notin R$). The same arguments show that

$$A_{\gamma,n^*}(1) \equiv -\frac{\gamma}{r^*}D_{\gamma,n^*} \mod (\mathcal{L}_{s_0-1}(v_0)J_1(O_{\text{sep}}) + J_{s_0+1}(O_{\text{sep}})). \qquad \square$$

7.14. Proposition. *Let* $\gamma \leqslant s_0 r^*$, $n^* \geqslant N_0$.
(a) *If γ is p-entier, then*

$$A_{\gamma,n^*}\left[t_1^{q\gamma}\right]^{p^{n^*}} \equiv A_{\gamma,n^*}(1) \sum_{1 \leqslant m < p} \frac{(-\gamma)^{m-1}}{m! r^{*m-1}} \left[t_1^{q\gamma - r^* m(q-1)}\right]^{p^{n^*}}$$

$$\mod \left(\mathcal{L}_{s_0-1}(v_0)J_1 + \sum_{1 \leqslant s \leqslant s_0} \left[t_1^{-r^*(p-s)p^{N_0}}\right]^{p^{n^*}} J_s(O_{\text{sep}}) + J_{s_0+1}\right).$$

(b) *If γ is not p-entier, then*

$$A_{\gamma,n^*}\left[t_1^{q\gamma}\right]^{p^{n^*}} \equiv A_{\gamma,n^*}(1)\left[t_1^{q\gamma - r^*(q-1)}\right]^{p^{n^*}}$$

$$\mod \left(\mathcal{L}_{s_0-1}(v_0)J_1 + \sum_{1 \leqslant s \leqslant s_0} \left[t_1^{-r^*(p-s)p^{N_0}}\right]^{p^{n^*}} J_s(O_{\text{sep}}) + J_{s_0+1}\right).$$

PROOF. We need several lemmas.

7.14.1. Lemma. *Let $n^* \geqslant N_0$. Then for any $\gamma \in M_{p-1}(R)$ (see 7.4.3) we have*

$$A_{\gamma,n^*}(1)\left[t_1^{q\gamma - r^*(q-1)}\right]^{p^{n^*}} \in \mathcal{L}_{s_0-1}(v_0)A_{\text{sep}} + \sum_{1 \leqslant s < s_0} \left[t_1^{-r^*(p-s)p^{N_0}}\right]^{p^{n^*}} J_s(O_{\text{sep}}) + J_{s_0}.$$

PROOF. The lemma from 7.5.4 (see also 7.11.5) implies

$$A_{\gamma,n^*}(1) = -\frac{1}{r^*}\mathcal{F}_{R,n^*}(\gamma, 0) \mod J_{s_0+1},$$

where the elements $\mathcal{F}_{R,n^*}(\gamma, 0)$ were defined in §5. If $\gamma \geqslant v_0$, then

$$A_{\gamma,n^*}(1) \in \mathcal{L}_{R,N}^{(v_0)} \otimes k \mod (C_{s_0+1} \otimes k).$$

Therefore, $A_{\gamma,n^*}(1) \in \mathcal{L}_{s_0-1}(v_0) \otimes k$ for $\gamma \geqslant v_0$ (see 7.7(a)). If $\gamma < v_0$, then

$$t_1^{q\gamma - r^*(q-1)} \in t_1^{-r^*(p-1)p^{N_0}} O'_{\text{tr}}$$

(see 7.4.3). Therefore,

$$A_{\gamma,n^*}(1)\left[t_1^{q\gamma - r^*(q-1)}\right]^{p^{n^*}} \in \left[t_1^{-r^*(p-1)p^{N_0}}\right]^{p^{n^*}} J_1(O_{\text{sep}})$$

$$\subset \sum_{1 \leqslant s < s_0}\left[t_1^{-r^*(p-s)p^{N_0}}\right]^{p^{n^*}} J_s(O_{\text{sep}}).$$

7.14.2. LEMMA. *If* $n^* \geqslant N_0$, *then*

$$B_{\gamma,n^*}(1)\left[t_1^{q\gamma - r^*(q-1)}\right]^{p^{n^*}} \in \mathcal{L}_{s_0-1}(v_0)A_{\text{sep}}$$
$$+ \sum_{1 \leqslant s_1, s_2 < s_0} \left[t_1^{-r^*(p-s_1)p^{N_0}}\right]^{p^{n^*}} \left[t^{s_2 \frac{r^*}{p}}\right]^{p^{n^*}} J_{s_1+s_2}(O_{\text{sep}}) + J_{s_0}.$$

PROOF. The arguments of the proof of Proposition 7.12 yield

$$C_{\gamma,n^*}^{(i)}[t_1^{q\gamma}]^{p^{n^*}} \in \mathcal{L}_{s_0-1}(v_0)A_{\text{sep}} + \sum_{1 \leqslant s < s_0}\left[t^{s\frac{r^*}{p}}\right]^{p^{n^*}} J_s(O_{\text{sep}}) + J_{s_0},$$

where $C_{\gamma,n^*}^{(i)}$, $i = 1, 2$, were defined in 7.11.4.

From Proposition 7.11.4 it follows that

$$A_{\gamma,n^*}(1) = \sum_{\substack{\gamma_1,\gamma_0,\gamma_2 \\ \gamma_1+\gamma_0+\gamma_2=\gamma}} C_{\gamma_1,n^*}^{(1)} B_{\gamma_0,n^*}(1) C_{\gamma_2,n^*}^{(2)} \mod J_{s_0+1}.$$

The set $\{\gamma \mid B_{\gamma,n^*} \neq 0\}$ is finite. Now our proposition can be proved by induction on γ using the above equality and Lemma 7.14.1.

7.14.3. LEMMA. *For any* $\gamma \in M_{p-1}(R)$, $1 \leqslant m < p$ *and* $n^* \geqslant N_0$, *we have:*

(a)
$$A_{\gamma,n^*}(m+1)\left[t_1^{q\gamma - (m+1)r^*(q-1)}\right]^{p^{n^*}}$$
$$\in \mathcal{L}_{s_0-1}(v_0)A_{\text{sep}} + \sum_{1 \leqslant s < s_0}\left[t_1^{-r^*(p-s)p^{N_0}}\right]^{p^{n^*}} J_s(O_{\text{sep}}) + J_{s_0};$$

(b)
$$B_{\gamma,n^*}(m)\left[t_1^{q\gamma - (m+1)r^*(q-1)}\right]^{p^{n^*}}$$
$$\in \mathcal{L}_{s_0-1}(v_0)A_{\text{sep}} + \sum_{1 \leqslant s < s_0}\left[t_1^{-r^*(p-s)p^{N_0}}\right]^{p^{n^*}} J_s(O_{\text{sep}}) + J_{s_0}.$$

PROOF. This statement will be proved by induction on m.

Assume that it is proved for some m such that $m + 1 < p$. Then Proposition 7.11.4 implies that

$$A_{\gamma,n^*}(m) = \sum_{\substack{\gamma_1,\gamma_0,\gamma_2 \\ \gamma_1+\gamma_0+\gamma_2=\gamma}} C_{\gamma_1,n^*}^{(1)} B_{\gamma_0,n^*}(m) C_{\gamma_2,n^*}^{(2)} \mod J_{s_0+1}.$$

By induction on γ, as in the above lemma, we obtain

$$B_{\gamma,n^*}(m)\left[t_1^{q\gamma - mr^*(q-1)}\right]^{p^{n^*}}$$
$$\in \mathcal{L}_{s_0-1}(v_0)A_{\text{sep}} + \sum_{1 \leqslant s_1, s_2 < s_0}\left[t_1^{-r^*(p-s_1)p^{N_0}}\right]^{p^{n^*}}\left[t^{s_2\frac{r^*}{p}}\right]^{p^{n^*}} J_{s_1+s_2}(O_{\text{sep}}) + J_{s_0}.$$

Multiplying both sides of this expression by $\left[t_1^{-r^*(q-1)}\right]^{p^{n^*}}$, we get formula (b) of our proposition. The formula (a) now follows from Proposition 7.11.5.

7.14.4. LEMMA. *Let $1 \leqslant m < p$, $n^* \geqslant N_0$, and $\gamma \in M_{p-1}(R)$ is p-entier, then*

$$r^*(m+1)A_{\gamma,n^*}(m+1)\left[t_1^{q\gamma-(m+1)r^*(q-1)}\right]^{p^{n^*}} \equiv -\gamma B_{\gamma,n^*}(m)\left[t_1^{q\gamma-(m+1)r^*(q-1)}\right]^{p^{n^*}}$$
$$\mod \left(\mathcal{L}_{s_0-1}(v_0)J_1 + \sum_{1\leqslant s\leqslant s_0}\left[t_1^{-r^*(p-s)p^{N_0}}\right]^{p^{n^*}}J_s(O_{\text{sep}}) + J_{s_0+1}\right).$$

PROOF. It follows from the above lemma, the relation from the proposition in 7.11.5, and the trivial fact that $A_{\gamma,n^*}(1)$, $C_{\gamma,n^*}^{(1)}$, $C_{\gamma,n^*}^{(2)} \in J_1$ for any $\gamma > 0$.

7.14.5. LEMMA. *Let $1 \leqslant m < p$ and $n^* \geqslant N_0$. Then*

$$A_{\gamma,n^*}(m)\left[t_1^{q\gamma-(m+1)r^*(q-1)}\right]^{p^{n^*}} \equiv B_{\gamma,n^*}(m)\left[t_1^{q\gamma-(m+1)r^*(q-1)}\right]^{p^{n^*}}$$
$$\mod \left(\mathcal{L}_{s_0-1}(v_0)J_1 + \sum_{1\leqslant s\leqslant s_0}\left[t_1^{-r^*(p-s)p^{N_0}}\right]^{p^{n^*}}J_s(O_{\text{sep}}) + J_{s_0+1}\right).$$

PROOF. This lemma can be deduced from the relation in the proposition from 7.11.4 in the same way as Lemma 7.14.4 was deduced from Proposition 7.11.5.

7.14.6. In order to finish the proof of our proposition, we note that

$$A_{\gamma,n^*}[t_1^{q\gamma}]^{p^{n^*}} \equiv \sum_{1\leqslant m<p} A_{\gamma,n^*}(m)\left[t_1^{q\gamma-mr^*(q-1)}\right]^{p^{n^*}} \mod \left[t_1^{q\gamma-pr^*(q-1)}\right]^{p^{n^*}} J_1(O_{\text{sep}}).$$

By the condition $\gamma \leqslant s_0 r^*$, we obtain

$$q\gamma - pr^*(q-1) \leqslant q(p-1)r^* - pr^*(q-1) = -r^*(q-p) \leqslant -r^*(p-1)p^{N_0}$$

(we have $q = p^N$ and $N > N_0$, see 7.4.3).

Therefore, the above equivalence is valid modulo

$$\mathcal{L}_{s_0-1}(v_0)J_1 + \sum_{1\leqslant s\leqslant s_0}\left[t_1^{-r^*(p-s)p^{N_0}}\right]^{p^{n^*}}J_s(O_{\text{sep}}) + J_{s_0+1}$$

and the previous lemmas yield the formula of our proposition.

7.15. PROPOSITION. *For any $\gamma \in M_{p-1}(R)$ we have*
(a) *if $\gamma > s_0 r^*$ or γ is not p-entier, then*

$$A(\gamma)_0[t_1^{q\gamma}]^q \equiv -\frac{1}{r^*}\mathcal{F}_{R,N}(\gamma,0)\left[t_1^{q\gamma-r^*(q-1)}\right]^q$$
$$\mod \left(\mathcal{L}_{s_0-1}(v_0)J_1 + \sum_{1\leqslant s\leqslant s_0} t^{-r^*(p-s)}J_s(O_{\text{sep}}) + J_{s_0+1}\right);$$

(b) *if $\gamma \leqslant s_0 r^*$ and is p-entier, then*

$$A(\gamma)_0[t_1^{q\gamma}]^q \equiv -\frac{1}{r^*}\mathcal{F}_{R,N}(\gamma,0)\sum_{1\leqslant m<p}\frac{(-\gamma)^{m-1}}{m!r^{*m-1}}\left[t_1^{q\gamma-mr^*(q-1)}\right]^q$$

$$\mod\left(\mathcal{L}_{s_0-1}(v_0)J_1 + \sum_{1\leqslant s\leqslant s_0}t^{-r^*(p-s)}J_s(O_{\text{sep}}) + J_{s_0+1}\right).$$

PROOF. The formulas

$$A_{\gamma_1,m_1}[t_1^{q\gamma}]^{p^{m_1}} \in \mathcal{L}_{s_0-1}(v_0)A_{\text{sep}} + \sum_{1\leqslant s<s_0}[t^{sr^*}]^{p^{N_0}}J_s(O_{\text{sep}}) + J_{s_0}$$

(see Proposition 7.12, 7.14 and Lemma 7.14.1),

$$A_{\gamma,N}[t_1^{q\gamma}]^q \in \mathcal{L}_{s_0-1}(v_0)A_{\text{sep}} + \sum_{1\leqslant s<s_0}t^{-r^*(p-s)p^{N_0}}J_s(O_{\text{sep}}) + J_{s_0}$$

(see Proposition 7.4 and Lemma 7.14.1) and the relation in Proposition 7.11.3 give

$$A(\gamma)_0[t_1^{q\gamma}]^q \equiv A_{\gamma,N}[t_1^{q\gamma}]^q \mod\left(\mathcal{L}_{s_0-1}(v_0)J_1 + \sum_{1\leqslant s\leqslant s_0}t^{-r^*(p-s)p^{N_0}}J_s(O_{\text{sep}}) + J_{s_0+1}\right).$$

Now Proposition 7.15 follows immediately from Propositions 7.13 and 7.14.

7.16. Let I be an ideal in $\mathcal{L}$ such that (see 7.10)

$$IA_{\text{sep}} \supset \mathcal{L}_{s_0-1}(v_0)J_1 + J_{s_0+1}$$

and let $X_1'' \in A_{\text{sep}}$ be the element from Proposition 7.11.1.

PROPOSITION. *The largest upper ramification number of the field of definition of X_1'' mod IA_{sep} over K is less then v_0 if and only if $\mathcal{F}_{R,N}(\gamma,n) \in I \otimes k$ for any $\gamma \geqslant v_0$ and $0 \leqslant n < N$.*

PROOF. By definition,

$$X_1''^{(q)} - X_1'' = \sum_{0\leqslant m<N} A(s_0)_0^{(p^m)},$$

where (see 7.11.2)

$$\sum_{0\leqslant m<N} A(s_0)_0^{(p^m)} = \sum_{\substack{\gamma\in\mathbb{Q}\\0\leqslant m<N}} A(\gamma)_0^{(p^m)}[t_1^{q\gamma}]^{qp^m}$$

$$\equiv -\sum_{\substack{0\leqslant m<N\\\gamma>s_0 r^*}}\frac{1}{r^*}\mathcal{F}_{R,N}(\gamma,m)\left[t_1^{q\gamma-r^*(q-1)}\right]^{qp^m}$$

$$-\sum_{\substack{0\leqslant m<N \\ \gamma\leqslant s_0 r^* \\ v_p(\gamma)<0}} \frac{1}{r^*}\mathcal{F}_{R,N}(\gamma,m)\left[t_1^{q\gamma-r^*(q-1)}\right]^{qp^m}$$

$$-\sum_{\substack{0\leqslant m<N \\ \gamma\leqslant s_0 r^* \\ v_p(\gamma)\geqslant 0}} \mathcal{F}_{R,N}(\gamma,m) \sum_{1\leqslant m_1<p} \frac{(-\gamma)^{m_1-1}}{m_1! r^{*m_1}}\left[t_1^{q\gamma-m_1 r^*(q-1)}\right]^{qp^m}$$

$$\mod\left(\mathcal{L}_{s_0-1}(v_0)J_1 + \sum_{1\leqslant s\leqslant s_0} t^{-r^*(p-s)}J_s(O_{\text{sep}}) + J_{s_0+1}\right).$$

Let $\gamma_0(I) = \max\{\gamma \mid \mathcal{F}_{R,N}(\gamma,m) \notin I \text{ for some } 0\leqslant m<N\}$. If $\gamma_0 < v_0$, then $q\gamma_0 - r^*(q-1) < 0$ and $X_1'' \mod (IA_{\text{sep}})$ defines a trivial extension of K_{tr}'.

If $\gamma_0 \geqslant v_0$, then $q\gamma_0 - r^*(q-1) > 0$ and the field of definition of $X_1'' \mod (IA_{\text{sep}})$, which we denote by $L''(I)$, has the largest upper ramification number equal to γ_0. Indeed, it follows from 6.3 that $v(L''(I)/K') = q\gamma_0 - r^*(q-1)$, hence

$$v(L''(I)/K) = \frac{q\gamma_0 - r^*(q-1) - r^*}{q} + r^* = \gamma_0.$$

Therefore,

$$I \supset \mathcal{L}_{s_0}(v_0) \iff \mathcal{F}_{R,N}(\gamma,m) \in I \otimes k \quad \text{for all } \gamma_0 \geqslant v_0 \text{ and } 0\leqslant m<N.$$

This gives the inductive assumption 7.7(a) for $s^* = s_0 + 1$. All the other assumptions are easy consequences of the above formulas.

References

[A1] V. A. Abrashkin, *The Galois modules of the period p group schemes over the ring of Witt vectors*, Math. USSR-Izv. **31** (1988), 1–46.

[A2] _____, *Group schemes over a discrete valuation ring with small ramification*, Leningrad Math. J. **1** (1990), no. 1, 57–97.

[A3] _____, *Modular representations of the Galois group of a local field and a generalisation of the Shafarevich conjecture*, Math. USSR-Izv. **35** (1990), no. 3, 469–518.

[A4] _____, *Ramification in étale cohomology*, Invent. Math. **101** (1990), 631–640.

[B] N. Bourbaki, *Lie groups and Lie algebras*, Part I: Ch. 1–3, Hermann, 1975.

[De] P. Deligne, *Les corps locaux de caractéristique p, limites de corps locaux de caractéristique 0*, Représentations des groupes réductifs (J.-N. Bernstein etc., eds.), Hermann, Paris, 1984.

[Di] B. Ditters, *Groupes formels*, Cours 3e cycle 1973–1974. Preprint Universite Paris XI, 149-75.42.

[F1] J.-M. Fontaine, *Il n'y a pas de varieté abélienne sur $\mathbb{Z}$*, Invent. Math. **81** (1985), 515–538.

[F2] _____, *Letter to W. Messing*, 15 Jan. 1986.

[F3] _____, *Representations p-adiques des corps locaux*, The Grotendieck Festschrift (P. Cartier etc., eds.), vol. 2, Birkhauser, Boston, 1990.

[Go] N. L. Gordeev, *Infinity of the number of relations in the Galois group of maximal p-extension with bounded ramification of a local field*, Dokl. Akad. Nauk SSSR **233** (1977), no. 6, 1031–1034; English transl. in Soviet Math. Dokl. **18** (1977).

[In1] E. Inaba, *On matrix equations for Galois extensions of fields with characteristic p*, Natur. Sci. Rep. Ochanomizu Univ. **12** (1961), 26–36.

[In2] _____, *On generalized Artin-Schreier equations*, Natur. Sci. Rep. Ochanomizu Univ. **13** (1962), 1–13.

[In3] _____, *Normal form of generalized Artin-Schreier equations*, Natur. Sci. Rep. Ochanomizu Univ. **14** (1963), 1–15.
[J-W] U. Janssen and K. Wingberg, *Die Struktur der absoluten Galoisgruppe $\wp$-adischer Zahlkorper*, Invent. Math. **70** (1982), 71–98.
[Jak] A. V. Jakovlev, *The Galois group of the algebraic closure of a local field*, Math. USSR-Izv. **2** (1968), 1231–1269.
[Ko] H. Koch, *Galoissche Theorie der p-Erweiterungen*, Springer-Verlag, New York, 1970.
[Ma] E. Maus, *Relationen in Verzweigungsgruppen*, J. Reine Angew. Math. **258** (1973), 23–50.
[R] M. Raynaud, *Schemas en groupes de type $(p, \ldots, p)$*, Bull. Soc. Math. France **102** (1974), 241–280.
[Se1] J.-P. Serre, *Sur les groupes de Galois attachés aux groupes p-divisibles*, Proc. Conf. Local Fields (Driebergen, 1966), Springer-Verlag, Berlin, 1967, pp. 118–131.
[Se2] _____, *Cohomologie Galoisienne*, Lecture Notes in Math., vol. 5, Springer-Verlag, Berlin, Heidelberg, New York, 1973.
[Se3] _____, *Local fields*, Graduate Texts in Math., vol. 67, Springer-Verlag, Berlin, Heidelberg, New York, 1979.
[Wtt] E. Witt, *Zyklische Korper und Algebren der Charakteristik p vom Grad p^n*, J. Reine Angew. Math **176** (1936), 126–140.
[Wnt] J.-P. Wintenberger, *Le corps des normes de certaines extensions infinies de corps locaux; applications*, Ann. Sci. Ecole Norm. Sup. **16** (1983), 59–89.

Translated by THE AUTHOR

Department of Mathematics, Moscow Institute of Engineers of the Civil Aviation, Kronshtadtskij bul., Moscow, 125490, Russia

Class Field Theory for Multidimensional Complete Fields with Quasifinite Residue Fields

B. M. BEKKER

In this paper we conclude the description of the abelian extensions of a complete multidimensional field of positive characteristic with quasifinite residue field that we started in [1]. In the forties and fifties Moriya, Nakayama, Schilling and Whaples generalized classical field theory to the case of one-dimensional complete fields with quasifinite residue fields. It turns out that, unlike the case of a finite residue field, it is not true anymore that every open subgroup of finite index in the multiplicative group of a given field is the norm group of some finite abelian extension. The norm groups were described by Whaples in [2, 3, 4]. In the present paper we generalize this result to the multidimensional case.

We preserve the terminology and notation of [1, 5].

§1. Let K be a multidimensional complete field of positive characteristic p with quasifinite residue field. Suppose that for every finite extension F/K we are given a nonempty set $\mathfrak{N}_F$ of subgroups of finite index in $A_F = K_n^{\text{top}}(F)$ that satisfy following conditions:

$1°$. If H and H' are subgroups of A_F such that $H \supset H'$ and $H' \in \mathfrak{N}_F$, then $H \in \mathfrak{N}_F$.

$2°$. If $H_1 \in \mathfrak{N}_F$ and $H_2 \in \mathfrak{N}_F$, then $H_1 \cap H_2 \in \mathfrak{N}_F$.

$3°$. If $H \in \mathfrak{N}_F$ and L/F is a cyclic totally ramified extension of prime degree, then $N_{L/F}^{-1}(H) \in \mathfrak{N}_L$.

$4°$. If $H \in \mathfrak{N}_F$ and F/K is a cyclic extension of prime degree, then $N_{F/K}(H) \in \mathfrak{N}_K$.

$5°$. If $H \in \mathfrak{N}_K$, H is of prime index in A_K, and H contains a prime $\pi_K \in A_K$, then there exists a finite totally ramified extension F/K such that $N_{L/K} A_F \subset H$.

We make several obvious remarks.

(a) $3°$ holds for all finite abelian totally ramified extensions.

(b) $4°$ holds for all finite abelian extensions.

1991 *Mathematics Subject Classification*. Primary 11S31; Secondary 12J10.

This research was partially supported by a grant from the Russian Foundation of Fundamental Research (project No. 94-01-00753-a).

© 1995, American Mathematical Society

(c) $A_F \in \mathfrak{N}_F$ and, therefore $N_{F/K} A_F \in \mathfrak{N}_K$ for all finite abelian extensions F/K.

(d) If $H \supset U_F$, then $H \in \mathfrak{N}_K$.

PROPOSITION 1.1. *Let $H \in \mathfrak{N}_K$ be of finite index ℓ^s in A_K, ℓ prime, and let H contain a prime π_K of A_K. Then there exists a finite extension F/K with $N_{F/K} A_F \subset H$.*

PROOF. If $s = 1$, we can use 5°. Let $s > 1$ and H_1 be a subgroup of index ℓ in A_K such that $H \subset H_1 \subset A_K$. Since $\pi_K \in H_1$, there exists a finite totally ramified extension F/K such that $N_{F/K} A_F \subset H_1$. Hence $H_1 \in \mathfrak{N}$. By 3°, we have $N_{F/K}^{-1}(H) \in \mathfrak{N}$. Moreover, the group $N_{F/K}^{-1}(H)$ contains a prime π_F of A_F such that $N_{F/K} \pi_F = \pi_K$. By induction on s, there exists an extension L/F of finite degree such that $N_{L/F} A_L \subset N_{F/K}^{-1}(H)$. Then

$$N_{L/F} A_L \subset N_{F/K}(N_{F/K}^{-1}(H)) \subset H.$$

PROPOSITION 1.2. *Every subgroup $H \in \mathfrak{N}_K$ of prime power index in A_K contains $N_{F/K} A_F$ for some finite F/K.*

PROOF. Let $(A_K : H) = \ell^s$, ℓ prime, and let π_K be a prime in A_K. By Proposition 1.1, we can assume $\pi_K \notin H$. Denote by H_1 the subgroup of A_K generated by π_K, $\ell^s U_K$, and $U_K \cap H$. Since $U_K \in \mathfrak{N}_K$, we have $U_K \cap H \in \mathfrak{N}_K$ and, therefore, $H_1 \in \mathfrak{N}_K$. Moreover, the index $U_K \cap H$ in U_K is an ℓ-power. Since $A_K = \langle \pi_K \rangle \times U_K$, the index H_1 in A_K is an ℓ-power. By Proposition 1.1, there exists a finite extension F/K such that $N_{F/K} A_F \subset H_1$. Denote by H_2 the subgroup of A_K generated by $\ell^s \pi_K$ and U_K. Fore some finite L/K we have $H_2 = N_{L/K} A_L$ and $N_{FL/K} A_{FL} \subset H_1 \cap H_2$. Finally, $H_1 \cap H_2 \subset H$. Indeed, if $m\pi_K + \ell^s u + u_1 = n\ell^s \pi_K + u_2$ with $m, n \in \mathbb{Z}$, $u_1, u_2 \in U_K$, and $u_1 \in U_K \cap H$, then $m = n\ell^s$. Since $\ell^s u \in H$, we have $u_2 \in H$ and $n\ell^s \pi_K + u_2 \in H$.

PROPOSITION 1.3. *Every subgroup $H \in \mathfrak{N}_K$ of finite index in A_K contains a subgroup $N_{F/K} A_F$ for some finite extension F/K.*

PROOF. Let $(A_K : H) = \ell^s m$, where ℓ is prime and m is prime to ℓ. Let H_1 be subgroup of A_K such that $H \subset H_1 \subset A_K$ and $(A_K : H_1) = m$. We have $H_1 \in \mathfrak{N}_K$. There exists a finite extension L/K such that $N_{L/K} A_L = H_1$. Then $N_{L/K}^{-1} H \in \mathfrak{N}_L$. Moreover, $(A_L : N_{L/K}^{-1}(H)) = \ell^s$. Using Proposition 1.2, we find an extension F/L such that $N_{F/L} A_F \subset N_{L/K}^{-1} H$. Hence $N_{F/K} A_F \subset H$.

It follows from Propositions 1.1–1.3 that $\mathfrak{N}_K$ coincides with the set of norm subgroups of A_K.

2. We shall construct a set of subgroups of A_K satisfying 1°–5° of §1. First we consider some subgroups of K^*.

An open subgroup H of K^* is said to be *analytic* if, for each multi-index $I = (i_n, \ldots, i_1) > 0$ there exists a polynomial $f_I(X) \in \mathcal{O}_K[X]$ such that the reduction $\overline{f}_I(X)$ is a nonzero additive polynomial, and $1 + f_I(\alpha) t_n^{i_n} \ldots t_1^{i_1} \in H$ for all $\alpha \in \mathcal{O}_K$, where $t_n, \ldots, t_1$ is a system of local parameters of K. This definition does not depend on the choice of $t_n, \ldots, t_1$. Moreover, a subgroup of K^* is analytic if it contain an analytic subgroup of K^*.

CLASS FIELD THEORY

PROPOSITION 2.1. *The intersection of analytic subgroups of K^* is an analytic subgroup of K^*.*

PROOF. Let H_1 and H_2 be analytic subgroups, and let $\{f_I(X)\}$ and $\{g_I(X)\}$ be the corresponding sets of polynomials. It follows from [6] that there exist polynomials $h(X), h_1(X), \ldots, h_s(X) \in \mathcal{O}[X]$ such that $\overline{h}_1(X) + \cdots + \overline{h}_s(X)$ is a nonzero additive polynomial, and $\overline{h}(X)$ is an additive polynomial satisfying the relation $\overline{g}_I(\overline{h}_1(X)) + \cdots + \overline{g}_I(\overline{h}_s(X)) = \overline{f}(\overline{h}(X))$. We have $1 + g_I(h_k(\alpha))t_n^{i_n} \ldots t_1^{i_1} \in H_2$ for all $\alpha \in \mathcal{O}_K$. Then

$$\prod_{k=1}^{s} \left(1 + g_I(h_k(\alpha))t_n^{i_n} \ldots t_1^{i_1}\right) \in H_2.$$

Therefore,

$$\prod_{k=1}^{s} \left(1 + g_I(h_k(\alpha))t_n^{i_n} \ldots t_1^{i_1}\right) = 1 + \sum_{k=1}^{s} g_I(h_k(\alpha))t_n^{i_n} \ldots t_1^{i_1} + \cdots$$

and $1 + f_I(h(\alpha))t_n^{i_n} \ldots t_1^{i_1} \in H_1$ for all $\alpha \in \mathcal{O}_K$. Replacing f_I and g_I by $f_I(h(X))$ and

$$\sum_{k=1}^{s} g_I(h_k(\alpha))t_n^{i_n} \ldots t_1^{i_1} + \cdots,$$

we can assume that the polynomials $f_I^{(1)}$ and $g_I^{(1)}$ from the definition of H_1 and H_2 satisfy the relation $\overline{f}_I^{(1)} = \overline{g}_I^{(1)}$. Further,

$$1 + g_I^{(1)}(\alpha)t_n^{i_n} \ldots t_1^{i_1} = (1 + f_I^{(1)}(\alpha)t_n^{i_n} \ldots t_1^{i_1})(1 + h_{I'}^{(1)}(\alpha)t_n^{i'_n} \ldots t_1^{i'_1}),$$

where $h_{I'}(X) \in \mathcal{O}_K[X]$, $\overline{h}_{I'}(X)$ is a nonzero additive polynomial, and $I' > I$. Applying the reasoning above to the polynomials $f_{I'}(X)$ and $h_{I'}(X)$, we obtain

$$1 + g_I^{(2)}(\alpha)t_n^{i_n} \ldots t_1^{i_1} = (1 + f_I^{(2)}(\alpha)t_n^{i_n} \ldots t_1^{i_1})(1 + h_{I''}(\alpha)t_n^{i''_n} \ldots t_1^{i''_1}).$$

Proceeding in this way, we find a polynomial $h_{I^{(k)}} \in \mathcal{O}_K[X]$ such that

$$1 + h_{I^{(k)}}(\alpha)t_n^{i_n^{(k)}} \ldots t_1^{i_1^{(k)}} \in H_1.$$

Since $1 + f_{I^{(\ell)}}^{(s)}(\alpha)t_n^{i_n} \ldots t_1^{i_1} \in H_1$ for all $\alpha \in \mathcal{O}_K$, we have

$$1 + g_I^{(k)}(\alpha)t_n^{i_n} \ldots t_1^{i_1} \in H_1 \cap H_2$$

for all $\alpha \in \mathcal{O}_K$. Thus $H_1 \cap H_2$ is an analytic subgroup.

PROPOSITION 2.2. *Let L/F be a totally ramified extension of prime degree, and let H be an analytic subgroup of F^*. Then $N_{L/F}^{-1}(H)$ is an analytic subgroup of L^*.*

PROOF. We have
$$N_{L/F}(1+\alpha t_n^{i_n}\ldots t_1^{i_1}) = 1+g_J(\alpha)t_n^{i_n}\ldots t_1^{i_1},$$
where $g_I(X) \in \mathcal{O}_F[X]$ and $\alpha \in \mathcal{O}_F$. On the other hand, for each $I > 0$, there exists $f_I(X) \in \mathcal{O}_F[X]$ such that $\overline{f}_I(X)$ is a nonzero additive polynomial and $1+f_I(\alpha)t_n^{i_n}\ldots t_1^{i_1} \in H$ for all $\alpha \in \mathcal{O}_F$. Applying the argument in the proof of Proposition 2.1 to $f_J(X)$ and $g_J(X)$, we can find a polynomial $h_I(X) \in \mathcal{O}_F[X]$ such that
$$N_{L/F}(1+h_I(\alpha)t_{n,L}^{i_n}\ldots t_{1,L}^{i_1}) = 1+g_J^{(1)}t_n^{j_n}\ldots t_1^{j_1},$$
where $\overline{g}_J^{(1)} = \overline{f}_J^{(1)}$, $f_J^{(1)} \in \mathcal{O}_F[X]$, $\overline{f}_J^{(1)}$ is a nonzero additive polynomial, and $1+f_J^{(1)}(\alpha)t_n^{j_n}\ldots t_1^{j_1} \in H$ for all $\alpha \in \mathcal{O}_F$. Proceeding as in the proof of Proposition 2.1, we find a polynomial $h_I^{(k)}(X) \in \mathcal{O}_F[X]$ such that $\overline{h}_I^{(k)}(X)$ is a nonzero additive polynomial and $N_{L/F}(1+h_I^{(k)}(\alpha)t_n^{i_n}\ldots t_1^{i_1}) \in H$ for all $\alpha \in \mathcal{O}_L$. Thus, $N_{L/F}^{-1}(H)$ is an analytic subgroup.

Now we construct the set $\mathfrak{N}_K$. An open subgroup H of finite index in $K_n^{\text{top}}(K)$ is said to be *analytic* if there exists an analytic subgroup H' of K^* such that $H'K_{n-1}^{\text{top}}(K) \subset H$. We must check that the set of analytic subgroups of $K_n^{\text{top}}(K)$ satisfies 1°–5° of §1.

Conditions 1° and 2° follow from similar properties of analytic subgroups in K^*. We shall prove 3°.

PROPOSITION 2.3. *Let $H \in \mathfrak{N}_F$ and let L/F be a cyclic totally ramified extension of prime degree. Then $N_{L/F}^{-1}(H) \in \mathfrak{N}_L$.*

PROOF. We may assume $[L:F]=p$. Moreover, we can choose the systems of local parameters $t_{n,F} = t_n, \ldots, t_{1,F} = t_1$ and $t_{n,L},\ldots,t_{1,L}$ for F and L respectively, so that $N_{L/F}t_{s,L} = t_s$ for some s and $t_{i,L} = t_i$ for all $i \neq s$. Let $H \in \mathfrak{N}_F$. Then there exists an analytic subgroup $H' \subset F^*$ such that $H'K_{n-1}^{\text{top}}(F) \subset H$. We shall check that, for each $I > 0$, there exists a polynomial $f_I(X)$ for which $\overline{f}_I(X)$ is an nonzero additive polynomial and
$$N_{L/F}\{1+f_I(\alpha)t_n^{i_n}\ldots t_{s,L}^{i_s}\ldots t_1^{i_1}, X\} \in H$$
for all $x \in K_{n-1}^{\text{top}}(L)$ and $\alpha \in \mathcal{O}_L$. It suffices to prove the displayed inclusion for $x = \{t_{i_1,L},\ldots,t_{i_{n-1}}\}$. If $i_1 \neq s,\ldots,i_{n-1} \neq s$, the assertion follows from Proposition 2.2. Moreover, it suffices to prove the inclusion for $\alpha \in \mathcal{O}_F$. Let $x = \{t_1,\ldots,t_{s,L},\ldots\widehat{t_k}\ldots t_n\}$. If i_s is divisible by p, the assertion reduces to a similar assertion for symbols of the form
$$\{1+f_J(\alpha)t_n^{j_n}\ldots t_1^{j_1}, x\} \quad \text{with } (j_n,\ldots,j_1) > (i_n,\ldots,i_1).$$

Now assume that i_s is not divisible by p. We have

$$A = i_s\{1 + at_n^{i_n} \ldots t_{s,L}^{i_s} \ldots t_1^{i_1}, t_1, \ldots, t_{s,L}, \ldots \widehat{t_k} \ldots, t_n\}$$
$$= \pm\{1 + at_n^{i_n} \ldots t_{s,L}^{i_s} \ldots t_1^{i_1}, t_1, \ldots, t_k^{i_k}, \ldots \widehat{t_s} \ldots, t_n\},$$

and

$$N_{L/F}A = \pm\{N_{L/F}(1 + at_n^{i_n} \ldots t_{s,L}^{i_s} \ldots t_1^{i_1}), t_1, \ldots, t_k^{i_k}, \ldots \widehat{t_s} \ldots, t_n\}.$$

The desired assertion is easily obtained from the fact that $VK_n^{\text{top}}(F)$ is uniquely i_s-divisible.

PROPOSITION 2.4. *Let H be an analytic subgroup of $K_n^{\text{top}}(F)$, and let F/K be a cyclic extension of degree p. Then $N_{F/K}$ is an cyclic subgroup of $K_n^{\text{top}}(K)$.*

PROOF. In the case of an unramified extension one can use the same argument as in the proof of similar statement in [2]. Let F/K be a ramified extension of degree p. Choose the systems of local parameters for F and K as in the proof of Proposition 2.3. For each $I > 0$ and $\{j_1, \ldots, j_{n-1}\} \subset \{1, 2, \ldots, n\}$, there exists an $f_{I, j_1, \ldots, j_{n-1}}(X) \in \mathcal{O}_K[X]$ such that $\overline{f}_{I, j_1, \ldots, j_{n-1}}(X)$ is a nonzero additive polynomial and

$$\{1 + f_{I, j_1, \ldots, j_{n-1}}(\alpha)t_n^{i_n} \ldots t_1^{i_1}, t_{j_1, L}, \ldots, t_{j_{n-1}, L}\} \in H$$

for all $\alpha \in \mathcal{O}_K$. Let $s \notin \{j_1, \ldots, j_{n-1}\}$. It suffices to consider the case when i_s is not divisible by p, and i_k is not divisible by p for some $k < s$.

Let $\varepsilon = 1 + at_n^{i_n} \ldots t_k^{i_k} \ldots t_s^{i_s} \ldots t_1^{i_1}$. We have

$$i_k\{\varepsilon, t_1, \ldots \widehat{t_s} \ldots, t_n\} = \{\varepsilon', t_1, \ldots \widehat{t_k} \ldots, t_n\},$$

where $\varepsilon' = 1 + a't_n^{i_n} \ldots t_1^{i_1} + \cdots$. Since

$$N_{F/K}\{\varepsilon', t_1, \ldots \widehat{t_k} \ldots, t_{s,L}, \ldots, t_n\} = \{\varepsilon', t_1, \ldots \widehat{t_k} \ldots, t_s, \ldots, t_n\},$$

there exists an $f_{I, j_1, \ldots, j_{n-1}}(X) \in \mathcal{O}_K[X]$, where $\{j_1, \ldots, j_{n-1}, s\} = \{1, 2, \ldots, n\}$, such that $\overline{f}_{I, j_1, \ldots, j_{n-1}}(X)$ is a nonzero additive polynomial and

$$\{1 + f_{I, j_1, \ldots, j_{n-1}}(\alpha)t_n^{i_n} \ldots t_1^{i_1}, t_1, \ldots \widehat{t_s} \ldots, t_n\} \in N_{F/K}(H)$$

for all $\alpha \in \mathcal{O}_K$.

Now by Proposition 2.1, we can find polynomials $f_I(X)$ with the required properties.

PROPOSITION 2.5. *Let $H \subset K_n^{\text{top}}(K)$ be an analytic subgroup of prime index. Let H contain a prime π_K of $K_n^{\text{top}}(K)$. Then there exists a finite totally ramified extension F/K such that $N_{F/K}K_n^{\text{top}}(F) \subset H$.*

PROOF. We may assume $(K_n^{\text{top}}(K):H) = p$. Since

$$K_n^{\text{top}}(K) \simeq \langle \pi_K \rangle \oplus \Theta \oplus VK_n^{\text{top}}(K), \quad \pi_K \in H, \quad \text{and} \quad \Theta \subset H,$$

it suffices to prove the existence of F/K such that $N_{F/K} VK_n^{\text{top}}(F) \subset H$. To that end, we show that there exists an $\alpha \in K$ such that the inclusion $\beta \in VK_n^{\text{top}}(K) \setminus H$ implies $[\beta, \alpha) \neq 0$ for all β. Then $N_{F/K} K_n^{\text{top}}(F) \subset H$ for $F = K(\gamma)$, where $\gamma^p - \gamma = \alpha$. Let A be an open (in the additive topology) subgroup of k such that $(k : A) = p$. Then $k = A \oplus \theta \mathbb{F}_p$ for some $\theta \in k$. Consider the element

$$x(\theta) = \{1 + \theta t_n^{i_n} \ldots t_1^{i_1}, t_{j_1}, \ldots, t_{j_{n-1}}\}.$$

Denote by $\widehat{x}(\theta)$ an element of $t_n^{-i_n} \ldots t_1^{-i_1} \mathcal{O}_K$ such that $[\widehat{x}(\theta), x(\theta)) = 1$ and

$$[\widehat{x}(\theta), \{1 + \eta t_n^{i_n} \ldots t_1^{i_1}, t_{j_1}, \ldots, t_{j_{n-1}}\}) = 0$$

for all $\eta \in A$. Let $(i_1, \ldots, i_n)$ be the largest multi-index such that there exists an element

$$x_0 = \{1 + \theta t_n^{i_n} \ldots t_1^{i_1}, t_{j_1}, \ldots, t_{j_{n-1}}\}$$

of the topological basis of $VK_n^{\text{top}}(K)$ that does not belong to H. We have $k = A \oplus \theta \mathbb{F}_p$, where A is an open subgroup of index p. We put $\alpha_0 = \widehat{x}(\theta, A)$. Let

$$x_1(\theta_1) = \{1 + \theta_1 t_n^{i'_n} \ldots t_1^{i'_1}, t_{j'_1}, \ldots, t_{j'_{n-1}}\}$$

be the next element of the topological basis of $VK_n^{\text{top}}(K)$ with the largest multi-index $(i'_n, \ldots, i'_1) \leq (i_n, \ldots, i_1)$ such that $[\alpha_0, x_1(\theta_1,)) \neq 0$ or $x_1(\theta_1) \in H$. For every $\theta \in k$ there exists a $\rho(\theta) \in \mathbb{F}_p$ such that $x_1(\theta) + \rho(\theta) x_0 \in H$. Consider the map $\lambda : k \longrightarrow \mathbb{F}_p$ defined by the formula

$$\lambda(\theta) = [\alpha_0, x_1(\theta) + \rho(\theta) x_0).$$

If the image of λ is nonzero, the kernel of λ is an open subgroup of index p, and $k = \ker \lambda \oplus \theta_2 \mathbb{F}_p$ for some $\theta_2 \in k$. We put $\alpha_1 = \widehat{x}(\theta_2, \ker \lambda)$. Proceeding in this way, we construct $\alpha_i \in V_K$ so that, for some $c_i \in \mathbb{Z}$, the element $\alpha = \sum c_i \alpha_i$ satisfies $[\alpha, x_0) \neq 0$, and if $x_i + a_i x_0 \in H$, then $[\alpha, x_i + a_i x_0) = 0$. This concludes the proof of the proposition.

Taking into account the remark at the end of §1, we see that the set of analytic subgroups coincides with the set of norm subgroups.

References

1. B. M. Bekker, *Abelian extensions of a complete discrete valuation field of finite height*, Algebra i Analiz **3** (1991), no. 6, 76–84; English transl. in Leningrad Math. J. **3** (1992), no. 6.
2. G. Whaples, *Generalized local class field theory. 2. Existence theorem*, Duke Math. J. **21** (1954), 247–255.
3. _____, *Generalized local class field theory. 3. Second form of existence theorem. Structure of analytic groups*, Duke Math. J. **21** (1954), 575–581.
4. _____, *Generalized local class field theory. 4. Cardinalities*, Duke Math. J. **21** (1954), 583–586.
5. I. B. Fesenko, *Multidimensional local class field theory*. 2, Algebra i Analiz **3** (1991), no. 5, 169–190; English transl. in Leningrad Math. J. **3** (1992), no. 5.
6. G. Whaples, *Additive polynomials*, Duke Math. J. **21** (1954), 55–65.

Translated by THE AUTHOR

On p-adic Representations Arising from Formal Groups

D. G. BENOIS

In this paper we investigate the action of the Galois group of a local field on torsion points of formal groups. The main result gives a partial solution of one of Fontaine's problems [F]. The proof will be given in another article. The author would like to thank Professor S. V. Vostokov for constant help and encouragement.

Let $\mathcal{O} = W(k)$ be the ring of Witt vectors over a perfect field of characteristic $p \neq 2$. We denote by K its quotient field and by K_s the algebraic closure of K. Let $v\colon K^* \to \mathbb{Z}$ be a discrete valuation of K such that $v(p) = 1$. Let $F = F(x, y)$ be a one-parameter formal group over $\mathcal{O}$ of finite height h. Then F induces a $\mathbb{Z}_p$-module structure on the maximal ideal $\mathfrak{M}_s$ of K_s:

$$\alpha +_F \beta = F(\alpha, \beta), \qquad a_F \alpha = [a]_F(\alpha).$$

We denote this module by $F(\mathfrak{M}_s)$. For every $n \in \mathbb{N}$ there exists an exact sequence

$$0 \to E_{F,n} \to F(\mathfrak{M}_s) \xrightarrow{p^n} F(\mathfrak{M}_s) \to 0,$$

where $E_{F,n}$ denotes the group of points of order p^n in $F(\mathfrak{M}_s)$. Then the Tate module $T_F = \varprojlim E_{F,n}$ is a free $\mathbb{Z}_p$-module of rank h. The Galois group $\mathrm{Gal}(K_s/K)$ acts on T_F and we obtain a p-adic representation

$$\rho_F \colon \mathrm{Gal}(K_s/K) \to GL(T_F).$$

J.-M. Fontaine [F] suggested the following description of $\mathrm{Im}\, \rho_F$. Put $q = p^h$ and assume that K contains all $(q-1)$th roots of unity. The group $G = GL_h(\mathbb{Z}_p)$ is a p-adic Lie group and we denote by G_m the subgroup $\{g \in G \mid g \equiv 1_h \pmod{p^m}\}$. Let $K_m = K(E_{F,m})$. Then $H_m = \rho_F(\mathrm{Gal}(K_s/K_m))$ is contained in G_m. On the other hand, $\mathrm{Gal}(K_1/K)$ is isomorphic to $\mathbb{F}_q^\times$ and acts on every quotient H_m/H_{m+1} by the conjugation:

$$\sigma * (g \pmod{H_{m+1}}) = \tilde{\sigma}^{-1} g \tilde{\sigma} \pmod{H_{m+1}},$$

where $\tilde{\sigma} \in G$ is an arbitrary lifting of σ.

1991 *Mathematics Subject Classification.* Primary 11S31.

For every $i \in \mathbb{Z}/h\mathbb{Z}$ define the $\mathbb{F}_q^\times$-module M_i in the following way. M_i is isomorphic to $\mathbb{F}_q^+$ as an abelian group and $\mathbb{F}_q^\times$ acts on it by the formula

$$\varepsilon * m = \varepsilon^{1-p^i} m, \qquad \varepsilon \in \mathbb{F}_q^\times, \quad m \in M_i.$$

Fontaine proved that the $\mathbb{F}_q^\times$-module H_m/H_{m+1} is isomorphic to a direct sum of M_i:

$$H_m/H_{m+1} \simeq \bigoplus_{i \in I_m} M_i,$$

where $I_m \subset \mathbb{Z}/h\mathbb{Z}$. He showed also that $0 \in I_m$ and $I_m \subset I_{m+1}$ for every $m \in \mathbb{N}$ and defined the function

$$v_F \colon \mathbb{Z}/h\mathbb{Z} \to \mathbb{N} \cup \{0\}, \qquad v_F(i) = \min\{m \mid i \in I_m\}.$$

Then

(1) $\qquad\qquad\qquad v_F(0) = 0,$

(2) $\qquad v_F(i+j) \leqslant v_F(i) + v_F(j) \quad \text{for every } i, j \in \mathbb{Z}/h\mathbb{Z}.$

Fontaine asked whether for an arbitrary function v satisfying conditions (1) and (2) there exists a formal group F such that $v_F = v$. In [F] he proved this conjecture for $h = 2$. T. Nakamura [N1, N2] also obtained some results on torsion points and checked the existence of F for some functions v. In the case when the residue field k is finite, S. V. Vostokov and the author [BV1, BV2] found a different proof of his main result using norm pairings associated to formal groups. In this paper we formulate some further applications of explicit formulas for norm pairings to Galois representations.

Assume that k is finite and denote by Δ the Frobenius operator on the ring $\mathcal{O}[\![x]\!]$ of formal power series:

$$\left(\sum_i a_i x^i\right)^\Delta = \sum_i a_i^{\mathrm{Fr}} x^{ip},$$

where Fr is the Frobenius automorphism of $K/\mathbb{Q}_p$. Let $\alpha_1, \alpha_2, \ldots, \alpha_h \in \mathcal{O}$. Suppose $\alpha_i \equiv 0 \pmod{p}$ for $1 \leqslant i \leqslant h-1$ and α_h is a unit. Put $\mathcal{A}(\Delta) = \sum_{i=1}^h \alpha_i \Delta^i$. Honda [H] proved that $\lambda_a(x) = (1 - \mathcal{A}(\Delta)/p)^{-1}(x)$ is the logarithm of a formal group over $\mathcal{O}$ and that every formal group over $\mathcal{O}$ of height h is strictly isomorphic to a unique group of this type.

Let F be a formal group over $\mathcal{O}$. Denote by $\mathcal{A}(\Delta)$ the corresponding Honda operator. Let F^* be a formal group with the logarithm $\lambda^*(x) = (1 - \alpha_h \Delta^h/p)^{-1}(x)$. Put

$$n = \min_{1 \leqslant i \leqslant h-1} v(\alpha_i) \quad \text{and} \quad \varphi(x) = \lambda^{-1}(\lambda^*(x)),$$

where λ^{-1} is the formal power series such that $\lambda^{-1}(\lambda(x)) = x$. Nakamura [N1] proved that the map

$$\varphi^* \colon E_{F^*,n} \to E_{F,n}, \qquad \varphi^*(\xi^*) = \varphi(\xi^*)$$

is an isomorphism. Choose $\pi \in E_{F^*, n}$. It follows immediately that π is a prime element of K_n. Let $\mathcal{H}_m$ be the multiplicative group of Laurent power series $\theta x^m + a_{m+1} x^{m+1} + \ldots$ where $a_i \in \mathcal{O}$, $m \in \mathbb{Z}$, and θ is a $(q-1)$th root of unity. Every element α of K_n^* can be expanded in a convergent series, i.e., $\alpha = f(\pi)$ for some $f(x) \in \mathcal{H}_m$. Let $\mathcal{O}_1 = \mathbb{Z}_p[\sqrt[q-1]{1}]$ and tr: $\mathcal{O} \to \mathcal{O}_1$ be the trace map. Denote by $[p^n]^*$ the isogeny of F^* and put

$$\tau_0(\alpha) = \operatorname{tr} \operatorname{res}\left\{\left(x \frac{df(x)}{f(x)} - \frac{\alpha_h}{p^{h+1}} \log(f(x)^q/f(x)^{\Delta^h}) d\lambda^*(x)^{\Delta^h}\right) / [p^n]^*\right\} \pmod{p^n},$$

$$\tau_i(\alpha) = \operatorname{tr} \operatorname{res}\left(\frac{\alpha_i}{p} \lambda^{*\Delta^i}(x) \frac{df(x)}{f(x)} / [p^n]^*\right) \pmod{p^n}$$

for $1 \leq i \leq h-1$. It may be shown that the elements $\tau_0(\alpha), \ldots, \tau_{h-1}(\alpha)$ of $\mathcal{O}_1/p^n \mathcal{O}_1$ do not depend on the choice of the expansion $f(x)$ and so we obtain homomorphisms

$$\tau_i: K_n^* \to \mathcal{O}_1/p^n \mathcal{O}_1, \qquad 0 \leq i \leq h-1.$$

Define the homomorphism

$$\tau: K_n^* \to \prod_{i=0}^{h-1} \mathcal{O}_1/p^n \mathcal{O}_1$$

by the formula $\tau(\alpha) = (\tau_0(\alpha), \tau_1(\alpha), \ldots, \tau_{h-1}(\alpha))$.

THEOREM 1. (i) *The kernel of τ coincides with the norm subgroup*

$$N_{K_{2n}/K_n}(K_{2n}^*).$$

(ii) *The reciprocity map $\theta_{K_n}: K_n^* \to \operatorname{Gal}(K_n^{ab}/K_n)$ induces an injective homomorphism*

$$\mathcal{H}: \operatorname{Gal}(K_{2n}/K_n) \to \prod_{i=0}^{h-1} \mathcal{O}_1/p^n \mathcal{O}_1,$$

which is compatible with p-adic filtrations.

Theorem 1 implies that for every $m = 1, 2, \ldots, n$ the homomorphism

$$\mathcal{H}_m: \operatorname{Gal}(K_{n+m}/K_{n+m-1}) \to \prod_{i=0}^{h-1} M_i,$$

$$\mathcal{H}_m(\alpha) = \frac{1}{p^{m-1}} \mathcal{H}(\alpha) \pmod{p}$$

is an injection.

THEOREM 2. (i) *For every $m \leq n$ $\mathcal{H}_m$ is an homomorphism of $\mathbb{F}_q^\times$-modules.*
(ii) $\operatorname{Im} \mathcal{H}_m = \prod_{i \in I_m} M_i$ *where* $I_m = \{i \mid v(\alpha_i) \leq n + m - 1\}$.

COROLLARY 1. *Let* $h = 2$. *Then* $v_F(0) = 0$, $v_F(1) = v(\alpha_1)$.

This result has been proved by Fontaine and Nakamura for an arbitrary perfect field k.

COROLLARY 2. *Let* $h = 3$. *Then*

$$v_F(0) = 0,$$
$$v_F(1) = \begin{cases} v(\alpha_1) & \text{if } v(\alpha_1) < 2v(\alpha_2), \\ 2v(\alpha_2) & \text{if } v(\alpha_1) \geqslant 2v(\alpha_2), \end{cases}$$
$$v_F(2) = \begin{cases} v(\alpha_2) & \text{if } v(\alpha_2) < 2v(\alpha_1), \\ 2v(\alpha_1) & \text{if } v(\alpha_2) \geqslant 2v(\alpha_1). \end{cases}$$

The proof of these theorems is based on the explicit formula for the Hilbert generalized norm pairing $(\ ,\ )_{F,m}$ which is defined in the following way. Let L be an arbitrary extension of K containing $E_{F,m}$. Then $(\alpha, \beta)_{F,m} = \gamma^{\theta_L(\alpha)} -_F \gamma$, where $[p^m]_F(\gamma) = \beta$ and $\theta_L \colon L^\times \to \text{Gal}(L^{ab}/L)$ is the reciprocity map. In the case when F is a Lubin-Tate formal group, an explicit formula for $(\ ,\ )_{F,m}$ was obtained by S. V. Vostokov [V2] (see also [V1]). The case $m = 1$ was studied in [BV1]. If $m = n$, the proof is similar but all computations are more complicated. On the other hand, all the main results of this paper may be proved for multidimensional local fields.

References

[BV1] D. G. Benois and S. V. Vostokov, *Norm pairing in formal groups and Galois representations*, Algebra i Analiz **2** (1990), no. 6, 69–97; English transl., Leningrad Math. J. **2** (1991), no. 6, 1221–1249.

[BV2] _____, *Galois representations associated to Honda's formal groups*, Trudy Sankt-Peterburg. Mat. Obshch. (to appear). (Russian)

[F] J.-M. Fontaine, *Points d'ordre fini d'un groupe formel sur une extension non ramifiée de* $\mathbb{Z}_p$, Bull. Soc. Math. France; Mémoire **37** (1974), 75–79.

[H] T. Honda, *On the theory of commutative formal groups*, J. Math. Soc. Japan **22** (1970), 213–246.

[N1] T. Nakamura, *On torsion points of formal groups over a ring of Witt vectors*, Math. Z. **193** (1986), 397–404.

[N2] _____, *Finite subgroups of formal A-modules over p-adic integer rings*, Trans. Amer. Math. Soc. **286** (1984), 765–769.

[V1] S. V. Vostokov, *An explicit form of the reciprocity law*, Izv. Akad. Nauk SSSR Ser. Mat. **42** (1978), 1288–1321; English transl., Math. USSR-Izv. **13** (1979), 557–588.

[V2] _____, *Symbols on formal groups*, Izv. Akad. Nauk SSSR Ser. Mat. **45** (1981), 985–1014; English transl., Math. USSR-Izv. **19** (1982), 261–284.

Translated by THE AUTHOR

The Pairing on K-Groups in Fields of Valuation of Rank n

S. V. VOSTOKOV

§1. Introduction

1°. The explicit formula for Hilbert's symbol, which was obtained in [V1], provided an impulse for the search of explicit formulas both in various fields and explicit expressions for various objects (for example, formal groups, see [V3, V4]). As local class fields theory developed, the scope of local fields was extended to those for which the methods of [V1] are applicable.

The present paper is devoted to n-dimensional complete fields of characteristic 0. In these fields we construct an explicit form of the pairing on topological K-groups for which the sum of dimensions is equal to $n+1$. In particular, the explicit formula for the Hilbert pairing in n-dimensional local fields is obtained.

Such fields may be represented as a sequence of complete discretely valued fields: $K = k^{(n)}, k^{(n-1)}, \ldots, k^{(0)}$, where the field $k^{(i-1)}$ is the residue field for the field $k^{(i)}$, $1 \leqslant i \leqslant n$. We suppose also that the first residue field $k^{(n-1)}$ is of characteristic p, and the last residue field $k^{(0)}$ is perfect.

In the standard terminology, the fields whose last residue field is finite are called the *multidimensional local fields*. The theory of such fields was developed in the papers of A. N. Parshin and K. Kato (see [P, K]).

The field of Laurent series $k((t))$ with coefficients from an ordinary (numerical or functional) local field k gives first examples of two-dimensional local fields. This case is called *equal characteristic*, because $\operatorname{char} k((t)) = \operatorname{char} \overline{k((t))}$. For a number local field k, the fields of Laurent series that are infinite on both sides may be defined as follows

$$K = k\{\{t\}\} = \left\{ \sum_{i=-\infty}^{\infty} a_i t^i, \ a_i \in k \right\},$$

where $a_i \to 0$ when $i \to -\infty$ and the set of coefficients is bounded from below in the norm of k. This is the second example of two-dimensional local fields and it is said to be the *mixed characteristic* one because $\operatorname{char} K = 0$ and $\operatorname{char} \overline{K} = p$.

1991 *Mathematics Subject Classification.* Primary 11S70; Secondary 11S31, 19F05.

Key words and phrases. Multidimensional local field, topological K-group, Hilbert pairing.

This work was completed in MPG-Arbeitsgruppe "Zahlentheorie" in Berlin.

© 1995, American Mathematical Society

A. N. Parshin gave a complete classification of multidimensional local fields. We shall present it here for the case of mixed characteristic fields.

The field $K = k\{\{t_1\}\}\ldots\{\{t_{n-1}\}\}$, where $t_1, \ldots, t_{n-1}$ are independent variables, will be called the *standard n-dimensional local field*. Parshin's classification is the following: every n-dimensional local field of characteristic 0 whose first residue field is of characteristic p (the case of mixed characteristics of an n-dimensional local field) contains a standard field as a finite subextension and moreover, such a field is contained in a finite standard extension, i.e., there are number local fields k and k', and systems of local parameters $t_1, \ldots, t_{n-1}$ and $t'_1, \ldots, t'_{n-1}$ such that

$$E = k\{\{t_1\}\}\ldots\{\{t_{n-1}\}\} \subset K \subset k'\{\{t'_1\}\}\ldots\{\{t'_{n-1}\}\} = K'$$

and K/E, K'/K are finite extensions.

In the general case the classification of complete discretely valuated fields of characteristic 0 whose residue field has characteristic p was found by I. B. Zhukov (see [Z]). Let k be such a field and F be its residue field. Suppose $F_0 = \bigcap_{i=1}^{\infty} F^{p^i}$ is the maximal perfect subfield in F and k_0 is the quotient field of the ring of Witt vectors $W(F_0)$. Obviously the field k_0 is isomorphically injected into the field K. Denote by k the algebraic closure of the field k_0 in K. The field k is said to be the *constant field for* K. The field K is said to be *standard* if the extension K/k is unramified. Further, the extension K'/k is said to be *constant* if there is a finite extension k'/k such that $K' = k'K$ (obviously, k' is the field of constants for K'). Moreover, it is clear that a constant extension of a standard field is standard.

REMARK. It is clear that not every complete field is standard. For example, the adjoining of $\sqrt[p]{-t\pi}$ (where π is a prime element of k) to the standard two dimensional local field $K = \{\{t\}\}$ gives a nonstandard field.

We introduce the basic notations.
- K is an n-dimensional complete field of characteristic 0;
- F is the first residue field of K, i.e., $F = k^{(n-1)}$;
- π is a prime element of K with respect to a discrete valuation of rank 1; $t_1, \ldots, t_n = \pi$ are local parameters of K (a lift of prime elements of the residue fields $k^{(i)}$ which define the field K);
- $\bar{v}_K = (v^{(1)}, \ldots, v^{(n)}): K^* \to \mathbb{Z}^{(n)}$ is a valuation of rank n corresponding to the chosen local parameters;
- $\mathbb{Z}^{(n)}$ is ordered as follows $(m_1, \ldots, m_n) < (m'_1, \ldots, m'_n)$ if $m_n = m'_n, \ldots, m_{i+1} = m'_{i+1}$, but $m_i < m'_i$;
- $\mathfrak{o}_K$ is the ring of valuation of rank n;
- $\mathfrak{M}_K$ is the unique maximal ideal of $\mathfrak{o}_K$;
- $v = v^{(n)}$ is the valuation of rank 1 of K as a discretely valuated field;
- $e = v(p)$ is the ramification index of K with respect to the valuation of rank 1;
- $\bar{e} = (e^{(1)}, \ldots, e^{(n)}) = \bar{v}_K(p)$;
- ζ is a fixed p^mth root of unity contained in K;
- F_0 is the maximal perfect subfield in the residue field F of K, which is assumed not to be algebraically closed;
- $\mathfrak{o}_0$ is the ring of Witt vectors for F_0;

- $\mathcal{R}$ is the Teichmüller system of the representatives of elements of the field F_0 in the ring $\mathfrak{o}$;
- k_0 is the quotient field of $\mathfrak{o}$;
- Δ is the Frobenius automorphism in k_0;
- $\wp(\alpha) = \alpha^\Delta - \alpha$ is the Cartier operator on the completion of the maximal unramified extension of the ring $\mathfrak{o}_0$;
- $\alpha \equiv \beta \mod (\wp, p^m)$ in the ring $\mathfrak{o}_0$ means that $\alpha = \beta + \wp(\gamma) + p^m \gamma'$, where $\gamma, \gamma' \in \mathfrak{o}_0$;
- $\alpha \approx \beta$ or $\alpha \equiv \beta \mod K^{*p^m}$ means that the elements α and β from K differ by a p^mth power in K;
- $K_{st} = k_0\{\{t_1\}\}\ldots\{\{t_n\}\}$ is a standard absolutely unramified field with a valuation of rank $n+1$ ($t_1, \ldots, t_n$ are independent variables);
- $\mathfrak{o} = \mathfrak{o}_0\{\{t_1\}\}\ldots\{\{t_{n-1}\}\}$ is a valuation ring of rank n in $k_0\{\{t_1\}\}\ldots\{\{t_{n-1}\}\} \subset K$;
- $\mathfrak{o}' = \mathfrak{o}((t_n))$ is a ring of Laurent series;
- $\mathcal{H}_m = (\mathfrak{o}')^*$ is the group of invertible elements of the ring $\mathfrak{o}'$.

The Frobenius operator Δ in the ring $\mathfrak{o}\{\{t_n\}\}$ is

$$\left(\sum_{\bar{r}} a_{\bar{r}} t_1^{r_1} \ldots t_n^{r_n}\right)^\Delta = \sum a_{\bar{r}}^\Delta t_1^{pr_1} \ldots t_n^{pr_n}, \qquad a_{\bar{r}} \in \mathfrak{o}_0$$

(note that Δ depends on the choice of local parameters $t_1, \ldots, t_n$).

$\deg f$ denotes the order of Laurent series $f(x) = a_m x^m + a_{m+1} x^{m+1} + \ldots$, i.e., $m = \deg f$.

The congruence $f \equiv g \mod (p^r, \deg s)$ means that the coefficients whose powers are less than s are congruent modulo p^r.

The congruence $f \equiv g \mod p^r$ means $f \equiv g \mod p^r \mathfrak{o}\{\{t_n\}\}$.

For a series $f(t_1, \ldots, t_n)$ from K, $\operatorname{res} f$ denotes the coefficient at $t_1^{-1} t_2^{-1} \ldots t_n^{-1}$ of f.

REMARK. For series in several variables, the notation above is defined in the sense of the lexicographic ordering.

2°. For any series f invertible in the ring $\mathfrak{o}\{\{t_n\}\}$ the function

$$(1.1) \qquad l(f) = \frac{1}{p} \log f^p / f^\Delta$$

is well defined, because $f^p \equiv f^\Delta \mod p$ (see also Proposition 1 from [V1] or [V5]). For any series g from the ideal $x\mathfrak{o}[\![x]\!]$, the Artin-Hasse function is well defined:

$$E(g) = \exp\left(\sum_{m=0}^\infty g^{\Delta^m} / p^m\right).$$

Besides, the following assertion holds (see [V5]).

PROPOSITION. *The functions l and E define reciprocally inverse isomorphisms between the multiplicative $\mathbb{Z}_p$-module $1 + t_n\mathfrak{o}[\![t_n]\!]$ and the additive $\mathbb{Z}_p$-module $t_n\mathfrak{o}[\![t_n]\!]$.*

In particular, any series $\varepsilon(t_n)$ *from* $1 + t_n \mathfrak{o}[\![t_n]\!]$ *may by uniquely represented in the form*

(1.2) $$\varepsilon(t_n) = E(\eta(t_n)),$$

where $\eta(t_n) = l(\varepsilon)$.

Let us expand an element α from the field K into a series with coefficients from the ring $\mathfrak{o}$ and denote this series by $\underline{\alpha}(t_n)$. Thus

(1.3) $$\underline{\alpha}(t_n)\Big|_{t_n=\pi} = \alpha.$$

Suppose

(1.4) $$s_r(t_n) = \underline{\zeta}(t_n)^{p^r} - 1;$$

(1.5) $$u_r(t_n) = s_r/s_{r-1} = p + \sum_{i=1}^{p-1} \binom{p}{i+1} s_{r-1}^i$$

(we shall omit an index for $r = m$ in the series $s_m(t_n)$ and $u_m(t_n)$).

3°. Series expansion. By π we denote the prime element of the field K with respect to a valuation of rank 1. Note that the local parameters $t_1, \ldots, t_{n-1}$ are algebraically independent while π may turn out to be algebraically connected with $t_1, \ldots, t_{n-1}$ over the field k where $E = k\{\{t_1\}\} \ldots \{\{t_{n-1}\}\}$ is a standard subextension in K (see 1°). As an example, one can consider the element $\pi = \sqrt[p-1]{-p} \cdot t$ in the field $K = \mathbb{Q}_p(\zeta_p)\{\{t\}\}$, where $\zeta_p^p = 1$, as the prime π. Then we obtain the algebraic relation: $\pi^{p-1} + pt^{p-1} = 0$.

The set of multi-indices $I \subset \mathbb{Z}^{(n)}$ is said to be *admissible* if for any fixed set of integers $i_{l+1}, \ldots, i_n$, $1 \leqslant l \leqslant n$, in the set of all multi-indices $\bar{r} = (r_1, \ldots, r_l, r_{l+1}, \ldots, r_n)$ from I for which the indices $r_{l+1}, \ldots, r_n$ coincide with the indices $i_{l+1}, \ldots, i_n$ respectively, the index r_l is bounded from below, i.e., there exists an integer i such that $r_l \geqslant i$ for any $\bar{r} = (r_1, \ldots, r_l, i_{l+1}, \ldots, i_n)$ from I.

Any element α from K can be uniquely represented as the following series

(1.6) $$\alpha = \sum_{\bar{r} \in I} \theta_{\bar{r}} t_1^{r_1} \ldots t_{n-1}^{r_{n-1}} \pi^{r_n},$$

where I is an admissible set and $\theta_{\bar{r}} \in \mathcal{R}$. This assertion may be proved as in Lemma 2 from the article [VZF] for a multidimensional local field, see also [MZ]. If the uniqueness of the expansion of the element α into series is not necessary, the elements of the Witt vectors ring $\mathfrak{o}_0 = W(F_0)$ may be taken as the coefficients of the series.

Together with the field K, we shall also consider the standard field $K_{st} = k_0\{\{t_1\}\} \ldots \{\{t_n\}\}$, where k_0 is the quotient field of the ring $\mathfrak{o}$ and $t_1, \ldots, t_n$ are independent variables.

The essence of the transition from the field K to the standard field K_{st} is as follows: the local parameters $t_1, \ldots, t_n$ are algebraically independent, while the parameters $t_1, \ldots, t_{n-1}, \pi$ of K may be algebraically dependent.

Let $\mathfrak{o}' = \mathfrak{o}((t_n))$ be a ring of Laurent series, where $\mathfrak{o} = \mathfrak{o}_0\{\{t_1\}\} \ldots \{\{t_{n-1}\}\}$. We shall write $T^{\bar{r}} = t_1^{r_1} \ldots t_n^{r_n}$ for the set of multi-indices $\bar{r} = (r_1, \ldots, r_n)$ and the set of variables $t_1, \ldots, t_n$. Any element $\alpha(x)$ of the ring $\mathfrak{o}'$ can be uniquely represented in the form

$$(1.7) \qquad \alpha(T) = \sum_{\bar{r} \in I} a_{\bar{r}} T^{\bar{r}}, \qquad a_{\bar{r}} \in \mathfrak{o}_0,$$

where I is an admissible set of multi-indices.

We shall denote the multiplicative group of the ring $\mathfrak{o}'$ by $\mathcal{H}_m$. There is the surjective homomorphism (uncanonical)

$$(1.8) \qquad \eta_m : \mathcal{H}_m \longrightarrow K^* \qquad \alpha(T) \longmapsto \alpha = \alpha(t_1, \ldots, t_{n-1}, \pi).$$

It is not hard to see that the multiplicative group $\mathcal{H}_m$ can be decomposed into the product of subgroups $\mathcal{H}_m = \langle t_1 \rangle \times \cdots \times \langle t_n \rangle \times \mathcal{R} \times U$, where U is the group of principal units consisting of the elements

$$(1.9) \qquad \varepsilon = 1 + \sum_{\bar{r} > 0} a_{\bar{r}} T^{\bar{r}}, \qquad a_{\bar{r}} \in \mathfrak{o}_0.$$

Using the Artin–Hasse function E, we can define $\mathbb{Z}_p$-generators of the group U in the form

$$(1.10) \qquad U = \langle E(aT^{\bar{r}}), \ a \in \mathfrak{o}_0, \ \bar{r} > 0 \rangle.$$

Moreover, if it is necessary to obtain the $\mathbb{Z}_p$-basis of the group U, one may choose the elements from the Teichmüller system $\mathcal{R}$ as coefficients.

The multiplicative group K^* of the field K has an analogous decomposition into the direct product

$$K^* = \langle t_1 \rangle \times \cdots \times \langle t_{n-1} \rangle \times \langle \pi \rangle \times \mathcal{R} \times U_K,$$

where U_K is the group of principal units of the field K.

However the group U_K has in general a more complicated structure (see §3 below).

§2. Preliminary results

1°. Let us consider the behavior of series in the rings $\mathfrak{o}$ and $\mathfrak{o}\{\{t_n\}\}$.

LEMMA 2.1. *Let*

$$f(t_n) = \sum_{\bar{i}} a_{\bar{i}} t_1^{i_1} \ldots t_n^{i_n}, \qquad a_{\bar{i}} \in \mathfrak{o}_0,$$

be a series from the ring $\mathfrak{o}\{\{t_n\}\}$. The series $f(t_n)$ is invertible if and only if there is at least one coefficient a_i which is invertible in $\mathfrak{o}_0$.

REMARK. The above condition is equivalent to the existence of coefficient $a_{(i)}$ which is prime to p.

PROOF. For $n = 1$ this lemma was proved in the article [V2, §1.3]. The general case can be verified by obvious induction. □

Further, we shall denote by $1/f$ the series which is inverse to $f(t_n)$ in the ring $\mathfrak{o}\{\{t_n\}\}$.

The following assertions are simple exercises (for the one-dimensional case see [V2, §1.3].

If $f(t_1, \ldots, t_n)$ is invertible in $\mathfrak{o}\{\{t_n\}\}$, then we have

(2.1) $\qquad f \equiv g \mod p^r \iff 1/f \equiv 1/g \mod p^r,$

(2.2) $\qquad \dfrac{\partial}{\partial t_i} f \equiv 0 \mod p^r \iff \dfrac{\partial}{\partial t_i}(1/f) \equiv 0 \mod p^r, \qquad 1 \leqslant i \leqslant n,$

(2.3) $\qquad g \equiv h \mod p^r \iff g/f \equiv h/f \mod p^r.$

LEMMA 2.2. *For any $r \geqslant 0$ the series $s_r(t_n)$ and $u_r(t_n)$ are invertible in the ring $\mathfrak{o}\{\{t_n\}\}$.*

PROOF. Consider the series $z_0(t_n) = \underline{\zeta}(t_n) - 1$. The order of the element $z_0(\pi) = \underline{\zeta}(\pi) - 1 = \zeta - 1$ in K (as in a discretely valuated field of rank 1) is less than e, thus not all the coefficients of $z_0(t_n)$ are divisible by p, i.e., there exists a coefficient of the series z_0 (element of $\mathfrak{o}$) which is invertible, according to Lemma 2.1, in $\mathfrak{o}$, but this means that the series $z_0(t_n)$ is invertible in $\mathfrak{o}\{\{t_n\}\}$.

Further, from the definition of $s_r(t_n)$, we have the congruence

$$s_r(t_n) \equiv z_0^{p^r}(t_n) \mod p,$$

and the invertibility of s_r follows from (2.1).

The invertibility of $u_r(t_n)$ follows from its definition and from the assertion proved above. So the lemma is proved. □

2°. The following congruences hold:

(2.4) $\qquad s_r \equiv s_{r-1}^\Delta \mod p^r, \qquad r \geqslant 1,$

(2.5) $\qquad 1/s_r \equiv 1/s_{r-1}^\Delta \mod p^r,$

(2.6) $\qquad \dfrac{\partial}{\partial t_i}(s_r) \equiv \dfrac{\partial}{\partial t_i}(1/s_r) \equiv 0 \mod p^r, \qquad 1 \leqslant i \leqslant n.$

Moreover, from the definition of u_m and (2.4), we have the congruence

(2.7) $\qquad u^\Delta \equiv u_{m+1} \mod p^m.$

Rewrite (2.4) in the form of the equality $u^\Delta = u_{m+1} + p^m h$ for some h from $\mathfrak{o}[\![t_n]\!]$. Further raise this equality to the power r and multiply it by the series $1/s$, obtaining

$$u^{r\Delta}/s = \sum_{i=0}^{r} \binom{r}{i} p^{mi} h^i u_{m+1}^{r-i}/s \equiv \sum_{i=0}^{r} \binom{r}{i} p^{mi} h^i p^{r-i}/s \mod \deg(0,\ldots,0).$$

From it follows that

(2.8) $\qquad u^{r\Delta}/s \equiv p^r/s \mod (p^{m+r-1}, \deg(0,\ldots,0)), \qquad r \geqslant 1.$

As earlier, using (2.4), we obtain

(2.9) $\qquad (u_{m-1}^{\Delta r} - s^r)/s \equiv 0 \mod (p^{mr}, \deg(0,\ldots,0)), \qquad r \geqslant 1.$

Finally, from the definition of the series u, we obtain the congruence

$$u^r/s_{m-1} \equiv p^r/s_{m-1} \mod \deg 0.$$

3°. Throughout this section the field K denotes an arbitrary complete discretely valued field of characteristic 0, whose residue field has characteristic p. We shall verify an assertion, which will be the basis of our proof of the independence of the pairing (see [V1]).

Let v be a valuation in K; $\mathfrak{o}_K$ be the ring of integers with respect to v; π be a prime in K. Let us fix a formal power series $e(x) \in \mathfrak{o}_K[\![x]\!]$, which is an Eisenstein series, i.e.,

$$e(x) = a_0 + a_1 x + \cdots + a_r x^r + \cdots,$$

where $a_0, a_1, \ldots, a_{r-1} \equiv 0 \mod \pi$, $a_0 \not\equiv 0 \mod \pi^2$, and the coefficient a_r is a unit of $\mathfrak{o}_K$.

PROPOSITION 2.1. *Let $f(x)$ be a series of the ring $\mathfrak{o}_K[\![x]\!]$. Then $f(x)$ is divisible by $e(x)$ if and only if both the series have a common root in the algebraic closure of the field K.*

PROOF. According to the preliminary Weierstrass lemma, there is a series $\varepsilon(x) = 1 + b_1 x + \cdots \in \mathfrak{o}_K[\![x]\!]$ such that $e_*(x) = e(x) \cdot \varepsilon(x)$ is the Eisenstein polynomial of degree r. From this it follows in particular that $e(x)$ has a root in the algebraic closure of K, and thus the necessity of the assertion is proved.

Let us verify the sufficiency of the assertion. Without loss of generality one can assume that the greatest common divisor of all the coefficients of $f(x)$ is equal to 1 (otherwise it could be taken out from $f(x)$). Again using Weierstrass lemma, one can find series $\eta(x) = 1 + c_1 x + \cdots \in \mathfrak{o}_K[\![x]\!]$ such that $f_*(x) = f(x)h(x)$ will be a polynomial. Let α be a common root of the series $f(x)$ and $e(x)$ in the algebraic closure of the field K. Then this root will be a common root for the polynomials f_* and e_*. Notice that the series ε and η are invertible in $\mathfrak{o}_K[\![X]\!]$.

The Eisenstein polynomial $e_*(x)$ is irreducible over K and has no multiple roots. From this it follows that $f_*(x)$ is divisible by $e_*(x)$, i.e., $f_*(x) = g(x)e_*(x)$, where $g(X) \in \mathfrak{o}_K[\![x]\!]$. But then $f = e \cdot \Psi$ where $\Psi = \varepsilon \cdot g \cdot \eta^{-1} \in \mathfrak{o}[\![x]\!]$. The proposition is proved. □

Now let K be the field specified in subsection 1°, i.e., complete discretely valued field of rank n. Then the series $u(t_n)$, constructed in (1.5), as it is easy to notice, is the Eisenstein series. We shall use the proposition proved above in the following form.

PROPOSITION 2.2. *Let $\varepsilon(t_n) = 1 + a_1 t_n + \cdots$ be an invertible series from the ring $\mathfrak{o}[\![t_n]\!]$. If $\varepsilon(\pi) = 1$ for a prime π in K, then there is a series $\Psi(t_n) \in \mathfrak{o}[\![t_n]\!]$ such that $\varepsilon(t_n) = 1 + u\Psi$.*

In the group $\mathcal{H}_m$ (see §1, 3°) we consider the subgroup

$$(2.10) \qquad \mathcal{U}_m = \langle 1 + u(t_n)\varphi(t_n) \rangle,$$

where $\varphi \in \mathfrak{o}[\![t_n]\!]$, $\varphi(0) = 0$.

PROPOSITION 2.3. *There exists an exact sequence $1 \to \mathcal{U}_m \to \mathcal{H}_m \xrightarrow{\eta_m} K^* \to 1$, where η_m is the homomorphism defined in (1.8).*

PROOF. From the definition of the epimorphism η_m it follows that η_m sends any elements of the subgroup $\mathcal{U}_m$ to 1. Conversely, if $\eta_m(h(t_n)) = 1$ for some series $h \in \mathcal{H}_m$, then $h(t_1, \ldots, t_{n-1}, \pi) - 1 = 0$. To complete the proof, Proposition 2.2 must be used. □

4°. We shall obtain a few congruences for the series $\varepsilon(t_n) \in \mathfrak{o}[\![t_n]\!]$ which satisfies the condition

$$(2.11) \qquad \varepsilon(t_n) \equiv 1 \mod \deg(1, 0, \ldots, 0), \qquad \varepsilon(\pi) = 1.$$

According to Proposition 2.2, this condition means that there exists a series $\Psi \in \mathfrak{o}[\![t_n]\!]$ such that

$$(2.12) \qquad \varepsilon(t_n) = 1 + u\Psi.$$

LEMMA 2.3. *Let $\varepsilon(t_n)$ be the series from (2.11) and η an arbitrary series from the ring $\mathfrak{o}[\![t_n]\!]$, whose free term is equal to 1. Then for $p \neq 2$ the following congruence holds:*

$$l(\eta) \log \varepsilon \frac{\partial}{\partial t_i}(1/s) \equiv 0 \mod (p^m, \deg(0, \ldots, 0)),$$

i.e., the coefficients of the series in the left side are divisible by p^m for powers less than $(0, 0, \ldots, 0)$.

PROOF. By definition

$$(2.13) \qquad u = s/s_{m-1}.$$

Besides, for $p \neq 2$, it follows from (1.5) that there are series φ_1 and φ_2 $\mathfrak{o}[\![t_n]\!]$ such that

$$(2.14) \qquad u = p\varphi_1 + s_{m-1}^2 \varphi_2.$$

Using the equality (2.13) we obtain

$$(\log \varepsilon)/s^2 = \sum_{r=1}^{\infty} \frac{(-1)^{r-1}}{r} u^r \Psi^r / s^2 = \Psi/(s_{m-1}s) + \sum_{r \geq 2} \frac{(-1)^{r-1}}{r} u^{r-1} \Psi^r/(s_{m-1}s)$$

$$= \Psi/(s_{m-1}s) + \sum_{r \geq 2} \frac{(-1)^{r-1}}{r} u^{r-2} \Psi^r/(s_{m-1}^2).$$

It is obvious, further, that (2.14) implies

$$u^{r-2}/s_{m-1}^2 \equiv p^{r-2}\varphi_1^{r-2}/s_{m-2}^2 \mod \deg(0,\ldots,0).$$

Hence,

$$l(\eta)\log\varepsilon \frac{\partial}{\partial t_i}(1/s) = -l(\eta)(\log\varepsilon)/s^2 \cdot \frac{\partial}{\partial t_i}s$$

$$\equiv -\left[\Psi \cdot l(\eta)/(s_{m-1}s) + (1/s_{m-1}^2)l(\eta)\sum_{r\geq 2}(-1)^{r-1}\frac{p^{r-2}}{r}\varphi_1^{r-2}\Psi^r\right]\frac{\partial}{\partial t_i}s \mod \deg(0,\ldots,0).$$

The number $\frac{p^{r-2}}{r}$ for $p \neq 2$ and $r \geq 2$ is a p-integer, so the series in brackets has integral coefficients. Besides,

$$\frac{\partial}{\partial t_i}s \equiv 0 \mod p^m$$

(see (2.6)). From this the congruence in lemma follows. □

LEMMA 2.4. *Let the series* $\varepsilon(t_n)$ *satisfy condition* (2.11). *Then for* $p \neq 2$ *the following congruence holds*

(2.15) $\qquad l(\varepsilon)/s \equiv (1-\Delta)((\log\varepsilon)/s) \mod (p^m, \deg(1,0,\ldots,0)).$

The series $\log\varepsilon/s$ *has integral coefficients for powers less than* $(1,0,\ldots,0)$.

PROOF. At first we shall verify the second assertion of the lemma. From the definition of u it follows that $u/s = 1/s_{m-1}$, hence

$$u^r/s = u^{r-1}(u/s) = u^{r-1}/s_{m-1} \equiv p^{r-1}/s_{m-1} \mod \deg(0,\ldots,0).$$

Then the following congruence holds:

(2.16) $\qquad \dfrac{u^r}{r}/s \equiv \dfrac{p^{r-1}}{r}/s_{m-1} \mod \deg(0,\ldots,0).$

Recall that $\varepsilon(t_n) = 1 + u\Psi$. Let us multiply both parts of congruence (2.16) by $(-1)^{r-1}\Psi^r$ and sum over r. Then we obtain

(2.17) $\qquad (\log\varepsilon)/s \equiv \dfrac{1}{p}(\log p\Psi)/s_{m-1} \mod \deg(1,0,\ldots,0)$

(since the series Ψ has no constant term, the order deg in the congruence increases). Finally we remark that p^{r-1}/r is p-integral coefficients. This proves the second part of the lemma.

Further from (2.8) when $r \geqslant 2$ we get the congruence

$$(2.18) \qquad \frac{u^{r\Delta}}{pr}/s \equiv \frac{p^{r-1}}{r}/s \mod (p^m, \deg(1, 0, \ldots, 0)).$$

This congruence holds also for $r = 1$ as

$$\frac{1}{p}(u^\Delta - u_{m+1})/s = \sum_{i=1}^{p-1} \left(\binom{p}{i-1}/p \right) (s_{m-1}^{\Delta i} - s^i)/s.$$

It remains to apply (2.9), taking into account (for $i = (p-1)$) the fact that $p \geqslant 3$.

Multiplying both sides of (2.18) by $(-1)^{r-1}\psi^{r\Delta}$ and summing it, we obtain

$$\left(\frac{\Delta}{p} \log \varepsilon \right)/s \equiv \left(\frac{\Delta}{p} \log p\Psi \right)/s \mod (p^m, \deg(1, 0, \ldots, 0)).$$

Applying the Frobenius operator Δ to both sides of (2.17) and using (2.5), we obtain the following congruence

$$((\log \varepsilon)/s)^\Delta \equiv \left(\frac{\Delta}{p} \log p\Psi \right)/s \mod (p^m, \deg(1, 0, \ldots, 0)).$$

Subtracting the two last congruences from each other, we obtain

$$\left(\frac{\Delta}{p} \log \varepsilon \right)(1 - p\Delta)(1/s) \equiv 0 \mod (p^m, \deg(1, 0, \ldots, 0)).$$

It remains to subtract the series $(\log \varepsilon)/s$ from both sides of the last congruence to obtain (2.15). The lemma is proved. $\square$

5°. The following lemmas are necessary for further calculations.

LEMMA 2.5. *Let $f(t_1, \ldots, t_n)$ be a series from the ring $\mathfrak{o}\{\{t_n\}\}$. Then the following relation holds*

$$(2.19) \qquad \operatorname{res} t_1^{-1} \ldots t_n^{-1} f = \operatorname{res} t_1^{-1} \ldots t_n^{-1} f^\Delta + \wp(\alpha),$$

where α is the free term of f. (Here res *is applied to all the variables $t_1, \ldots, t_n$.)*

The proof is obvious.

REMARK. If the last residue field $F_0 = k^{(0)}$ of the field K is finite, then we have

$$(2.20) \qquad \operatorname{tr} \operatorname{res} t_1^{-1} \ldots t_n^{-1} f = \operatorname{tr} \operatorname{res} t_1^{-1} \ldots t_n^{-1} f^\Delta,$$

where tr is the trace operator in the extension $k_0/\mathbb{Q}_p$.

The relation (2.20) follows immediately from (2.19), because $\operatorname{tr} \wp(\alpha) = \operatorname{tr}(\alpha^\Delta - \alpha) = 0$.

Let $\bar{e} = \bar{v}_K(p) = (e_1, \ldots, e_n)$ and $\bar{e}_m = \bar{e}/(p^{m-1}(p-1))$. Then $\bar{v}_K(\zeta - 1) = \bar{e}_m$ and hence one can choose from the all expansions of the element $\zeta - 1$ the one satisfying $\deg(\underline{\zeta}(t_n) - 1) = \bar{e}_m$. From this and the definition of s_r it immediately follows that

$$(2.21) \qquad \deg s_r^{\Delta^i} \geqslant \bar{e}_m, \qquad r \geqslant m, \ i \geqslant 0.$$

LEMMA 2.6. *For any $i \geq 1$ the following inequality holds*:

$$\bar{v}_K \left(s^{\Delta^i}(t_n) \big|_{t_n=\pi} \right) \geq \bar{e}(1 + \max(i, m)). \tag{2.22}$$

PROOF. From the congruence $s_{r-1}^{\Delta} \equiv s_r \bmod p^r$ (see (2.4)), we have

$$s_{m+\alpha-1}^{\Delta} = s_{m+\alpha} + p^{m+\alpha} f_\alpha, \qquad \alpha \geq 1,$$

$\deg f_\alpha \geq \bar{e}_m$. Applying the operator $\Delta^{i-\alpha}$ to both parts of the last relation and summing from 1 to i, we obtain

$$s^{\Delta^i} = s_{m+i} + p^{m+i} f_i + p^{m+i-1} f_{i-1}^{\Delta} + \cdots + p^{m+1} f_1^{\Delta^{i-1}},$$

where $\deg f_\alpha \geq \bar{e}_m$. Substitute $t_n = \pi$, then

$$v\left(p^{m+i-\alpha} f_{i-\alpha}^{\Delta^\alpha} \big|_{t_n=\pi} \right) \geq (m+i-\alpha)\bar{e} + p^\alpha \bar{e}_m \geq \bar{e}(1 + \max(i, m)).$$

This implies (2.22), since $s_{m+1}(\pi) = \zeta^{p^{m+i}} - 1 = 0$. □

Using relations (2.21) and (2.22) and arguing as in Proposition 1 of the paper [V2], we obtain the following result.

LEMMA 2.7. *Let $a \in \mathfrak{o}_0$, then for $p \neq 2$*,

$$E(as(t_n))\big|_{t_n=\pi} \approx E(p^m a \log \underline{\zeta}(t_n))\big|_{t_n=\pi}.$$

The following relations, which may be easily verified, will be used further. Let $\alpha(t_1, \ldots, t_n)$ be a series in n variables and let

$$\delta_i(\alpha) = \alpha^{-1} \frac{\partial}{\partial t_i} \alpha$$

be ith logarithmic derivative, then

$$\delta_i(\alpha) = \frac{1}{p} \delta_i(\alpha^\Delta) = \frac{\partial}{\partial t_i} l(\alpha). \tag{2.23}$$

Indeed, according to the definition of $l(\alpha)$ (see (1.1)), we have

$$\frac{\partial}{\partial t_i} l(\alpha) = \frac{1}{p} \frac{\partial}{\partial t_i} \log \alpha^p/\alpha^\Delta = \frac{1}{p} (\alpha^p/\alpha^\Delta)^{-1} \frac{\partial}{\partial t_i} (\alpha^p/\alpha^\Delta)$$

$$= \alpha^{-1} \frac{\partial}{\partial t_i} \alpha - \frac{1}{p} \alpha^{-\Delta} \frac{\partial}{\partial t_i} \alpha^\Delta = \delta_i(\alpha) - \frac{1}{p} \delta_i(\alpha^\Delta)$$

and (2.23) is proved.

§3. Arithmetic of the field K

1°. A unit ε of K is said to be *principal* if its image in the last residue field is equal to 1. We shall denote the group of principal units by U_1. Then we have

$$K^* \cong \mathbb{Z}^n \times \mathcal{R} \times U_1,$$

and any principal unit ε possesses a unique expansion into the convergent product

$$\varepsilon = \prod_{i_n \geqslant 0} \prod_{i_{n-1} \geqslant I_{n-1}(i_n)} \cdots \prod_{i_1 \geqslant I_1(i_n, \ldots, i_2)} (1 + \theta_{(i)} \pi^{i_n} \cdot t_{n-1}^{i_{n-1}} \ldots t_1^{i_1}),$$

where $I_{n-1}(0) \geqslant 0, \ldots, I_1(0, \ldots, 0) > 0$ and the elements $\theta_{(i)} \in \mathcal{R}$.
Let $\bar{v}_K(p) = \bar{e} = (e_1, \ldots, e_n) > 0$ and

$$\bar{e}' = \frac{\bar{e}}{p-1} = \left(\frac{e_1}{p-1}, \ldots, \frac{e_n}{p-1}\right),$$

$p = \theta_p t_1^{e_1} \ldots t_{n-1}^{e_{n-1}} \pi^{e_n} + \ldots$, where $\theta_p \in \mathcal{R}$. Let

$$(3.1) \qquad I = \{\bar{r} \mid \bar{0} < \bar{r} < p\bar{e}',\ p \nmid \bar{r}\},$$

i.e., not all the indices $r_1, \ldots, r_n$ are divisible by p.

PROPOSITION 3.1. *If the field K does not contain any nontrivial pth roots of unity, then the group of principal units has the following set of generators over $\mathbb{Z}_p$:*

$$(3.2) \qquad \varepsilon_{\bar{r}}(\theta) = 1 - \theta t_1^{r_1} \ldots t_{n-1}^{r_{n-1}} \pi^{r_n},$$

where $\bar{r}$ runs through the set I and θ runs through the Teichmüller system $\mathcal{R}$.

PROOF. From the relation $(1-\alpha)^p = 1 - \alpha^p + p\alpha - \ldots$ it follows that

$$(3.3) \qquad \varepsilon_{\bar{r}}^p(\theta) = \begin{cases} 1 - \theta^p t_1^{pr_1} \ldots t_{n-1}^{pr_{n-1}} \pi^{pr_n} + \ldots, & \text{if } \bar{r} < \bar{e}', \\ 1 - \theta_p \theta t_1^{r_1+e_1} \ldots t_{n-1}^{r_{n-1}+e_{n-1}} \pi^{r_n+e_n} + \ldots, & \text{if } \bar{r} > \bar{e}', \\ 1 - (\theta_p \theta + \theta^p) t_1^{pe_1'} \ldots t_{n-1}^{pe_{n-1}'} \pi^{pe_n'} + \ldots, & \text{if } \bar{r} = \bar{e}' \end{cases}$$

(where the dots denote terms of higher powers).

REMARK 1. We shall use the following system of generators of the group U_1 more often

$$(3.4) \qquad \mathcal{E}_{\bar{r}}(\theta) = E(\theta t_1^{r_1} \ldots t_{n-1}^{r_{n-1}} \pi^{r_n})$$

(under the same condition for the multi-index $\bar{r}$ and the element $\theta \in \mathcal{R}$). This is possible due to the following congruence

$$E(x) \equiv 1 - x \mod x^2.$$

REMARK 2. It is easy to note that the group of principal units U_1 is l-divisible for $(l, p) = 1$.

From (3.2) and (3.3) it follows in particular that any principal unit $\varepsilon = 1 + \theta t_1^{r_1} \ldots t_{n-1}^{r_{n-1}} \pi^{r_n} + \ldots$ for which the multi-index $\bar{r} = (r_1, \ldots, r_n)$ is greater than $p\bar{e}'$ is a pth power. Besides, the map $U_1 \xrightarrow{\uparrow p} U_1$ is an injection, because the field K has no nontrivial pth roots of unity. Hence any unit $\varepsilon = 1 + \theta t_1^{pe_1'} \ldots t_{n-1}^{pe_{n-1}'} \pi^{pe_n'} + \ldots$ is a pth power. (The proof of the necessity is similar to that for number local fields, and is described in detail in [H, II, §15]). The assertion of the proposition now follows from all the above. $\square$

If the field K contains nontrivial pth roots of unity, the problem is more difficult, because the map that raises to the power p then has a nontrivial kernel. In this case the principal units of order $p\bar{e}'$ are not all pth powers. Below we shall find a minimal system of generators (canonical Shafarevich basis, see Theorem 5.7), but now we shall restrict ourselves to the following assertion.

PROPOSITION 3.2. *Let the field K contain nontrivial pth roots of unity. Then the group of principal units has the following set of generators over $\mathbb{Z}_p$:*

(3.5) $$\{\varepsilon_{\bar{r}}(\theta), \ \bar{r} \in I\} \& \{\varepsilon_*(\theta) = 1 - \theta t_1^{pe_1'} \ldots t_{n-1}^{pe_{n-1}'} \pi^{pe_n'} + \ldots\}$$

or

(3.6) $$\{\mathcal{E}_{\bar{r}}(\theta), \ \bar{r} \in I\} \& \{\mathcal{E}_*(\theta)\},$$

where θ runs through the Teichmüller system $\mathcal{R}$.

REMARK. The system of units $\varepsilon_*(\theta)$ is redundant since it contains pth powers.

2°. Primary elements. Now we consider the complementary generators $\varepsilon_*(\theta)$ for the system (3.5) in detail. It will be shown that these generators give primary units in the field K. Recall that an element $\omega \in K$ is said to be p^m-*primary* if the extension $K(\sqrt[p^m]{\omega})/K$ is the unique unramified one (i.e., may be obtained when the residue field $k^{(0)}$ is extended). These elements, both in a local number field, and in a multidimensional local field have been constructed in the articles [V1, V5].

Let $\widetilde{K}$ be the maximal abelian purely unramified p-extension of the field K. If for the maximal complete subfield $F_0 = k^{(0)}$ in the residue field $\overline{K}$ we have

$$F_0/\wp(F_0) \cong \bigoplus_{\varkappa} \mathbb{Z}/p\mathbb{Z},$$

then the Galois group of the maximal abelian p-extension of the field F_0 is isomorphic to $\prod_{\varkappa} \mathbb{Z}_p$ and hence

(3.6) $$\mathrm{Gal}(\widetilde{K}/K) \cong \mathrm{Gal}(F_0^{p,\,\mathrm{ab}}/F_0) \cong \prod_{\varkappa} \mathbb{Z}_p.$$

Let $\mathfrak{o}_0 = W(F_0)$ be the Witt vectors ring of the field F_0 and $\mathfrak{o}_0^{\mathrm{nr}}$ be the ring of integers of the completion of the maximal unramified extension k_0 (i.e., the quotient field for the ring $\mathfrak{o}_0$).

LEMMA 3.1. *For any element $a \in \mathfrak{o}_0$ there is an element $A \in \mathfrak{o}_0^{nr}$ that satisfies the following equation*

(3.7) $$\wp(A) = A^\Delta - A = a.$$

Besides, for any automorphism φ from $\mathrm{Gal}(\widetilde{K}/K)$, the element $a_\varphi = A^\varphi - A$ is a p-adic integer that does not depend of the choice of the root A.

PROOF. The equation $\wp(x) = \bar{a}$, where $\bar{a}$ is the residue of a in the field F_0, is solvable in the extension $F_0^{p,\mathrm{ab}}$. Using the method of successive approximations we obtain the solution of equation (3.7).

If $\varphi \in \mathrm{Gal}(\widetilde{K}/K)$, then $a_\varphi^\Delta - a_\varphi = a_\varphi - a$. The element a belongs to $\mathfrak{o}_0$, hence $a^\varphi = a$. Therefore $a_\varphi^\Delta = a_\varphi$, so $a_\varphi \in \mathbb{Z}_p$. Obviously, a_φ does not depend on the choice of A. The lemma is proved.

Let ζ be a primitive p^mth root of unity contained in the field K, and let $\underline{\zeta}(x)$ be a series from the ring $\mathfrak{o}[\![x]\!]$ obtained from the expansion the element ζ into a power series with respect to element π. As before, $s(x) = \underline{\zeta}(x)^{p^m} - 1$.

THEOREM 3.1. *Any p^m-primary element p^m of the field K may be represented (up to a p th power) as follows:*

(3.8) $$\alpha \approx \omega(a) = E(as(x))\Big|_{x=\pi}$$

for some $a \in \mathfrak{o}_0$; vice versa, for any $a \in \mathfrak{o}_0$ the element $\omega(a)$ of the form (3.8) is p^m-primary. Besides, the element $\omega(a)$ does not depend on the expansion of the root ζ into a series in the prime element π. Further,

(3.9) $$\sqrt[p^m]{\omega(a)}^{\varphi-1} = \zeta^{A^\varphi - A} = \zeta^{a_\varphi}$$

for any automorphism φ of the Galois group $\mathrm{Gal}(\widetilde{K}/K)$. Finally,

(3.10) $$\omega(a) \approx 1 \iff a \equiv \wp(a_0) \mod p^m, \quad a_0 \in \mathfrak{o}_0.$$

COROLLARY. *Let Ω be the group of p^m-primary elements of the field K and μ_{p^m} be the group of p^mth roots of unity. If $F_0 = k^{(0)}$ is a finite field and φ is the Frobenius automorphism in $\widetilde{K}/K$, then*

(3.11) $$\sqrt[p^m]{\omega(a)}^{\varphi-1} = \zeta^{\mathrm{tr}\, a},$$

and there is an isomorphism χ such that

(3.12) $$\chi: \Omega/\Omega \cap K^{*p^m} \xrightarrow{\sim} \mu_{p^m}, \quad \chi(\omega(a)) = \zeta^{\mathrm{tr}\, a},$$

where tr is the trace operator in $k^{(0)}/\mathbb{Q}_p$.

PROOF OF THE COROLLARY. Let $q = p^f$ denote the number of elements of the field F_0. Then $\varphi = \Delta^f$ and hence $a_\varphi = A^\varphi - A = A^{\Delta^f} - A = a + a^\Delta + \cdots + a^{\Delta^{f-1}} =$

tr a. Hence formula (3.8) follows from equality (3.12) of the theorem. The second statement of the corollary follows from formula (3.10) of the theorem. □

PROOF OF THEOREM 3.1. We shall first verify that if α is a p^m-primary element of K, then it may be represented in the form (3.8). The following argument is due to I. B. Fesenko, to whom the author is very grateful. Indeed, for the series $s_1(x) = \underline{\zeta}_1^p(x) - 1$, where $\zeta_1^p = 1$, the following congruence is obviously holds:

$$s_1(x) \equiv (\underline{\zeta}_1(x) - 1)^p = c^p t_1^{pe'_1} \ldots t_{n-1}^{pe'_{n-1}} x^{pe'_n} + \ldots \mod p,$$

where $\underline{\zeta}(x) = 1 + ct_1^{e'_1} \ldots t_{n-1}^{e'_{n-1}} x^{e'_n} + \ldots$. Hence, using the properties of the function E, we have

(3.13) $$E(s_1(x)) \underset{p}{\approx} E(c^p t_1^{pe'_1} \ldots t_{n-1}^{pe'_{n-1}} x^{pe'_n}).$$

The element α is p^m-primary, hence p-primary, so $\alpha = 1 + a_* t_1^{pe'_1} \ldots t_{n-1}^{pe'_{n-1}} \pi^{pe'_n} + \ldots$ for some $a_* \in \mathfrak{o}_0$ (see (3.5)). Then, there is an $a_0 \in \mathfrak{o}_0$ such that

$$E(a_0 s_1(x))\Big|_{x=\pi} \equiv E(c^p a_0 t_1^{pe'_1} \ldots t_{n-1}^{pe'_{n-1}} x^{pe'_n})\Big|_{x=\pi} \equiv \alpha \mod K^{*p},$$

i.e.,

$$\alpha = E(a_0 s_1(x))\Big|_{x=\pi} \cdot \beta^p$$

for some β, $\bar{v}_K(\beta - 1) > \bar{e}_1$.

Now Lemma 2.8 yields the following equality:

$$E(a_0 s_1(x))\Big|_{x=\pi} = E(p a_0 \log \underline{\zeta}_1(x))\Big|_{x=\pi} \cdot \gamma^p.$$

Further, for $\zeta = \sqrt[p^{m-1}]{\zeta_1}$ we have

$$E(p a_0 \log \underline{\zeta}_1(x))\Big|_{x=\pi} = E(p a_0 \log \underline{\zeta}(x)^{p^{m-1}})\Big|_{x=\pi} = E(p^m a_0 \log \underline{\zeta}(x))\Big|_{x=\pi}$$
$$= E(a_0 s(x))\Big|_{x=\pi} (\gamma')^p, \quad \gamma' \in K^*.$$

Hence we obtain $\alpha = \omega(a_0) \alpha_1^p$, where the element α_1 is p^{m-1}-primary. By induction one can assume that

$$\alpha_1 \equiv E(p^{m-1} a_1 \log \underline{\zeta}_{m-1}(x))\Big|_{x=\pi} \equiv E(a_1 s(x))\Big|_{x=\pi} \mod K^{*p^{m-1}}.$$

Then $\alpha \equiv \omega(a_0 + a_1) \mod K^{*p^m}$ and the decomposition (3.8) is proved.

Now consider an element of the form

$$\omega(a) = E(as(x))\Big|_{x=\pi}.$$

We shall verify that this element is p^m-primary. Suppose

$$H(a) = E(p^m A^\Delta l(\underline{\zeta}(x)))\Big|_{x=\pi},$$

where A is a root of the equation $\wp(x) = a$. The equality

(3.14) $$H(a) = E(p^m a \log(\underline{\zeta}(x)))\Big|_{x=\pi}$$

is verified similarly to Lemma 9 from [V1]. To obtain the relation

(3.15) $$\omega(a) \underset{p^m}{\approx} H(a)$$

one can now use Lemma 2.8.

It is easier to verify that $H(a)$ is primary, then to do that for $\omega(a)$. Namely, it follows from (3.14) that the element $H(a)$ belongs to the field K. But $H(a)$ is p^m th power of the element

$$E(A^\Delta l(\underline{\zeta}(x)))\Big|_{x=\pi},$$

which obviously belongs to an unramified extension $\widetilde{K}$ of the field K, because the element A is in $\widetilde{K}$. Thus, the p^m-primarity of $H(a)$ (and also of $\omega(a)$ at the same time) is proved.

Now we shall verify the element (3.8) does not depend on the choice of the expansion of the root ζ. Since (3.15) holds, it is sufficient to verify a similar independence of the element $H(a)$. Let $\zeta^{(1)}$ be the series corresponding to some another expansion of ζ. Then we have

(3.16) $$H(a) = E(p^m A^\Delta l(\underline{\zeta}(x)))\Big|_{x=\pi} = E(p^m A^\Delta l(\underline{\zeta}^{(1)}(x)))\Big|_{x=\pi} \cdot \eta^{p^m},$$

where $\eta = E(A^\Delta l(\underline{\zeta}(x)/\underline{\zeta}^{(1)}(x)))\Big|_{x=\pi}$. It remains to verify that $\eta \in K$. Let φ be an arbitrary automorphism of the Galois group $\mathrm{Gal}(\widetilde{K}/K)$. Then we have (see Lemma 3.1) $A^{\Delta^\varphi} - A^\Delta = a_\varphi^\Delta = a_\varphi \in \mathbb{Z}_p$. Since the function E is $\mathbb{Z}_p$-multiplicative and the functions E and l are reciprocally inverse, we obtain (see Proposition 1.1)

$$\eta^\varphi = \eta \cdot E(l(\underline{\zeta}(x)/\underline{\zeta}^{(1)}(x))^{a_\varphi})\Big|_{x=\pi} = \eta \left(\frac{\underline{\zeta}(\pi)}{\underline{\zeta}^{(1)}(\pi)}\right)^{a_\varphi} = \eta \left(\frac{\zeta}{\zeta}\right)^{a_\varphi} = \eta.$$

Hence $\eta \in K$. The independence of the element $H(a)$ (and at the same time of $\omega(a)$) from the expansion in a series with respect to the element π follows from the last relation and formula (3.16).

We shall verify (3.9) for the element $H(a)$. Using formula (3.15), the definition of $H(a)$ and Lemma 3.1, we obtain

$$\sqrt[p^m]{\omega(a)}^{\varphi-1} = \sqrt[p^m]{H(a)}^{\varphi-1} = E((a^{\Delta^\varphi} - A^\Delta)l(\underline{\zeta}(x)))\Big|_{x=\pi} = E(a_\varphi^\Delta l(\underline{\zeta}(x)))\Big|_{x=\pi}.$$

But $a_\varphi \in \mathbb{Z}_p$, hence $a_\varphi^\Delta = a_\varphi$, besides we can use Proposition 1.1 for the functions E and l:
$$E(a_\varphi^\Delta l(\underline{\zeta}(x)))\Big|_{x=\pi} = E(l(\underline{\zeta}(x)))\Big|_{x=\pi}^{a_\varphi} = \underline{\zeta}(\pi)^{a_\varphi} = \zeta^{a_\varphi}.$$

Thus relation (3.9) is obtained.

It remains to verify condition (3.10). From formula (3.15) and the definition of the element $H(a)$ we get
$$\omega(a) \approx 1 \iff H(a) \approx 1 \iff h = H(A^\Delta l(\underline{\zeta}(x)))\Big|_{x=\pi} \in K.$$

The last condition is equivalent to the following: for any automorphism φ of the Galois group $\mathrm{Gal}(\widetilde{K}/K)$ we have $\eta^\varphi = \eta$. But
$$\eta^{\varphi-1} = E((A^{\Delta^\varphi} - a^\Delta)l(\underline{\zeta}(x)))\Big|_{x=\pi} = \zeta^{a_\varphi},$$

where $A^{\Delta^\varphi} - A^\Delta = a_\varphi^\Delta = a_\varphi \in \mathbb{Z}_p$. Hence
$$\eta \in K \iff a_\varphi \equiv 0 \mod p^m \iff A^\varphi - A \equiv 0 \mod p^m,$$

and the element A may be represented as $A = a_0 + p^m a_1$, where $a_0 \in \mathfrak{o}_0$, $A_1 \in \mathfrak{o}_0^{\mathrm{nr}}$. It immediately follows that
$$a = \wp(A) = \wp(a_0) + p^m \wp(A_1) \equiv \wp(a_0) \mod p^m$$

and (3.10) is proved. Thus the proof of the theorem is complete. □

§4. Maps $\gamma_{(T)}$ and $\Gamma_{(T)}$

We shall construct a mapping from the product of $m+1$ copies of the group $\mathcal{H}_m$ to the ring $\mathfrak{o}_0$, afterwards we shall interpret this map as that from the product of $m+1$ copies of the group K^* into the group of p^m-primary elements Ω.

Let $(T) = (t_1, \ldots, t_n)$ be a system of local parameters in the field $k_0\{\{t_1\}\} \ldots \{\{t_n\}\}$. In the ring $\mathfrak{o}\{\{t_n\}\}$, where $\mathfrak{o} = \mathfrak{o}_0\{\{t_1\}\} \ldots \{\{t_{n-1}\}\}$, the function
$$l(\alpha) = \frac{1}{p} \log \alpha^p / \alpha^\Delta$$

is defined for any invertible series $\alpha(T)$. Further, for a series $\alpha(t_1, \ldots, t_n)$ from the ring $\mathfrak{o}\{\{t_n\}\}$, we shall denote the ith logarithmic derivative by $\delta_i(\alpha)$, i.e.,

(4.1) $$\delta_i(\alpha) = \frac{\partial}{\partial t_i}\alpha, \quad 1 \leq i \leq n.$$

Recall that the series $\zeta(t_n)^{p^m} - 1$ was denoted by $s(t_n)$ (see (1.4)).

For $p \neq 2$ we shall define the mapping

(4.2) $$\gamma_{(T)}: \mathcal{H}_m \times \cdots \times \mathcal{H}_m \to \mathfrak{o} \mod (\wp, p^m),$$
$$\gamma_{(T)}(\alpha_1, \ldots, \alpha_{n+1}) = \mathrm{res}\,\Phi(\alpha_1, \ldots, \alpha_{n+1})/s,$$

where

$$\Phi(\alpha_1, \ldots, \alpha_{n+1}) = l(\alpha_{n+1})D_{n+1} - l(\alpha_n)D_n + \cdots + (-1)^n l(\alpha_1)D_1,$$

$$D_i = \frac{1}{p^{n-i+1}} \begin{vmatrix} \delta_1(\alpha_1) & \ldots & \delta_n(\alpha_1) \\ \cdots & & \cdots \\ \delta_1(\alpha_{i-1}) & \ldots & \delta_n(\alpha_{i-1}) \\ \delta_1(\alpha_{i+1}^\Delta) & \ldots & \delta_n(\alpha_{i+1}^\Delta) \\ \cdots & & \cdots \\ \delta_1(\alpha_{n+1}^\Delta) & \ldots & \delta_n(\alpha_{n+1}^\Delta) \end{vmatrix}, \quad 1 \leqslant i \leqslant n+1.$$

REMARK 1. If we suppose that the first term for the series $\alpha_1, \ldots, \alpha_n$ is not necessarily chosen from the Teichmüller system, the constant $1/2$ must be subtracted from the series $1/s$.

We note, further, that if one of the elements $\alpha_1, \ldots, \alpha_{n+1}$ (α_i for example) is chosen from the Teichmüller system $\mathcal{R}$, then, since the group $\mathcal{R}$ is infinitely p-divisible, we have

(4.3) $$\gamma_{(T)}(\alpha_1, \ldots, \alpha_i, \ldots, \alpha_{n+1}) \equiv 0 \mod p^m.$$

In the sequel the following particular case will be used. Let the sets of indices $I = (i_1 < i_2 < \cdots < i_{u-1})$ and $J = (j_1 < j_2 < \cdots < j_{v-1})$, where $u + v = n - 1$, be such that they give some permutation of $(1, 2, \ldots, n)$, together with some k, i.e., $I \cup \{k\} \cup J = \{1, 2, \ldots, n\}$. Then we have

(4.4) $$\Phi(t_{i_1}, \ldots, t_{i_{u-1}}, \alpha, t_{j_1}, \ldots, t_{j_{v-1}}, \beta)$$
$$= (-1)^\varkappa (t_1^{-1} \ldots \hat{t}_k^{-1} \ldots t_n^{-1}) l(\beta) \delta_k(\alpha) + (-1)^{n-k+v} l(\alpha) \frac{1}{p} \delta_k(\beta^\Delta),$$

where $\varkappa$ is the number of inversions in the permutation (I, k, J). Notice that $\delta_i(t_j)$ equals 0 for $i \neq j$ and it is equal to t_i^{-1} when $i = j$. Moreover, if $\omega(a)$ is a p^m-primary element of K, then from the definition we immediately get

(4.5) $$\gamma_{(T)}(t_1, \ldots, t_n, \omega(a)) \equiv a \mod (\wp, p^m).$$

PROPOSITION 4.1. *The mapping $\gamma_{(T)}$ is well defined (i.e., is invariant relative to the choice of local parameters $t_1, \ldots, t_n$ of the field K). It is also multiplicative in all of its arguments, skew-symmetric, and proportional, i.e.,*

$$\gamma_{(T)}(\ldots, \alpha_i, \ldots, -\alpha_i, \ldots) \approx 1.$$

Besides, it satisfies the Steinberg relation

$$\gamma_{(T)}(\ldots, \alpha_i, \ldots, 1 - \alpha_i, \ldots) \approx 1.$$

PROOF. At first we shall verify the multiplicative property, proportionality, and skew-symmetry, as well as the Steinberg relation for every fixed set of local parameters $t_1, \ldots, t_n$. Afterwards we shall prove that $\gamma_{(T)}$ is well defined.

The multiplicative property in all its arguments follows, obviously, from the additivity of the functions l and δ_i and from the linearity of the determinant with respect to its rows.

To verify the other properties, the following assertion is necessary.

LEMMA 4.1. *Let $f_{ij}(\alpha)$, $1 \leqslant i, j \leqslant n$, be differentiable functions in n arguments from the ring $\mathfrak{o}$ whose values are in the same ring and the following conditions are satisfied*

$$\frac{\partial}{\partial t_k} f_{ij} = \frac{\partial}{\partial t_j} f_{ik}, \qquad 1 \leqslant k, j \leqslant n.$$

Then for any differentiable function $\varphi(t_1, \ldots, t_n)$ the following equality holds:

$$\Delta_1 \frac{\partial}{\partial t_1} \varphi - \Delta_2 \frac{\partial}{\partial t_2} \varphi + \cdots + (-1)^{n-1} \Delta_n \frac{\partial}{\partial t_n} \varphi$$
$$= \frac{\partial}{\partial t_1}(\Delta_1 \varphi) - \frac{\partial}{\partial t_2}(\Delta_2 \varphi) + \cdots + (-1)^{n-1} \frac{\partial}{\partial t_n}(\Delta_n \varphi),$$

where Δ_k is the determinant of the matrix obtained from the matrix

$$\begin{pmatrix} f_{11} & \cdots & f_{1n} \\ \cdots \cdots \cdots \cdots \cdots \cdots \\ f_{n-1,1} & \cdots & f_{n-1,n} \end{pmatrix}$$

by striking out the kth column. □

Besides we shall use the following notation for the function $\varphi(t_1, \ldots, t_n)$:

(4.6) $$\varphi \equiv 0 \mod \partial,$$

if φ is the sum of partial derivatives of a the series from the ring $\mathfrak{o}\{\{t_n\}\}$. We shall need the following lemma.

LEMMA 4.2. *Let φ be a series from the ring $\mathfrak{o}$ that satisfies the congruence $\varphi \equiv 0 \mod \partial$, then $\operatorname{res} \varphi/s \equiv 0 \mod p^m$, where the residue res is taken in all the variables $t_1, \ldots, t_n$.*

PROOF. The above condition means that

$$\varphi = \frac{\partial}{\partial t_1} \varphi_1 + \cdots + \frac{\partial}{\partial t_n} \varphi_n,$$

where $\varphi_i \in \mathfrak{o}\{\{t_n\}\}$. Hence, using the congruence (2.6), we obtain

$$\operatorname{res} \varphi/s = \sum_{i=1}^{n} \operatorname{res} \frac{\partial \varphi_i}{\partial t_i}/s \equiv \sum_{n=1}^{n} \operatorname{res} \frac{\partial}{\partial t_i}(\varphi_i/s) = 0 \mod p^m,$$

and the lemma is proved. □

For simplicity we shall suppose that $\alpha_1 = \alpha$, $\beta_1 = \beta$, when the skew-symmetry of the mapping will be verified. Then, according to Lemma 4.2, it sufficices to verify the congruence

(4.7) $$\varphi = \Phi(\alpha, \beta, \alpha_3, \ldots, \alpha_{n+1}) + \Phi(\beta, \alpha, \alpha_3, \ldots, \alpha_{n+1}) \equiv 0 \mod \partial.$$

From the definition of the series Φ and the skew-symmetry property of determinants (the first row and the second one are interchanged) it follows that only the two last terms remain when the series $\Phi(\alpha, \beta, \alpha_3, \ldots, \alpha_{n+1})$ and $\Phi(\beta, \alpha, \alpha_3, \ldots, \alpha_{n+1})$ are summed:

$$\varphi = ((-1)^{n-1}l(\underline{\beta})D_2 + (-1)^n l(\underline{\alpha})D_1) + ((-1)^{n-1}l(\underline{\alpha})D_2' + (-1)^n l(\underline{\beta})D_1').$$

Using the relation $\delta_i(\alpha) - \frac{1}{p}\delta_i(\alpha^\Delta) = \partial_i l(\alpha)$ (see [V5]), we obtain

$$\varphi = \frac{(-1)^{n-1}}{p^{n-1}} \begin{vmatrix} \partial_1 l(\underline{\alpha})l(\underline{\beta}) & \ldots & \partial_n l(\underline{\alpha})l(\underline{\beta}) \\ \delta_1(\underline{\alpha}_3^\Delta) & \ldots & \delta_n(\underline{\alpha}_3^\Delta) \\ \ldots & \ldots & \ldots \\ \delta_1(\underline{\alpha}_{n+1}^\Delta) & \ldots & \delta_n(\underline{\alpha}_{n+1}^\Delta) \end{vmatrix}.$$

If we expand the determinant in the first row, then

$$\varphi = (-1)^{n-1} \sum_{i=1}^{n} (-1)^{i-1} (\partial_i l(\underline{\alpha})l(\underline{\beta})) \Delta_i,$$

where

(4.8) $$\Delta_i = \frac{1}{p^{n-1}} \begin{vmatrix} \delta_1(\underline{\alpha}_3^\Delta) & \ldots & \widehat{\delta_i(\underline{\alpha}_3^\Delta)} & \ldots & \delta_n(\underline{\alpha}_3^\Delta) \\ \ldots & \ldots & \ldots & \ldots & \ldots \\ \delta_1(\underline{\alpha}_{n+1}^\Delta) & \ldots & \widehat{\delta_i(\underline{\alpha}_{n+1}^\Delta)} & \ldots & \delta_n(\underline{\alpha}_{n+1}^\Delta) \end{vmatrix}$$

is the determinant when the ith column is struck out. The series $l(\underline{\alpha})l(\underline{\beta})$ belongs to the ring $\mathfrak{o}\{\{t_n\}\}$. Hence, using Lemma 4.1, we get (4.7).

Now let us verify the proportionality of the mapping $\gamma_{(T)}$. Since it is skew-symmetric, we suppose that $\alpha_1 = \alpha$, $\alpha_2 = -\alpha$. According to Lemma 4.2, it suffices to verify the congruence

(4.9) $$\Phi(\alpha, -\alpha, \alpha_3, \ldots, \alpha_{n+1}) \equiv 0 \mod \partial.$$

It is obvious that only the two last terms remain in the definition of Φ, because the two first rows of the corresponding determinants are the same. Further, the functions $l(\underline{\alpha})$ and $l(-\underline{\alpha})$ are equal when p is odd. Hence

$$\Phi(\alpha, -\alpha, \alpha_3, \ldots, \alpha_{n+1}) = (-1)^{n-1} l(\underline{\alpha})(D_2 - D_1).$$

We note that $\delta_i(\alpha) - \frac{1}{p}\delta_i(\alpha^\Delta) = \partial_i l(\alpha)$, so

$$(-1)^{n-1} l(\underline{\alpha})(D_2 - D_1) = (-1)^{n-1} \sum_{i=1}^{n} (-1)^{i-1} l(\underline{\alpha}) \partial_i l(\underline{\alpha}) \Delta_i$$

$$= (-1)^{n-1} \sum_{i=1}^{n} (-1)^{i-1} \left(\partial_i \frac{l(\underline{\alpha})^2}{2} \right) \Delta_i,$$

where Δ_i is the same as in (4.8).

The series $l(\underline{\alpha})^2/2$ belong to the ring $\mathfrak{o}\{\{t_n\}\}$. Hence we can apply Lemma 4.1 to obtain (4.9). The proportionality of the mapping $\gamma_{(T)}$ is proved.

To verify the Steinberg relations, we can use the skew-symmetry property, so we suppose that $\alpha_1 = \alpha$, $\alpha_2 = 1 - \alpha$. Let α be an element from the maximal ideal $\mathfrak{M}$ of the ring of integers of K. Due to Lemma 4.2, it suffices to verify the congruence

$$\Phi(\alpha, 1 - \alpha, \alpha_3, \ldots, \alpha_{n+1}) \equiv 0 \mod \partial. \tag{4.10}$$

We note that all items in the definition of Φ, except the last two, are equal to 0, because it is obvious that

$$\delta_i(1 - \underline{\alpha}) = \left(-\sum_{r \geq 1} \underline{\alpha}^r\right) \delta_i(\underline{\alpha}),$$

and hence the corresponding determinants have proportional first rows. So

$$\Phi(\alpha, 1 - \alpha, \alpha_3, \ldots, \alpha_{n+1}) = (-1)^{n-1}\left(l(1 - \underline{\alpha})D_2 - l(\underline{\alpha})D_1\right).$$

Let us expand the determinants D_2 and D_1 in the first row, then we obtain

$$\Phi(\alpha, 1 - \alpha, \alpha_3, \ldots, \alpha_{n+1})$$
$$= (-1)^{n-1}\sum_{i=1}^n (-1)^{i-1}\left(l(1 - \underline{\alpha})\delta_i(\underline{\alpha}) - l(\underline{\alpha})\frac{1}{p}\delta_i(1 - \underline{\alpha}^\Delta)\Delta_i\right),$$

where Δ_i is a determinant of the form (4.8).

In [V1, (11)] (see also [V5, proof of Proposition 1]) it was shown that

$$l(1 - \underline{\alpha})\delta_i(\underline{\alpha}) - l(\underline{\alpha})\frac{1}{p}\delta_i(1 - \underline{\alpha}^\Delta) = \partial_i \Psi,$$

where Ψ is a series from the ring $\mathfrak{o}\{\{t_n\}\}$. Note, further, that

$$\partial_i(\delta_j(\underline{\alpha})) = \partial_j(\delta_i(\underline{\alpha})),$$

and hence we can apply Lemma 4.1 to obtain (4.10).

Further, if $\alpha \notin \mathfrak{M}$ but $\alpha^{-1} \in \mathfrak{M}$, then from the above and proportionality it follows that

$$1 \approx \gamma_{(T)}(\alpha^{-1}, 1 - \alpha^{-1}, \alpha_3, \ldots, \alpha_{n+1})$$
$$= \gamma_{(T)}(\alpha^{-1}, (1 - \alpha)(-\alpha^{-1}), \alpha_3, \ldots, \alpha_{n+1})$$
$$= \gamma_{(T)}(\alpha^{-1}, 1 - \alpha, \alpha_3, \ldots, \alpha_{n+1}) \cdot \gamma_{(T)}(\alpha^{-1}, -\alpha^{-1}, \alpha_3, \ldots, \alpha_{n+1})$$
$$= \gamma_{(T)}^{-1}(\alpha, 1 - \alpha, \alpha_3, \ldots, \alpha_{n+1}),$$

which proves the Steinberg relation in this case.

If $\alpha \notin \mathfrak{M}$ but $1 - \alpha \in \mathfrak{M}$, then it suffices to use the fact that $\alpha' = 1 - \alpha \in \mathfrak{M}$ and $1 - \alpha'$ is a principal unit and then to apply the relations proved above and skew-symmetry.

If, finally, α and $\alpha - 1 \in \mathfrak{M}$, then $\alpha = \theta\varepsilon$, where θ is in the Teichmüller system $\mathcal{R}$, ε is a principal unit. Let $x = (1-\varepsilon)/(1-\alpha)$, then $x \in \mathfrak{M}$ and hence as it was proved above

$$\gamma_{(T)}(x, 1-x, \alpha_3, \ldots, \alpha_{n+1}) \approx 1.$$

Further, $1 - \varepsilon \in \mathfrak{M}$, hence according to the relations proved above we have

$$\gamma_{(T)}(1-\varepsilon, \varepsilon, \alpha_3, \ldots, \alpha_{n+1}) \approx 1.$$

From the last two relations we obtain

$$\gamma_{(T)}\left(1-\varepsilon, \frac{1-\alpha}{1-\theta}, \alpha_3, \ldots, \alpha_{n+1}\right) \cdot \gamma_{(T)}(1-\alpha, \varepsilon, \alpha_3, \ldots, \alpha_{n+1})$$
$$\times \gamma_{(T)}(1-\alpha, 1-\theta, \alpha_3, \ldots, \alpha_{n+1}) \approx 1.$$

Taking into account (4.3), we get

(4.11)
$$\gamma_{(T)}\left(1-\varepsilon, \frac{1-\alpha}{1-\theta}, \alpha_3, \ldots, \alpha_{n+1}\right) \cdot \gamma_{(T)}\left(1-\alpha, \frac{1-\theta}{\theta}, \alpha_3, \ldots, \alpha_{n+1}\right)$$
$$\times \gamma_{(T)}(1-\alpha, \alpha, \alpha_3, \ldots, \alpha_{n+1}) \approx 1.$$

Further, for $y = (1-\alpha)/(1-\theta)$ the element $1 - y$ may be represented as $(\varepsilon - 1)\theta/(1-\theta) \in \mathfrak{M}$, hence $\gamma_{(T)}(1-y, y, \alpha_3, \ldots, \alpha_{n+1}) \approx 1$ as it was proved above. Hence

$$\gamma_{(T)}\left(1-\varepsilon, \frac{1-\alpha}{1-\theta}, \alpha_3, \ldots, \alpha_{n+1}\right) \approx \gamma_{(T)}\left(\frac{1-\theta}{\theta}, \frac{1-\alpha}{1-\theta}, \alpha_3, \ldots, \alpha_{n+1}\right).$$

Substituting the last relation in (4.11), we obtain

$$\gamma_{(T)}\left(\frac{1-\theta}{\theta}, \frac{1-\alpha}{1-\theta}, \alpha_3, \ldots, \alpha_{n+1}\right) \cdot \gamma_{(T)}\left(1-\alpha, \frac{1-\theta}{\theta}, \alpha_3, \ldots, \alpha_{n+1}\right)$$
$$\times \gamma_{(T)}(1-\alpha, \alpha, \alpha_3, \ldots, \alpha_{n+1}) \approx 1.$$

Taking into account the skew-symmetry and (4.5), we immediately obtain the required relation

$$\gamma_{(T)}(1-\alpha, \alpha, \alpha_3, \ldots, \alpha_{n+1}) \approx 1.$$

The Steinberg relation is completely proved.

It remains to verify the invariance of the mapping $\gamma_{(T)}$ under changes of variables. Let us consider at first the special case of $\gamma_{(T)}(t_1, \ldots, t_n, \alpha)$, where α is an arbitrary series from $\mathcal{H}_m$ and there is only one change of variable. One must verify the congruence

(4.12) $\gamma_{(t_1, \ldots, t_n)}(t_1, \ldots, t_n, \alpha) \equiv \gamma_{(t_1, \ldots, t_i', \ldots, t_n)}(t_1, \ldots, t_n, \alpha) \mod (\wp, p^m).$

For simplicity we suppose that the variable t_n changes: $t_n = g(t'_n)$. We denote the set $(t_1, \ldots, t_{n-1}, t'_n)$ by (T'). Further, we can suppose that α is a principal unit, since if $\alpha = t_1^{r_1} \ldots t_n^{r_n} \theta \varepsilon$, where $\theta \in \mathcal{R}$ and ε is a principal unit, then due to (4.3) and to the proportionality of the mapping $\gamma_{(T)}$, we have

$$\gamma_{(T)}(t_1, \ldots, t_n, \alpha) \equiv \gamma_{(T)}(t_1, \ldots, t_n, \varepsilon) \mod (\wp, p^m).$$

The group of principal units in $\mathcal{H}_m$ is generated by the units $E(aT^{\bar{r}})$, $a \in \mathfrak{o}_0$, $\bar{r} > 0$ (see (1.10)). Hence it suffices to verify the congruence

(4.13) $\quad \gamma_{(T)}(t_1, \ldots, t_n, E(aT^{\bar{r}})) \equiv \gamma_{(T')}(t_1, \ldots, t_n, E(aT^{\bar{r}})) \mod (\wp, p^m).$

In further calculations the definition of the functions E and l depends on the choice of variables. Hence we shall denote them by $E_{(T)}$, $l_{(T)}$ for each set (T) of variables.

From (4.2) we get

(4.14) $\quad \gamma_{(T)}(t_1, \ldots, t_n, E_{(T)}(aT^{\bar{r}})) = \operatorname{res}_{(T)} aT^{\bar{r}-1}/s(T),$

where $T^{\bar{r}-1} = t_1^{r_1-1} \ldots t_n^{r_n-1}$. On the other hand, according to the same formula, we obtain

(4.15)
$$\gamma_{(T')}(t_1, \ldots, t_n, E(aT^{\bar{r}}))$$
$$= \operatorname{res}_{(T')}\{l_{(T')}(E_{(T)}(aT^{\bar{r}}))\delta_{t'_n}(g) - l_{(T')}(g)\delta_{t'_n}(E_{(T)}(aT^{\bar{r}}))^{\Delta'}\}/s(T').$$

Let us denote the product $at_1^{r_1} \ldots t_{n-1}^{r_{n-1}}$ by α, then $aT^{\bar{r}} = \alpha t_n^{r_n} = \alpha g^{r_n}$, where $t_n = g(t'_n)$. From the definitions of the functions l and E it follows (in our notation) that

$$l_{(T')}(E_{(T)}(aT^{\bar{r}})) = ag^{r_n} + \sum_{k \geq 1} \alpha^{p^k} \frac{g^{p^k r_n} - g^{p^{k-1} r_n \Delta'}}{p^k} - \frac{1}{p}\delta_{t'_n}(E_{(T)}(aT^{\bar{r}})^{\Delta'})$$

$$= \delta_{t'_n} \sum_{k \geq 1} \frac{\alpha^{p^k}(g^{p^{k-1} r_n})^{\Delta'}}{p^k}.$$

Hence

$$\gamma_{(T')}(t_1, \ldots, t_n, E_{(T)}(aT^{\bar{r}}))$$
$$= \operatorname{res}_{(T')}\left(aT^{\bar{r}}\delta_{t'_n}(g) + \sum_{k \geq 1} \frac{\alpha^{p^k}}{p^k}\right.$$
$$\left. \times \{(g^{p^k r_n} - g^{p^{k-1} r_n \Delta'})\delta_{t'_n}(g) - l_{(T')}(g)\delta_{t'_n}(g^{p^{k-1} r_n \Delta'})\}\right)/s(T').$$

As in the paper [V1, (11)] it can be proved that

(4.16) $\quad \displaystyle\sum_{k \geq 1} \frac{\alpha^{p^k}}{p^k}\{\ldots\} = \frac{\partial}{\partial t'_n}\varphi(T'),$

where the series $\varphi(T')$ for $p \neq 2$ has integral coefficients.

The series $1/s(T)$ satisfies the congruence

$$\frac{\partial}{\partial t_n}(1/s) \equiv 0 \mod p^m$$

(see (2.6)), hence if we change t_n to t'_n, we obtain a similar congruence. Therefore

(4.17)
$$\operatorname{res}_{(T')} \frac{\partial \varphi}{\partial t'_n}/s(T') = \operatorname{res}_{t_1,\ldots,t_{n-1}} \left(\operatorname{res}_{t'_n} \frac{\partial \varphi}{\partial t'_n}/s(T') \right)$$
$$= -\operatorname{res}_{t_1,\ldots,t_{n-1}} \left(\operatorname{res}_{t'_n} \varphi \cdot \frac{\partial}{\partial t'_n}(1/s) \right) \equiv 0 \mod p^m.$$

Further,

$$\operatorname{res}_{(T')} \left(aT^{\bar{r}-1} \frac{\partial g}{\partial t'_n} \right)/s(T') = \operatorname{res}_{t_1,\ldots,t_{n-1}} \left(\operatorname{res}_{t'_n}(aT^{\bar{r}-1}/s(T')) \right) \cdot \frac{\partial g}{\partial t'_n}$$
$$= \operatorname{res}_{t_1,\ldots,t_{n-1}} \left(\operatorname{res}_{t_n}(aT^{\bar{r}-1}/s(T)) \right)$$
$$= \operatorname{res}_{(T)}(aT^{\bar{r}-1}/s(T)).$$

From this relation, congruence (4.17) and (4.15), and from (4.16), we get

$$\gamma_{(T')}(t_1, \ldots, t_n, E_{(T')}(aT^{\bar{r}})) = \operatorname{res}_{(T')} \left(aT^{\bar{r}-1} \frac{\partial g}{\partial t'_n} + \frac{\partial}{\partial t'_n} \varphi \right)/s(T')$$
$$\equiv \operatorname{res}_{(T)}(aT^{\bar{r}-1}/s(T)) \mod p^m.$$

Comparing the last result with (4.14), we obtain (4.13), which proves the invariance of the mapping $\gamma_{(T)}$ in our case.

The invariance proved may be applied to the case of $\gamma_{(T)}(t_1, \ldots, t_n, \alpha)$ using induction on the number of variables, changing the variables $(T) = (t_1, \ldots, t_n)$ to $(t'_1, \ldots, t'_n)$. We omit the corresponding straightforward technical calculations since they are rather complicated.

The general case is proved by induction on the number of elements, assuming that

(4.18)
$$\gamma_{(T)}(\alpha_1, \ldots, \alpha_{r-1}, t_r, \ldots, t_n, \alpha)$$
$$\equiv \gamma_{(T')}(\alpha_1, \ldots, \alpha_{r-1}, t_r, \ldots, t_n, \alpha) \mod (\wp, p^m),$$

where $(T) = (t_1, \ldots, t_n)$, $(T') = (t'_1, \ldots, t'_n)$. Let t''_r be another rth local parameter, then by the induction assumption we obtain

(4.19)
$$\gamma_{(t'_1,\ldots,t''_r,\ldots,t'_n)}(\alpha_1, \ldots, \alpha_{r-1}, t''_r, t_{r+1}, \ldots, t_n, \alpha)$$
$$\equiv \gamma_{(T)}(\alpha_1, \ldots, \alpha_{r-1}, t''_r, t_{r+1}, \ldots, t_n, \alpha) \mod (\wp, p^m).$$

On the other hand

$$\gamma_{(t'_1,\ldots,t''_r,\ldots,t'_n)}(\alpha_1, \ldots, \alpha_{r-1}, t''_r, t_{r+1}, \ldots, t_n, \alpha)$$
$$\equiv \gamma_{(t'_1,\ldots,t'_r,\ldots,t'_n)}(\alpha_1, \ldots, \alpha_{r-1}, t''_r, t_{r+1}, \ldots, t_n, \alpha) \mod (\wp, p^m).$$

Hence

(4.20) $$\gamma_{(T)}(\alpha_1, \ldots, \alpha_{r-1}, t_r'', t_{r+1}, \ldots, t_n, \alpha)$$
$$\equiv \gamma_{(T')}(\alpha_1, \ldots, \alpha_{r-1}, t_r'', t_{r+1}, \ldots, t_n, \alpha) \mod (\wp, p^m).$$

One should remark that any element α_r from $\mathcal{H}_m$ can be represented as $\alpha_r = t_r^a (t_r'')^b t_{r+1}^{i_{r+1}} \ldots t_n^{i_n}$ for some local parameter t_r'' and integers a, b. Hence from (4.18) and (4.20), using the multiplicativity and proportionality of the mapping $\gamma_{(T)}$, we obtain

$$\gamma_{(T)}(\alpha_1, \ldots, \alpha_r, t_{r+1}, \ldots, t_n, \alpha)$$
$$\equiv \gamma_{(T')}(\alpha_1, \ldots, \alpha_r, t_{r+1}, \ldots, t_n, \alpha) \mod (\wp, p^m).$$

This verifies the induction step. The invariance of $\gamma_{(T)}$ is completely proved.

2°. Consider the subgroup $\mathcal{U}_m = \langle 1 + u(t_n)\Psi(t_n) \rangle$ in the group $\mathcal{H}_m$, where the series $u(t_n) \in \mathfrak{o}[\![t_n]\!]$ is the Eisenstein series from Proposition 2.2, and the series Ψ is an arbitrary one without free term from the ring $\mathfrak{o}[\![t_n]\!]$.

PROPOSITION 4.2. *The subgroup $\mathcal{U}_m$ belongs to the kernel of the mapping $\gamma_{(T)}$ in all its arguments, so for $p \neq 2$ $\gamma_{(T)}$ induces a nontrivial mapping (denoted by the same letter)*

$$\gamma_{(T)}(\mathcal{H}_m/\mathcal{U}_m \mathcal{H}_m^{p^m})^{n+1} \to \mathfrak{o}_0 \mod (\wp, p^m).$$

PROOF. First we verify that

(4.21) $$\gamma_{(T)}(t_1, \ldots, t_n, \varepsilon) \equiv 0 \mod (\wp, p^m)$$

for any series $\varepsilon(t_n)$ from $\mathcal{U}_m$. By the definition of $\gamma_{(T)}$, we have

$$\gamma_{(T)}(t_1, \ldots, t_n, \varepsilon) = \operatorname{res}_{(T)} t_1^{-1} \ldots t_n^{-1} l(u\Psi)/s,$$

where $\varepsilon = 1 + u\Psi$. Taking into account congruence (2.15), we obtain (4.21).

Further, any element α of $\mathcal{H}_m$ can be represented as the product of local parameters (each α has its own set of local parameters $t_1, \ldots, t_n$). Hence the relations proved above, congruence (4.21), the multiplicativity and invariance of $\gamma_{(T)}$ imply the congruence

(4.22) $$\gamma_{(T)}(\alpha_1, \ldots, \alpha_n, \varepsilon) \equiv 0 \mod (\wp, p^m),$$

where $\alpha_1, \ldots, \alpha_n \in \mathcal{H}_m$ and $\varepsilon \in \mathcal{U}_m$. From this congruence and the skew-symmetry of $\gamma_{(T)}$ the assertion of the proposition follows. □

The exact sequence (see Proposition 2.3)

$$1 \to \mathcal{U}_m \to \mathcal{H}_m \xrightarrow{\eta_m} K^* \to 1$$

determines the homomorphism

$$\eta_m^* \colon K^* \xrightarrow{\sim} \mathcal{H}_m/\mathcal{U}_m$$

which is defined as follows: every element α from K^* is expanded into the series in the local parameters $t_1, \ldots, t_{n-1}, \pi$, and then the prime element π of the field K is replaced by the variable t_n. According to Proposition 4.2, the homomorphism η_m^* is well defined and is an isomorphism. With the help of this isomorphism, the mapping $\gamma_{(T)}$ can be defined on the field K. Namely, for $p \neq 2$, we define the mapping

(4.23)
$$\Gamma_{(T)} \colon (K^*)^{n+1} \to \Omega/(\Omega \cap K^{*p^m})$$
$$\Gamma_{(T)}(\alpha_1, \ldots, \alpha_{n+1}) = \omega(\gamma_{(T)}),$$

where $\gamma_{(T)} = \gamma_{(T)}(\eta_m^*(\alpha_1), \ldots, \eta_m^*(\alpha_{n+1}))$.

PROPOSITION 4.3. *The mapping $\Gamma_{(T)}$ for $p \neq 2$ is $\mathbb{Z}_p$-multiplicative in all its arguments, invariant, independent of the choice of local parameters $t_1, \ldots, t_{n-1}, \pi$ of K, independent of the expansion of the elements $\alpha_1, \ldots, \alpha_{n+1}$ into series with respect to the system $t_1, \ldots, t_{n-1}, \pi$. Besides, it satisfies the Steinberg relation*

$$\Gamma(\ldots, \alpha_i, \ldots, 1 - \alpha_i, \ldots)_{(T)} \approx 1$$

for any $\alpha_i \neq 1$ from K^*.

PROOF. The $\mathbb{Z}_p$-multiplicativity, invariance, and Steinberg relation follow from the corresponding properties of the mapping $\gamma_{(T)}$ (see Proposition 4.1). Independence from the expansion of elements into series follows from Proposition 4.2 and the isomorphism η_m^*. □

§5. The pairing on topological K-groups

1°. Let $K_u(K)$ be the uth Milnor group of the field K, $u \geq 0$. We shall choose the strongest topology on $K_u(K)$, under which the map

$$(K^*)^u \to K_u(K)$$

is sequentially continuous in every argument, i.e., if $x_n \to x$, $y_n \to y$, then $x_n + y_n \to x + y$ and $-x_n \to -x$ in $K_u(K)$. Let $\Lambda_u(K)$ be the subgroup in $K_u(K)$ which is the intersection of all the neighborhoods of 0 in $K_u(K)$. Then we define the topological Parshin K-group as follows:

$$K_u^{\mathrm{top}}(K) = K_u(K)/\Lambda_u(K)$$

(see [P]).

In this section we shall define a nondegenerate pairing on K-groups with values in the group of p^m-primary elements Ω. Let K contain all the p^mth roots of the unity. The mapping $\Gamma_{(T)}$ constructed in the preceding section induces pairing in the K-groups $K_u^{\mathrm{top}}(K)$ and $K_v^{\mathrm{top}}(K)$, where $u + v = n + 1$:

$$\langle \,,\, \rangle_{(T)}^{(u)} \colon K_u^{\mathrm{top}}(K)/p^m K_u^{\mathrm{top}}(K) \times K_v^{\mathrm{top}}(K)/p^m K_v^{\mathrm{top}}(k) \to \Omega/(\Omega \cup K^{*p^m}),$$

which is defined on the symbols $\alpha = \{\alpha_1, \ldots, \alpha_u\}$ and $\beta = \{\beta_1, \ldots, \beta_v\}$ as follows:

(5.1)
$$\langle \alpha, \beta \rangle_{(T)}^{(u)} = \Gamma_{(T)}(\alpha_1, \ldots, \alpha_u, \beta_1, \ldots, \beta_v).$$

The pairing $\langle \,,\, \rangle_{(T)}^{(u)}$ can be extended to the remaining elements by linearity. To begin with, we shall establish some auxiliary results.

Let $K_u^{\mathrm{top}}(K)$ be the topological Milnor K-group of the field K, $1 \leq i \leq n$; as above, we denote by $t_1, \ldots, t_n$ a system of local parameters of K.

LEMMA 5.1. *Any element x of the K-group $K_u^{\text{top}}(K)$ may be represented as follows*:

(5.2)
$$x \equiv \sum_I a_I \{t_{i_1}, \ldots, t_{i_{u-1}}, \mathcal{E}_I\} + \sum_I b_I \{t_{i_1}, \ldots, t_{i_{u-1}}, \omega_I\}$$
$$+ \sum_{I'} c_{I'} \{t_{i'_1}, \ldots, t_{i'_u}\} \mod p,$$

where a_I, b_I, $c_{I'}$ are certain integers, the sets $I = (i_1, \ldots, i_{u-1})$ and $I' = (i'_1, \ldots, i'_u)$ run through $1 \leq i_1 < \cdots < i_{u-1} \leq n$, $1 \leq i'_1 < \cdots < i'_u \leq n$; ω_I is some p^m-primary element in K and the unity $\mathcal{E}_I$ is

(∗)
$$\mathcal{E}_I = E\left(\sum_{(\alpha)} \theta_{(\alpha)} t_1^{\alpha_1} \ldots t_n^{\alpha_n}\right), \qquad \theta_{(\alpha)} \in \mathcal{R},$$

where the sum is taken over any set of multi-indices $(\alpha_1, \ldots, \alpha_n)$ not divisible by p.

PROOF. As in Proposition 1, §2, [P],
$$x = \sum_I a_I \{t_{i_1}, \ldots, t_{i_{u-1}}, \varepsilon_I\} + \sum_{I'} c_{I'} \{t_{i_1}, \ldots, t_{i_u}\},$$

where ε_I is a unit of K, is verified.

The unit ε_I may be regarded the principal one, since $\varepsilon_I = \theta \varepsilon_1$, where $\theta \in \mathcal{R}$ and ε_1 is a principal unit, and besides the Teichmüller system is infinitely p-divisible.

To obtain (5.2), it is sufficient to represent the principal unit ε_I in the canonical form (5.11).

REMARK. For our purpose it is only necessary to have the canonical decomposition (which is proved in Proposition 3.2), but not uniqueness (which follows from the nondegeneracy of pairing $\langle \, , \, \rangle$).

LEMMA 5.2. *Suppose the power α_k is not divisible by p for the unit*
$$\mathcal{E} = E(\theta t_1^{\alpha_1} \ldots t_n^{\alpha_n}).$$

Then we have

(5.3)
$$x = \{t_{i_1}, \ldots, t_k, \ldots, t_{i_{u-1}}, \mathcal{E}\}$$
$$\equiv \sum_{I'} a_{I'} \{t_{i'_1}, \ldots, t_k, \ldots, t_{i'_{u-1}}, \mathcal{E}\} \mod p,$$

where the sets $I' = (i'_1, \ldots, i'_{u-1})$ do not contain k.

PROOF. We use the expansion of the Artin–Hasse function E into the product
$$E(\alpha) = \prod_{(r, p)=1} (1 - \alpha^r)^{\frac{\mu(r)}{r}}.$$

Then the symbol x may be represented as follows:

$$x = -\sum_{(r,p)=1} \frac{\mu(r)}{r}\{t_{i_1}, \ldots, t_k, \ldots, t_{i_{u-1}}, 1 - \theta^r t_1^{r\alpha_1} \ldots t_n^{r\alpha_n}\},$$

where the power $r\alpha_k$ is not divisible by p.

Using the Steinberg relation in the K-groups we obtain

$$0 = \{t_{i_1}, \ldots, \theta^r t_1^{r\alpha_1} \ldots t_k^{r\alpha_k} \ldots t_n^{r\alpha_n}, \ldots, t_{i_{u-1}}, 1 - \theta^r t_1^{r\alpha_1} \ldots t_n^{r\alpha_n}\}$$
$$\equiv r\alpha_k\{t_{i_1}, \ldots, t_k, \ldots, t_{i_{u-1}}, 1 - \theta^r t_1^{r\alpha_1} \ldots t_n^{r\alpha_n}\}$$
$$+ \sum_{s \ne k} r\alpha_s\{t_{i_1}, \ldots, t_s, \ldots, t_{i_{u-1}}, 1 - \theta^r t_1^{r\alpha_1} \ldots t_n^{r\alpha_n}\} \mod p,$$

thus proving (5.3).

Let the element β from $K_v^{\text{top}}(K)$ have the form $\beta = \{t_{j_1}, \ldots, t_{j_{v-1}}, \omega\}$, where ω is p^m-primary element of K. Let $\alpha = \{t_{i_1}, \ldots, t_{i_{u-1}}, t_{i_u}\varepsilon\}$, where ε is a principal unit. From the definition of the pairing $\langle\,,\,\rangle$ and formula (4.2) we have the following lemma.

LEMMA 5.3. *If the intersection of the sets* $I = \{i_1, \ldots, i_u\}$ *and* $J = \{j_1, \ldots, j_{v-1}\}$ *is nonempty, then* $\langle\alpha, \beta\rangle \approx 1$. *If* $I \cup J = \{1, 2, \ldots, n\}$, *then*

(5.4) $$\langle\alpha, \beta\rangle \approx \omega.$$

Now let $\beta = \{t_{j_1}, \ldots, t_{j_v}\}$ and $\alpha = \{t_{i_1}, \ldots, t_{i_{u-1}}, \varepsilon\}$, $\varepsilon = \mathcal{E}\omega$ being the Shafarevich decomposition (5.11).

LEMMA 5.4. *If the intersection* $I = \{i_1, \ldots, i_{u-1}\}$ *with* $J = \{j_1, \ldots, j_v\}$ *is nonempty, then*

(5.5) $$\langle\alpha, \beta\rangle \approx 1.$$

If $I \cup J = \{1, 2, \ldots, n\}$, then

(5.6) $$\langle\alpha, \beta\rangle \approx \omega.$$

The formulas of Lemma 5.4 are obtained from the definition of the pairing $\langle\,,\,\rangle$. □

Let the sets $I = \{i_1, \ldots, i_{u-1}\}$ and $J = \{j_1, \ldots, j_{v-1}\}$ be disjoint and the number k be the complement $I \cup J$ to the set $\{1, 2, \ldots, n\}$. Let the unit

$$\mathcal{E} = E(\theta_{(\alpha)} t_1^{\alpha_1} \ldots t_n^{\alpha_n} + \ldots), \quad \theta_{(\alpha)} \in \mathcal{R}$$

be such that α_k is not divisible by p and the powers of the all monomials are not divisible by p. Denote by $\mathcal{E}^\perp$ the unit

$$\mathcal{E}^\perp = E(\theta_* t_1^{pe_1' - \alpha_1} \ldots t_n^{pe_n' \alpha_n})$$

for some $\theta_* \in \mathcal{R}$.

LEMMA 5.5. For $\alpha = \{t_{i_1}, \ldots, t_{i_{u-1}}, \mathcal{E}_I\} \in K_u^{\text{top}}(K)$, $\beta = \{t_{j_1}, \ldots, t_{j_{v-1}}, \mathcal{E}_I^\perp\}$
$\in K_v^{\text{top}}(K)$, we have

$$\langle \alpha, \beta \rangle \not\equiv 1 \mod k^{*p}$$

for a suitable choice of θ^*.

PROOF. By the definition of the series $\Phi_{\alpha,\beta}$, we obtain in our case (see (4.4)):

$$\Phi_{\alpha,\beta}/s = (-1)^{v-1}\left\{\left(\frac{\partial}{\partial t_k}\log\mathcal{E}_I^\perp\right)l(\mathcal{E}_I) - l(\mathcal{E}_I^\perp)\frac{\partial}{\partial t_k}\frac{\Delta}{p}\log\mathcal{E}_I\right\}/s$$

$$\equiv (-1)^{v-1}\{t_k^{-1}(-\alpha_k)(\theta_* t_1^{pe_1'-\alpha_1}\ldots t_n^{pe_n'-\alpha_n}$$
$$+ (\theta_* t_1^{pe_1'-\alpha_1}\ldots t_n^{pe_n'-\alpha_n})^p + \ldots)(\theta_{(\alpha)}t_1^{\alpha_1}\ldots t_n^{\alpha_n} + \ldots)$$
$$- \theta_* t_1^{pe_1'-\alpha_1}\ldots t_n^{pe_n'-\alpha_n}\alpha_k t_k^{-1}((\theta_{(\alpha)}t_1^{\alpha_1}\ldots t_n^{\alpha_n})^p + \ldots)\}/s \mod p,$$

where the dots denote terms of higher degrees. Next, we note that the series $1/s$, according to (2.5), consists only of those monomials whose powers are divisible by p. Since the power $(pe_1' - \alpha_1, \ldots, pe_n' - \alpha_n)$ is not divisible by p, we have

$$\text{res}_{(T)}\, t_1^{-1}\ldots t_n^{-1}\{((\theta_* t_1^{pe_1'-\alpha_1}\ldots t_n^{pe_n'-\alpha_n})^p + \ldots)$$
$$\times (\theta_{(\alpha)}t_1^{\alpha_1}\ldots t_n^{\alpha_n} + \ldots)\}/s \equiv 0 \mod p^m.$$

Similarly,

$$\text{res}_{(T)}\, t_1^{-1}\ldots t_n^{-1}\{\theta_* t_1^{pe_1'-\alpha_1}\ldots t_n^{pe_n'-\alpha_n})((\theta_{(\alpha)}t_1^{\alpha_1}\ldots t_n^{\alpha_n})^p + \ldots)\}/s \equiv 0 \mod p^m.$$

Hence

$$\text{res}_{(T)}\,\Phi_{\alpha,\beta}/s \equiv \text{res}_{(T)}(-1)^v t_1^{-1}\ldots t_n^{-1}\alpha_k$$
$$\times \theta_* t_1^{pe_1'-\alpha_1}\ldots t_n^{pe_n'-\alpha_n}(\theta_{(\alpha)}t_1^{\alpha_1}\ldots t_n^{\alpha_n} + \ldots)/s \mod p.$$

Further, from (2.5) it follows that the series $1/s$ has the form (mod p)

$$1/s \equiv z_0 t_1^{-pe_1'}\ldots t_n^{-pe_n'} + \ldots \mod p,\ p \nmid z_0.$$

Therefore

$$\text{res}_{(T)}\,\Phi_{\alpha,\beta}/s \equiv \text{res}_{(T)}(-1)^v \alpha_k \theta_* \theta_{(\alpha)} t_1^{pe_1'-1}\ldots t_n^{pe_n'-1}/s$$
$$= (-1)^v \alpha_k \theta_* \theta_{(\alpha)} z_0 \mod p,$$

and for some $\theta_* \in \mathcal{R}$ we have $\text{res}_{(T)}\,\Phi_{\alpha,\beta}/s \not\equiv 0 \mod p$. The lemma is proved. □

The proof of this lemma gives us the following result.

COROLLARY. *If for the unit*

$$\mathcal{E} = E(\theta_{(\alpha')}t_1^{\alpha'_1}\ldots t_n^{\alpha'_n} + \ldots)$$

the power $(\alpha'_1, \ldots, \alpha'_n)$ *is greater than* $(\alpha_1, \ldots, \alpha_n)$, *then for the element* $\alpha = \{t_{i_1}, \ldots, t_{i_{u-1}}, \mathcal{E}'\}$ *of* $K_u^{\text{top}}(K)$ *the following relation holds*:

$$\langle \alpha, \beta \rangle \approx 1.$$

THEOREM 5.1. *The pairing* $\langle\,,\,\rangle_{(T)}^{(u)}$ *for* $p \neq 2$ *is well defined and nondegenerate in both arguments.*

REMARK. From the skew-symmetry of $\Gamma_{(T)}$ we get the following property of $\langle\,,\,\rangle_{(T)}$:

(5.7) $$\langle \alpha, \beta \rangle_{(T)} = \langle \beta, \alpha \rangle_{(T)}^{(-1)^{uv}}.$$

PROOF OF THE THEOREM. From the properties of $\Gamma_{(T)}$ (see Proposition 4.3), it follows that $\Gamma_{(T)}$ defines the pairing $\langle\,,\,\rangle_{(T)}$ on K-groups.

Nondegeneracy. One must verify that for any α from $K_u^{\text{top}}(K)$ which is not divisible by p there is an element β from $K_v^{\text{top}}(K)$ such that the value $\langle \alpha, \beta \rangle_{(T)}$ generates the group of p^mth primary elements Ω. This is equivalent to the relation

(5.8) $$\gamma_{(T)}(\alpha, \beta) \not\equiv 0 \mod (\wp, p)$$

(see (5.1) and (4.23)).

Let us represent the element α from $K_u^{\text{top}}(K)$ as

$$\alpha = \sum_I a_I\{t_{i_1}, \ldots, t_{i_{u-1}}, \mathcal{E}_I\} + \sum_I b_I\{t_{i_1}, \ldots, t_{i_{u-1}}, \omega_I\} + \sum_{I'} b_{I'}\{t_{i'_1}, \ldots, t_{i'_u}\},$$

where a_I, b_I, $c_{I'}$ are defined modulo p, the unit $\mathcal{E}_I$ has the form $(*)$ and ω_I is the p^mth primary unit.

1. If there is a coefficient $c_{I'}$ for some I', which is not divisible by p, then we consider $\beta = \{t_{j_1}, \ldots, t_{j_{v-1}}, \omega_*\}$ to be an element β from $K_v^{\text{top}}(K)$, where the set of indices $J = (j_1, \ldots, j_{v-1})$ together with I' forms a permutation of n numbers, ω_* is the generator of the group of p^m-primary elements Ω.

Then by Lemma 5.3 $\langle \alpha, \beta \rangle_{(T)} = \omega_*$, and the nondegeneracy is proved.

2. If all the coefficients $c_{I'}$ are divisible by p, but there is a coefficient b_I which is relatively prime to p, then we choose $\beta = \{t_{j_1}, \ldots, t_{j_v}\}$ as an element β from $K_v^{\text{top}}(K)$. Here the set $J = (j_1, \ldots, j_v)$ together with I is the permutation of $u + v - 1 = n$ elements.

Using Lemma 5.4 again, we obtain $\langle \alpha, \beta \rangle_{(T)} = \omega_I$ and the nondegeneracy in this case is also proved.

3. Finally, if all the coefficients b_I and $c_{I'}$ are divisible by p, then there exists a coefficient a_I which is not divisible by p (otherwise the element α would be divisible by p), i.e.,

$$\alpha \equiv \sum_I a_I\{t_{i_1}, \ldots, t_{i_{u-1}}, \mathcal{E}_I\} \mod p,$$

where only those summands for which $a_I \not\equiv 0 \mod p$ are written.

Among all the sets I we shall find those for which the unit $\mathcal{E}_I = E(\theta t_{i_1}^{\alpha_1} \ldots t_{i_{u-1}}^{\alpha_{u-1}})$ has minimal rank, i.e., the power $(\alpha_1, \ldots, \alpha_{u-1})$ is minimal (in the lexicographical sense). Among the powers $\alpha_1, \ldots, \alpha_{u-1}$ not divisible by p at least one can be found, for example α_k. From the relation (5.3) of Lemma 5.2 it follows that the symbol $\{t_{i_1}, \ldots, t_{i_{u-1}}, \mathcal{E}_I\}$ is the linear combination of those symbols $\{t_{i'_1}, \ldots, t_{i'_{u-1}}, \mathcal{E}_{I'}\}$ for which set I' does not contain k. Hence $\alpha \equiv \alpha' + \alpha'' \mod p$, where

$$\alpha' = \sum_{\substack{I \\ k \notin I}} a'_I \{t_{i_1}, \ldots, t_{i_{u-1}}, \mathcal{E}_I\},$$

$$\alpha'' = \sum_{I'} a'_{I'} \{t_{i'_1}, \ldots, t_{i'_{u-1}}, \mathcal{E}_{I'}\},$$

and the rank of the units $\mathcal{E}_{I'}$ in the summand α'' is greater than that of $\mathcal{E}_I$ in the summand α.

Let us fix any set $I^{(0)} = (i_1^{(0)}, \ldots, i_{u-1}^{(0)})$, $k \notin I^{(0)}$ of the first sum and consider the set $J = (j_1, \ldots, j_{v-1}) = (1, 2, \ldots, n) \setminus (I^{(0)} \cup (k))$. Then, obviously, $I \cap J \neq \varnothing$ for $I \neq I^{(0)}$ of the first sum, hence any summand of this sum α', except $\alpha_{I^{(0)}} = \{t_{i_1^{(0)}}, \ldots, t_{i_{u-1}^{(0)}}, \mathcal{E}_I\}$ is orthogonal to $\beta = \{t_{j_1}, \ldots, t_{j_{v-1}}, \mathcal{E}_I^\perp\}$, since at least one of indices $j_1, \ldots, j_{v-1}$ coincides with $i_1, \ldots, i_{u-1}$.

Next, the element β is orthogonal (modulo p) to any summand of the second sum (see the Corollary to Lemma 5.5).

Hence

(5.9) $$\gamma_{(T)}(\alpha, \beta) \equiv \gamma_{(T)}(\alpha' + \alpha'', \beta) \equiv \gamma_{(T)}(\alpha, \beta) = \gamma_{(T)}(\alpha_{I^{(0)}}, \beta)$$
$$\not\equiv 0 \mod (\wp, p),$$

and for the element α from $K_u^{\text{top}}(K)$ we have found an element β from $K_v^{\text{top}}(K)$ that satisfies the congruence (5.8). So the nondegeneracy of the pairing is completely proved. $\square$

2°. Shafarevich canonical decomposition. Let $\bar{v}_k(p) = \bar{e} = (e_1, \ldots, e_n)$ be the ramification index of the field K, and $\bar{e}' = \overline{e/(p-1)} = (e'_1, \ldots, e'_n)$. Let I be the set of multi-indices satisfying the condition

(5.10) $$I = \{\bar{i} \mid \bar{0} < \bar{i} < p\bar{e}', \ p \nmid \bar{i}\},$$

i.e., not all indices $i_1, \ldots, i_n$ in the set $\bar{i}$ are divisible by p.

Next, let $t_1, \ldots, t_{n-1}, \pi$ be the system of local parameters of the field K. We suppose that K contains all the p^mth roots of unity.

THEOREM 5.2. *Any principal unit ε of K can be uniquely represented in the form (up to p^mth powers)*

(5.11) $$\varepsilon = \omega(a_\varepsilon) E(w_\varepsilon(T))\Big|_{(T)=(t_1, \ldots, t_{n-1}, \pi)},$$

where
$$w_\varepsilon(T) = \sum_{\bar{i} \in I} c_{\bar{i}} T^{\bar{i}},$$

$$\varepsilon \in K^{*p^m} \iff a_\varepsilon \equiv 0 \mod (\wp, p^m), \quad w_\varepsilon(T) \equiv 0 \mod p^m.$$

PROOF. The existence of decomposition (5.11) is proved in Proposition 3.2.

The verification of the uniqueness in (5.11) is based on the nondegeneration of the pairing $\langle\,,\,\rangle_{(T)} = \langle\,,\,\rangle_{(T)}^{(n)}$, which was proved in Theorem 5.1.

If $\varepsilon \approx 1$, then $\langle t_1, \ldots, t_n, \varepsilon\rangle_{(T)} \approx 1$. From the definition of the pairing $\langle\,,\,\rangle_{(T)}$ we have
$$\langle t_1, \ldots, t_n, E(w_\varepsilon(T))\rangle_{(T)} = \omega(\gamma),$$
where
$$\gamma = \text{res}_{(T)} t_1^{-1} \ldots t_n^{-1} w_\varepsilon(T)/s \equiv 0 \mod p^m$$

(the last congruence follows from the fact that all the monomials of the series $w_\varepsilon(T)$ contain powers not divisible by p, and all the powers of monomials of the series $1/s$, according to (2.5), are divisible by p). So
$$\langle t_1, \ldots, t_n, E(w_\varepsilon(T))\rangle_{(T)} \approx 1,$$
hence
$$1 \approx \langle t_1, \ldots, t_n, \varepsilon\rangle_{(T)} \approx \langle t_1, \ldots, t_n, \omega(a_\varepsilon)\rangle_{(T)} \approx \omega(a_\varepsilon)$$

(see (4.5)). From this congruence it follows that $a_\varepsilon \equiv 0 \mod (\wp, p^m)$.

Further, let the greatest common p-divisor of the coefficients $c_{\bar{i}}$ in the series $w_\varepsilon(T)$ be equal p^r. If $r \geq m$, then everything is proved. Otherwise, let $\bar{i}^{(0)}$ be the minimal multi-index for which $v_p(c_{\bar{i}^{(0)}}) = r$. So, $c_{\bar{i}}$ is divisible by p^{r+1}, if $\bar{i} < \bar{i}^{(0)}$. Consider the unit
$$\eta = E(dt_1^{pe_1' - i_1^{(0)}} \ldots t_{n-1}^{pe_{n-1}' - i_{n-1}^{(0)}} \pi^{pe_n' - i_n^{(0)}})$$

for some $d \in \mathfrak{o}$. Let k be the minimal number for which $i_k^{(0)}$ is not divisible by p. Consider the pairing $\langle t_1, \ldots, \varepsilon, \ldots, t_n, \eta\rangle_{(T)}$, where the unit ε is at the kth position. Then condition $\varepsilon \approx 1$ implies the congruence
$$cd \equiv 0 \mod (\wp, p^m)$$

(see Lemma 5.4), which cannot be true for all $d \in \mathfrak{o}$ if $r < m$. Hence $r \geq m$ and the uniqueness of the canonical decomposition is proved. □

§6. The pairing $\langle\,,\,\rangle_{(T)}$ and the Hilbert symbol

Let us consider the K-groups $K_n(K)$ and $K_1(K) \cong K^*$ in the context of preceding section. So,

(6.1) $\quad \langle\,,\,\rangle_{(T)} = \langle\,,\,\rangle_{(T)}^n : K_n^{\text{top}}(K)/p^m K_n^{\text{top}}(K) \times K^*/K^{*p^m} \to \Omega/(\Omega \cap K^{*p^m}).$

If the field K is a multidimensional local field of characteristic 0 (i.e., the last residue field $k^{(0)}$ of K is finite), then the character χ, defined on the group of p^m-primary elements as

$$\chi(\omega) = \sqrt[p^m]{\omega}^{\Delta-1} = \zeta^{\operatorname{tr} a},$$

where Δ is the Frobenius automorphism in the unramified extension $K(\sqrt[p^m]{\omega})/K$ and tr is the trace operator in $k_0/\mathbb{Q}_p$, gives the isomorphism

(6.2) $$\chi: \Omega/\Omega^{p^m} \xrightarrow{\sim} \langle \zeta \rangle = \mu_{p^m}.$$

Thus, the pairing (6.1) induces the map

(6.3) $$\chi(\langle\, ,\, \rangle_{(T)}): K_n^{\operatorname{top}}(K)/p^m K_n^{\operatorname{top}}(K) \times K^*/K^{*p^m} \to \mu_{p^m}.$$

Similarly to the map (6.3), in the multidimensional local field the Hilbert norm residue symbol is defined on the same set

$$(\, ,\,)_m: K_n^{\operatorname{top}}(K)/p^m K_n^{\operatorname{top}}(k) \times K^*/K^{*p^m} \to \mu_{p^m},$$
$$(\alpha, \beta)_m = \sqrt[p^m]{\beta}^{\psi(a)-1},$$

where ψ is the canonical reciprocity homomorphism

$$\psi: K_n(K) \to \operatorname{Gal}(K^{ab}/K)$$

(see [P, K, F1]).

THEOREM 6.1. *For any $\alpha \in K_n(K)$ and any $\beta \in K^*$, where $p \neq 2$, the following relation holds*

(6.4) $$(\alpha, \beta)_m = \chi(\langle \alpha, \beta \rangle_{(T)}),$$

which determines the Hilbert symbol explicitly by means of the (4.2).

PROOF. According to the isomorphism (6.2), the mapping $\chi(\langle\, ,\, \rangle_{(T)})$ coincides with the pairing $\langle\, ,\, \rangle_\Gamma$ constructed in Theorem 3 of the article [V5]. Then Theorem 4 from this article implies (6.4). □

2°. Now we consider the complete discretely valued field K of charactistic 0 with perfect residue field F of characteristic $p > 0$. The field F is not algebraically closed. In the context of this article, the field K is a 1-dimensional complete field.

The reciprocity map for such a field is constructed as follows (see [F2]). Let L/K be a totally ramified Galois extension of power p^m, then

$$(\, , L/K): U_{1,K}/N_{L/K}U_{1,L} \xrightarrow{\sim} \operatorname{Hom}(\operatorname{Gal}(\widetilde{K}/K), \operatorname{Gal}(L/K)^{ab}),$$

where $U_{1,K}$ is the group of principal units, $\widetilde{K}$ is the maximal abelian unramified p-extension of K, and Hom means continuous $\mathbb{Z}_p$-homomorphisms. We note that

$$\operatorname{Gal}(\widetilde{K}/K) \cong \prod_\varkappa \mathbb{Z}_p,$$

where $F/\wp F \cong \oplus \mathbb{Z}/p\mathbb{Z}$ is the direct sum of $\varkappa$ copies of $\mathbb{Z}/p\mathbb{Z}$. Moreover, it is clear that the group Hom is noncanonically isomorphic to

$$\bigoplus_{\varkappa} \operatorname{Gal}(L/K)^{ab}.$$

We define the Hilbert symbol of degree p^m in the field K, which contains all the p^mth roots of the unity

$$(\,,\,)_m\colon U_{1,K} \times K^* \to \operatorname{Hom}(\operatorname{Gal}(\widetilde{K}/K), \mu_{p^m}),$$
$$(\alpha, \beta)_m\colon \varphi \longmapsto \sqrt[p^m]{\beta}^{\sigma_\alpha(\varphi)-1},$$

where $\varphi \in \operatorname{Gal}(\widetilde{K}/K)$, $\sigma_\alpha(\varphi)$ is the extension of $(\alpha, M/K)(\varphi)$ to the group $\operatorname{Gal}(\widetilde{K}(\sqrt[p^m]{\beta})/\widetilde{K})$, M/K is a totally ramified extension such that

$$\begin{array}{ccc} K & \longrightarrow & M \\ | & & | \\ \widetilde{K} & \longrightarrow & \widetilde{K}(\sqrt[p^m]{\beta}) \end{array}$$

It is easy to see that the symbol $(\,,\,)_m$ is multiplicative in both arguments, skew-symmetric, i.e., $(\alpha, \beta)_m = (\beta, \alpha)_m^{-1}$ if $\alpha, \beta \in U_{1,K}$, and satisfies the Steinberg relation $(1-\alpha, \alpha)_m = 1$ for any $\alpha \equiv 0 \mod \pi$.

Besides, it is clear that the symbol $(\,,\,)_m$ induces the nondegenerate pairing

$$U_{1,K}/U_{1,K}^{p^m} \times K^*/\Omega K^{*p^m} \to \operatorname{Hom}(\operatorname{Gal}(\widetilde{K}/K), \mu_{p^m}),$$

where Ω is the group of p^m-primary elements of K.

Using the skew-symmetry property, one can extend the symbol $(\,,\,)_m$ to $K^* \times K^*$: for (π, α), $\alpha \in U_{1,K}$, we set $(\pi, \alpha) = (\alpha, \pi)^{-1}$ and $(\pi, \pi) = (-1, \pi)$. Then the pairing $(\,,\,)_m$ induces the nondegenerate Hilbert pairing

$$(\,,\,)_m\colon K^*/K^{*p^m} \times K^*/K^{*p^m} \to \operatorname{Hom}(\operatorname{Gal}(\widetilde{K}/K), \mu_{p^m}).$$

Define the mapping

$$\lambda\colon \Omega/\Omega \cap K^{*p^m} \to \operatorname{Hom}(\operatorname{Gal}(\widetilde{K}/K), \mu_{p^m}),$$
$$\lambda(\omega)(\varphi) = \sqrt[p^m]{\omega}^{\varphi-1}.$$

Then the mapping λ is an isomorphism and the Hilbert symbol $(\,,\,)_m$ coincides with the composition

$$_{p^m}\operatorname{Br}(K) \xrightarrow{\mu} \Omega/\Omega \cap K^{*p^m} \xrightarrow{\lambda} \operatorname{Hom}(\operatorname{Gal}(\widetilde{K}/K), \mu_{p^m}),$$

where μ is defined as follows. Any cyclic pair (α, β) can be uniquely represented as $(\alpha, \beta) = (\pi, \omega)$, where ω is p^m-primary element (see [W]). Then $\mu(\alpha, \beta) = \omega$.

The pairing $\langle\,,\,\rangle_{(T)}^{(u)}$, constructed in §5 in our case has the following form. Let $\alpha,\beta \in K^*$, then

(6.5)
$$\Gamma_\pi \colon K^*/K^{*p^m} \times K^*/K^{*p^m} \to \Omega/\Omega \cap K^{*p^m},$$
$$\Gamma_\pi(\alpha,\beta) = \omega(\gamma_{\alpha,\beta}),$$

where

(6.6) $$\gamma_{\alpha,\beta} = \mathrm{res}_x\bigl(l(\underline{\alpha})l(\underline{\beta})' - l(\underline{\alpha})\underline{\beta}'/\beta = l(\underline{\beta})\underline{\alpha}'/\alpha\bigr)/s$$

($\underline{\alpha}(x)$ denotes the series $\eta_m^*(\alpha)$ see (4.23)).

Along with the pairing Γ_π that defines the mapping into the group of p^m-primary elements, it is natural to consider the pairing

$$\langle\,,\,\rangle_\pi \colon K^* \times K^* \to \mathrm{Hom}(\mathrm{Gal}(\widetilde{K}/K), \mu_{p^m}),$$

which is the composition of Γ_π and of the isomorphism

(6.7)
$$\lambda \colon \Omega/\Omega \cap K^{*p^m} \to \mathrm{Hom}(\mathrm{Gal}(\widetilde{K}/K), \mu_{p^m}),$$
$$\lambda(\omega)(\varphi) = \sqrt[p^m]{\omega}^{\varphi-1},$$

i.e., $\langle \alpha,\beta\rangle_\pi = \lambda(\Gamma_\pi(\alpha,\beta))$ (here $\widetilde{K}$ is a maximal abelian unramified p-extension of K) (see [F2]).

THEOREM 6.2. *The pairing Γ_π, as well as $\langle\,,\,\rangle_\pi$, is bilinear, skew-symmetric, and well defined (i.e., is invariant relative to the choice of the prime element π and does not depend on the expansion of α and β, and of the root ζ into series in π). Both pairings satisfy the Steinberg relation, i.e.,*

$$\Gamma_\pi(\alpha, 1-\alpha) \approx 1, \quad \langle\alpha, 1-\alpha\rangle_\pi = 1, \qquad \alpha \neq 0, 1.$$

Besides, if the field F is not algebraically closed, then both pairings define mappings nondegenerate up to p^m th powers. Both pairings have the norm property.

PROOF. All the properties of the pairings Γ_π and $\langle\,,\,\rangle_\pi$ are proved in Proposition 4.4. For the norm property, see Theorem 7 in [V5]. □

Let us note the following properties of the pairing $\langle\,,\,\rangle_\pi$.
1. If $w(x) = \sum_{i \geq 1} a_i x^i$, $a_i \in \mathfrak{o}$ and $(i, p) = 1$, then

(6.8) $$\langle \pi, E(w(x))\big|_{x=\pi}\rangle_\pi = 1.$$

In fact, one can use the equality

(6.9) $$E(x) = \prod_{m \geq 1}(1 - x^m)^{-\mu(m)/m}, \quad (m, p) = 1,$$

where $\mu(m)$ is the Möbius function and expand every element a_i from the ring $\mathfrak{o}$ into series in powers of p with coefficients from the Teichmüller system $\mathcal{R}$. Then to prove (6.8) means to verify the relation

$$(6.10) \qquad \langle \pi, 1 - \theta\pi^i \rangle_\pi = 1, \quad \theta \in \mathcal{R}, \ (i, p) = 1.$$

Since k_0 is perfect, we have $\mathcal{R}^p = \mathcal{R}$, hence $\langle \theta, \alpha \rangle_\pi = 1$ for any $\alpha \in K^*$, so

$$1 = \langle \theta\pi^i, 1 - \theta\pi^i \rangle_\pi = \langle \pi, 1 - \theta\pi^i \rangle^i.$$

Relation (6.10) immediately follows from the last equality since $(i, p) = 1$. Thus (6.8) is proved.

2. If $\omega(a)$, $a \in \mathfrak{o}$ is some p^m-primary element of the field K, then

$$(6.11) \qquad \langle \pi, \omega(a) \rangle_\pi = 1 \iff a \equiv 0 \mod (\wp, p^m).$$

In fact, according to the definition of the pairings, we have

$$\langle \alpha, \beta \rangle_\pi = 1 \iff \Gamma_\pi(\alpha, \beta) \approx 1.$$

Next, we obtain $\Gamma(\pi, \omega(a)) = \omega(a)$, hence

$$\langle \pi, \omega(a) \rangle_\pi = 1 \iff \omega(a) \approx 1 \iff a \equiv 0 \mod (\wp, p^m),$$

and condition (6.11) is established.

4°. Now we verify the main result of this section. Let K be a complete discretely valuated field whose residue field is perfect, not algebraically closed, and is of characteristic p. Let the field K contain all the p^mth roots of unity and $(\ ,\ )_m$ be the Hilbert symbol in the field K. We shall consider the isomorphism

$$\lambda: \Omega/\Omega \cap K^{*p^m} \to \mathrm{Hom}(\mathrm{Gal}(\widetilde{K}/K), \mu_{p^m}),$$
$$\lambda(\omega) = \sqrt[p^m]{\omega}^{\varphi-1},$$

where Ω is the group of p^m-primary elements of K, and $\widetilde{K}$ is a maximal abelian unramified p-extension of K. In the group of principal units we shall use the Shafarevich canonical decomposition, which in our case may be represented as follows.

Let e be the ramification index of K and $e_1 = e/(p-1)$. Let, further, K contain all the p^mth roots of unity and let π be a prime element of K.

PROPOSITION 6.1. *Any principal unit ε of the field K can be uniquely (up to p^mth powers) represented as*

$$(6.12) \qquad \varepsilon = \omega(a_\varepsilon) E(w_\varepsilon(x)) \Big|_{x=\pi},$$

where $w_\varepsilon(x) = \sum c_i x^i$, $1 \leqslant i \leqslant pe_1$, $(p, i) = 1$. Moreover,

$$\varepsilon \in K^{*p^m} \iff a_\varepsilon \equiv 0 \mod (\wp, p^m), \ w_\varepsilon(x) \equiv 0 \mod p^m.$$

THEOREM 6.3. *Let* $p \neq 2$, *then*
$$(\alpha, \beta)_m = \lambda(\omega(\gamma_{\alpha,\beta})),$$

where
$$\gamma_{\alpha,\beta} = \operatorname{res}_x(l(\underline{\alpha})l(\underline{\beta})' - l(\underline{\alpha})\underline{\beta}'/\underline{\beta} + l(\underline{\beta})\underline{\alpha}'/\underline{\alpha})/s$$

(*see* (6.6)).

PROOF. First consider the case $\alpha = \pi$, $\beta = \varepsilon$, where ε is a principal unit. Notice that since the pairing
$$\langle \alpha, \beta \rangle_\pi = \lambda(\Gamma_\pi(\alpha, \beta)) = \lambda(\omega(\gamma_{\alpha,\beta}))$$

has the independence property, any expansion of the unit ε into power series in the prime π may be used. Hence, we shall represent ε in the canonical form
$$\varepsilon = \omega(a_\varepsilon) E(w_\varepsilon(x)) \Big|_{x=\pi},$$

where $w_\varepsilon(x) = \sum c_i x^i$, $p \nmid i$. Then from (6.8), (6.11) and the properties of the pairing $\langle\,,\,\rangle_\pi$ we obtain
$$\langle \pi, \varepsilon \rangle_\pi = \lambda(\omega(a_\varepsilon)).$$

On the other hand, the unity $E(c\pi^i)$, $p \nmid i$, is the norm in the extension $K(\sqrt[p^m]{\pi})$ (this follows from the expansion (6.9), see also [Sh, p. 128]). Hence for the Hilbert symbol we obtain
$$\left(\pi, E(w_\varepsilon(x))\Big|_{x=\pi}\right)_m = 1.$$

So, $(\pi, \varepsilon)_m = (\pi, \omega(a_\varepsilon)) = \lambda(\omega(a_\varepsilon))$, and the following relation

(6.13) $$(\pi, \varepsilon)_m = \langle \pi, \varepsilon \rangle_\pi$$

is established.

If ε, η are principal units of the field K, then, supposing $\tau = \pi\varepsilon$ and using the invariance property of the pairing $\langle\,,\,\rangle_\pi$ we obtain
$$\langle \varepsilon, \eta \rangle_\pi = \langle \pi\varepsilon, \eta \rangle_\pi \cdot \langle \pi, \eta \rangle_\pi^{-1} = \langle \tau, \eta \rangle_\tau \cdot \langle \pi, \eta \rangle_\pi^{-1}.$$

Relation (6.13) proved above implies
$$(\varepsilon, \eta)_m = \langle \varepsilon, \eta \rangle_\pi.$$

The general case follows from the statement first proved and from the properties of bilinearity and skew-symmetry. These properties are possessed both by the pairing and the Hilbert symbol. The theorem is proved. □

The author is very grateful to I. B. Fesenko for valuable remarks.

References

[BV] D. G. Benois and S. V. Vostokov, *A norm pairing in formal groups and Galois representations*, Algebra i Analiz **2** (1990), no. 6, 69–97; English transl. in Leningrad Math. J. **2** (1991), no. 6.

[Ha] H. Hasse, *Zahlentheorie*, Akademie-Verlag, Berlin, 1963.

[Ho] T. Honda, *On the theory of commutative formal groups*, J. Math. Soc. Japan **22** (1970), 213–243.

[F1] I. B. Fesenko, *Class field theory of multidimensional local fields of characteristic zero with residue fields of positive characteristic*, Algebra i Analiz **3** (1991), no. 3; English transl. in Leningrad Math. J. **3** (1992).

[F2] _____, *On abelian totally ramified p-extensions of local fields* (to appear).

[K] K. Kato, *A generalization of local class field theory by using K-groups*. II, J. Fac. Sci. Univ. Tokyo Sect. IA Math. **27** (1980), 603–683.

[P] A. N. Parshin, *Local class field theory*, Trudy Mat. Inst. Steklov **165** (1984), 143–170; English transl. in Proc. Steklov Inst. Math. **1985**, no. 3.

[Sh] I. R. Shafarevich, *A general reciprocity law*, Mat. Sb. **26 (68)** (1950), no. 1, 113–146; English transl., Amer. Math. Soc. Transl. **(2) 4** (1956), 73–106.

[V1] S. V. Vostokov, *Explicit form of the law of reciprocity*, Izv. Akad. Nauk SSSR Ser. Mat. **42** (1978), no. 6, 1288–1321; English transl. Math. USSR-Izv. **13** (1979), no. 3, 557–588.

[V2] _____, *The Hilbert symbol in a discretely valuated field*, Zap. Nauch. Semin. Leningrad. Otdel. Mat. Inst. Steklov (LOMI) **94** (1979), 50–69; English transl. in J. Soviet Math. **19** (1982).

[V3] _____, *A norm pairing in formal modules*, Izv. Akad. Nauk SSSR Ser. Mat. **43** (1979), no. 4, 765–794; English transl, Math. USSR-Izv. **15** (1980), no. 1, 25–52.

[V4] _____, *Symbols on formal groups*, Izv. Akad. Nauk SSSR Ser. Math. **45** (1981), no. 5, 985–1014; English transl., Math. USSR-Izv. **19** (1982), no. 2, 261–284.

[V5] _____, *Explicit construction of class field theory for a multidimensional local field*, Izv. Akad. Nauk SSSR. Ser. Mat. **49** (1985), no. 2, 283–308; English transl, Math. USSR-Izv. **26** (1986), no. 2, 263–288.

[VZF] S. V. Vostokov, I. B. Zhukov, and I. B. Fesenko, *On the theory of higher-dimensional local fields. Methods and constructions*, Algebra i Analiz **2** (1990), no. 4, 91–118; English transl., Leningrad Math. J. **2** (1991), no. 4, 775–800.

[W] E. Witt, *Schiefkörper über diskret bewerteten Körpern*, J. Reine und Angew. Math. **176** (1936), 153–156.

[Z] I. B. Zhukov, *A structure theorem for complete fields*, this volume.

Translated by THE AUTHOR

St. Petersburg University, Department of Mathematics and Mechanics, Bibliotechnaya pl. 2, 198904 St. Petersburg, Russia

Artin-Hasse Exponentials and Bernoulli Numbers

S. V. VOSTOKOV

§1. Dimension of the cyclotomic units space

1°. In the cyclotomic field $\mathbb{Q}(\zeta)$, where ζ is a pth root of unity, any cyclotomic unit can be represented as follows

$$u = \zeta^a \prod_i (\zeta^{\alpha_i} - 1)^{n_i}, \tag{1}$$

where $\{\alpha_i\}$ and $\{n_i\}$ are finite sets of integers such that
  (i) $(\alpha_i, p) = 1$,
  (ii) $\prod_i \alpha_i^{n_i} \equiv 1 \mod p$,
  (iii) $\sum n_i = 0$.

Let $\mathcal{U}$ be the closure of the set of cyclotomic units in the field $\mathbb{Q}_p(\zeta)$ and s be the number of Bernoulli integers B_i, $2 \leqslant i \leqslant p - 3$, that are not divisible by p, $p \neq 2$.

The main goal of this article is to prove the following theorem.

THEOREM 1. *The dimension of the space $\mathcal{U}/\mathcal{U}^p$ over a finite field $\mathbb{F}_p$ is equal to $s + 1$.*

2°. At first we shall calculate the logarithmic derivatives of the cyclotomic units. We consider the Lubin-Tate formal groups (see [LT]). Any such group is known to be uniquely defined by its endomorphism.

So, for the polynomial $[p](x) = px + x^p$ there is a unique Lubin-Tate formal group $F(x, y)$ for which this polynomial is an endomorphism, i.e., $F \circ [p] = [p] \circ F$. All the formal groups constructed from a series of the form $px + x^p + \sum_{i \geqslant 2} a_i x^i$, $a_i \in p\mathbb{Z}_p$, are isomorphic over $\mathbb{Z}_p$ (see [LT]). In particular the group $F(x, y)$ is isomorphic to the multiplicative formal group $F_m(x, y) = x + y + xy$. For any formal group $G(x, y)$ over $\mathbb{Z}_p$ there is a series, called the *logarithm* of the formal group, which establishes an isomorphism between the formal group $G(x, y)$ and

---

1991 *Mathematics Subject Classification.* Primary 11S31; Secondary 11B65, 11B68.

Key words and phrases. Local field, Hilbert symbol, cyclotomic units, Bernoulli numbers.

These studies were supported by the Russian Fund of Fundamental Studies (project No. 94-01-00753-a).

the additive formal group $G_a = x + y$. We denote it by $\log_G(x)$ or $\lambda_G(x)$. Then
$$\log_G(G(x, y)) = \log_G(x) + \log_G(y).$$

The general theory of formal groups states that the series $\lambda_1^{-1}(\lambda_2(x))$ determines an isomorphism between the formal groups $F_1(x, y)$ and $F_2(x, y)$ with the logarithms $\lambda_1(x)$ and $\lambda_2(x)$ respectively.

In our case, if $\lambda(x)$ is the logarithm of the formal group $F(x, y)$, then the series $f(x) = \exp \lambda(x)$ determines the isomorphism over $\mathbb{Z}_p$ between the formal group F and the multiplicative formal group F_m, i.e., $F_m \circ f = f \circ F$.

In particular,
$$f(x) = \exp \lambda(x) \in \mathbb{Z}_p[\![x]\!].$$

It should be noted that the series $\lambda(x)$ can be represented as follows (see [La])
$$\lambda(x) = \lim_{n \to \infty} \frac{[p^n](x)}{p^n},$$

where $[p^n]$ denotes the nth iteration of the series $[p](x)$. The last relation implies the following congruence

(2) $$\lambda(x) \equiv x \mod x^p.$$

Since the formal groups $F(x, y)$ and $F_m(x, y)$ are isomorphic over $\mathbb{Z}_p$, the field $\mathbb{Q}_p(w)$ coincides with the field $\mathbb{Q}_p(\zeta)$, where w is a root of the equation $[p](x) = 0$. We shall normalize this root as follows:

(3) $$\exp \lambda(x)\big|_{x=w} = \zeta.$$

Then we have

(4) $$\zeta^a - 1 = \exp(a\lambda(x))\big|_{x=w} - 1.$$

From the definition of Bernoulli numbers we get

(5) $$(\exp(ax) - 1)^{-1} \frac{d}{dx}(\exp(ax) - 1) = a + \sum_{k \geq 0} \frac{1}{k!} B_k a^k x^{k-1}.$$

Taking into account congruence (2), we easily obtain the following congruence
$$(\exp(a\lambda(x)) - 1)^{-1} \frac{d}{dx}(\exp(a\lambda(x)) - 1)$$
$$\equiv (\exp(ax) - 1)^{-1} \frac{d}{dx}(\exp(ax) - 1) \mod x^{p-1}.$$

From this and (5) for the cyclotomic unit $u = \prod_i (\zeta^{\alpha_i} - 1)^{n_i}$, we obtain

(6) $$\frac{d}{dx} \log u \equiv \sum_i n_i \alpha_i + \sum_{k \geq 0} \frac{B_k}{k!} \left(\sum n_i \alpha_i^k\right) x^{k-1}$$
$$= \frac{1}{2} \sum n_i \alpha_i + \sum_{k \geq 2} \frac{B_k}{k!} \left(\sum n_i \alpha_i^k\right) x^{k-1} \mod x^{p-1}$$

as $B_1 = -\frac{1}{2}$ and for $k = 0$ we have $\sum_i n_i = 0$ (by assumption).

Now we shall calculate the logarithmic derivative for cyclotomic units of the following special form:

$$u_k = \prod_{a=1}^{p-1} \left(\zeta^{a\frac{1-g}{2}} \frac{\zeta^{ag} - 1}{\zeta^a - 1}\right)^{a^{p-1-k}}, \qquad 2 \leqslant k \leqslant p - 3,$$

where g is a root of unity modulo p (see [Wa]). Note that the unit u_k is the principal unit in the field $\mathbb{Q}_p(\zeta)$ since

$$\frac{\zeta^{ag} - 1}{\zeta^a - 1} \equiv g \mod (\zeta - 1)$$

and hence

$$\prod_{a=1}^{p-1} \left(\frac{\zeta^{ag} - 1}{\zeta^a - 1}\right)^{a^{p-1-k}} \equiv \prod_{a=1}^{p-1} g^{a^{p-1-k}} \equiv g^{\sum_{a=1}^{p-1} a} \equiv 1 \mod (\zeta - 1).$$

For the cyclotomic unit u_k the following congruence holds

(7) $$\frac{d}{dx} \log u_k \equiv -\frac{B_k}{k!} g^k x^{k-1} \mod (p, x^{p-1}).$$

Indeed, using (6) and the congruence $\sum_{a=1}^{p-1} a^{p-k} \equiv 0 \mod p$, we obtain

$$\frac{d}{dx} \log u_k = \sum_{a=1}^{p-1} a^{p-1-k} \left(a \cdot \frac{1-g}{2}\right) \frac{d}{dx} \log \zeta$$

$$+ \sum_{a=1}^{p-1} a^{p-1-k} \left(\frac{d}{dx} \log(\zeta^{ag} - 1) - \frac{d}{dx} \log(\zeta^a - 1)\right)$$

$$= \frac{1-g}{2} \left(\frac{d}{dx} \log \zeta\right) \sum_{a=1}^{p-1} a^{p-k}$$

$$+ \sum_{a=1}^{p-1} a^{p-1-k} \left(a(g - 1) + \sum_{m \geqslant 1} \frac{B_m}{m!}(a^m - 1) g^m x^{m-1}\right)$$

$$\equiv \sum_{m \geqslant 1} \frac{B_m}{m!} g^m x^{m-1} \left(\sum_{a=1}^{p-1} a^{p-1-k+m} - \sum_{a=1}^{p-1} a^{p-1-k}\right)$$

$$\equiv \sum_{m \geqslant 1} \frac{B_m}{m!} g^m x^{m-1} \sum_{a=1}^{p-1} a^{p-1-k+m} \mod (p, x^{p-1})$$

since $2 \leqslant k \leqslant p - 3$. The congruence (7) follows immediately.

3°. Now we consider cyclotomic units with respect to the Hilbert pairing. Let (,) be a pth power Hilbert symbol in the field $\mathbb{Q}_p(\zeta)$. Its explicit form was obtained in the articles [V1, V2]. For the principal units ξ, η, this formula has the form:

$$(8) \qquad (\xi, \eta) = \zeta^\gamma,$$

where

$$\gamma = \mathrm{res}_x \left(\log \varepsilon \frac{d}{dx} \log \eta \right) x^{-p}$$

and the units ξ, η are expanded into series in the prime element w, and then w is replaced by the variable x. In particular, the following statement, proved in a different way in [Wa, Theorem 8.16], can be obtained from this formula.

PROPOSITION 1. *If u_k is pth power, then B_k is divisible by p.*

PROOF. Let u_k be a pth power. Consider the unit $\varepsilon_k = 1 + w^{p-k}$. Then $(\varepsilon_k, u_k) = 1$ since u_k is a pth power and we can calculate the Hilbert symbol using (8)

$$\gamma_k = \mathrm{res}_x \left(\log \varepsilon_k \frac{d}{dx} \log u_k \right) x^{-p} \equiv -\frac{B_k}{k!} g^k \quad \mod p.$$

Hence $B_k g^k / k!$ must be divisible by p and the proposition is proved.

PROPOSITION 2. *The cyclotomic units group is isotropic, i.e., for any pair u, $u' \in \mathcal{U}$ we have $(u, u') = 1$.*

PROOF. The calculation of γ by formula (8) using (6) gives

$$\gamma = \left(\frac{1}{2} \sum n_i \alpha_i \right) \left(\frac{B_{p-1}}{(p-1)!} \sum_j n'_j (\alpha'_j)^{p-1} \right)$$
$$+ \sum_{k=2}^{p-1} \frac{B_k}{k \cdot k!} \frac{B_{p-k}}{(p-k)!} \left(\sum_i n_i \alpha_i^k \right) \left(\sum_j n'_j (\alpha'_j)^{p-k} \right).$$

The first sum is divisible by p since $(\alpha'_j)^{p-1} \equiv 1 \mod p$ and by the condition $\sum_j n'_j = 0$. In the second sum either B_k or B_{p-k} is equal 0 (since k or $p-k$ is even). Hence this sum is also divisible by p. The proposition is proved.

4°. In this section we shall prove Theorem 1, which was formulated in the Introduction. First we shall calculate the orthogonal complement to the cyclotomic units group $\mathcal{U}$ in the space E/E^p, where E is the group of principal units of the field $\mathbb{Q}_p(\zeta)$ with respect to the Hilbert pairing (,).

Let

$$(9) \qquad \eta_k = \exp \lambda(w^k), \qquad k = 1, 2, \ldots, p$$

(in particular, $\eta_1 = \zeta$, see (3)).

LEMMA 1. *The elements* $\eta_1, \ldots, \eta_p$ *form an orthogonal basis (with respect to the Hilbert pairing) from proper vectors in the space* E/E^p, *i.e.,*

$$(10) \qquad (\eta_i, \eta_j) = \begin{cases} 1 & \text{if } i+j \neq p, \\ \zeta^j & \text{if } i+j = p, \end{cases}$$

and $\eta_k^\sigma = \eta_k^{g^k}$ *for the automorphism* $\sigma \in \mathrm{Gal}(\mathbb{Q}_p(\zeta)/\mathbb{Q}_p)$ *satisfying the equality*

$$\zeta^\sigma = \zeta^g; \quad g \bmod p.$$

PROOF. We use formula (8) for the Hilbert symbol and the congruence $\lambda(x) \equiv x \bmod x^p$. Then $(\eta_i, \eta_j) = \zeta^{\gamma_{ij}}$, where

$$\begin{aligned}
\gamma_{ij} &= \mathrm{res}_x \left(\log \eta_i \frac{d}{dx} \log \eta_j \right) x^{-p} = \mathrm{res}_x \left(\lambda(x^i) \frac{d}{dx} \lambda(x^j) x^{-p} \right) \\
&= \mathrm{res}_x \left(x^i + a x^{pi} + \ldots \right) \left(j x^j + a p j x^{pj} + \ldots \right) x^{-1-p} \\
&= \mathrm{res}_x \, x^{-1} \left(j x^{i+j} + j a x^{pi+j} + \ldots \right) x^{-p} \\
&= \mathrm{res}_x \, x^{-1} j x^{i+j-p} = \begin{cases} 0 & \text{if } i+j \neq p, \\ j & \text{if } i+j = p. \end{cases}
\end{aligned}$$

Condition (10) is proved.

The second statement may be obtained as follows. From $\zeta^\sigma = \zeta^g$ and $\zeta = \exp \lambda(w)$ it follows that

$$\exp \lambda(w^\sigma) = (\exp \lambda(w))^g = \exp(g \lambda(w)) = \exp \lambda([g]w).$$

Hence $w^\sigma = [g]w$ and $(w^k)^\sigma = [g^k]w$. Then

$$\eta_k^\sigma = \exp \lambda(w^k)^\sigma = \exp \lambda([g^k]w) = \exp g^k \lambda(w) = \eta_k^{g^k}.$$

The lemma is proved.

PROOF OF THEOREM 1. Let $i_1, \ldots, i_s$ be even numbers, $2 \leqslant i_\alpha \leqslant p-3$, for which B_{i_α} is not divisible by p. Consider the subgroup of principal units V generated by the elements

$$(11) \qquad V = \langle \eta_j \mid 1 \leqslant j \leqslant p, \; j \neq p - i_1, \ldots, p - i_s, p - 1 \rangle.$$

The units $\eta_1, \ldots, \eta_p$ form a basis in the space E/E^p, hence

$$(12) \qquad \dim_{\mathbb{F}_p} V / V \cap E^p = p - s - 1.$$

So, the dimension of the orthogonal complement $V^\perp$ to the subgroup V in the space E/E^p is equal to $s+1$.

Let us prove that

(13) $$\mathcal{U} \subset V^\perp,$$

which would imply the following inequality:

(14) $$\dim_{\mathbb{F}_p} \mathcal{U}/\mathcal{U} \cap E^p \leqslant s + 1.$$

To do this, we must verify that any cyclotomic unit $u \in \mathcal{U}$ is orthogonal to any generator η_j from the subgroup V.

Indeed, if $\eta_j \in V$, $j \neq 1$, then from formula (8) and congruence (6) we obtain

$$(\eta_j, u) = \zeta^\gamma,$$

where

$$\gamma = \operatorname{res}_x \left(\log \eta_j \frac{d}{dx} \log u \right) x^{-p} = \operatorname{res}_x x^j \left(\frac{1}{2} \sum_i n_i \alpha_i + \sum_{k \geqslant 2} \frac{B_k}{k!} \left(\sum_i n_i \alpha_i^k \right) x^{k-1} \right) x^{-p}$$

$$= \operatorname{res}_x x^{-1} \left(\sum_{k \geqslant 2} \frac{B_k}{k!} \left(\sum n_i \alpha_i^k \right) x^{j+k-p} \right) = \frac{B_{p-j}}{(p-j)!} \sum_j n_i \alpha_i^{p-j}.$$

Further, since $j \neq p - i_1, \ldots, p - i_s, p - 1$ we have $p - j \neq 1, i_1, \ldots, i_s$. Hence $B_{p-j} \equiv 0 \mod p$ and so $\gamma \equiv 0 \mod p$. Hence

(15) $$(\eta_j, u) = 1$$

for $j \neq 1, p - i_1, \ldots, p - 1$.

If $j = 1$, then the unit η_1 is equal to ζ. Since the cyclotomic unit consists of multipliers of the form $\zeta^a - 1$, where $p \nmid a$ (see (1)), it follows that $(\eta_1, \zeta^a - 1) = (\zeta^a, \zeta^a - 1)^{1/a} = 1$ and so

(16) $$(\eta_1, u) = 1$$

for any cyclotomic unit $u \in \mathcal{U}$. Now (15) and (16) imply (13) and hence (14). Similar calculations using (7) give the following result for the units u_{i_α}:

$$\left(\eta_{p-i_\alpha}, u_{i_\alpha} \right) = \zeta^\gamma, \quad \alpha = 1, 2, \ldots, s,$$

where

$$\gamma \equiv -\frac{B_{i_\alpha}}{i_\alpha!} g^{i_\alpha} \not\equiv 0 \mod p.$$

By Lemma 1, the units $\eta_{p-1}, \eta_{p-i_1}, \ldots, \eta_{p-i_s}$ are linearly independent, hence the units $\zeta, u_{i_1}, \ldots, u_{i_s}$, which belong to the group of cyclotomic units $\mathcal{U}$, are also linearly independent. Therefore $\dim_{\mathbb{F}_p} \mathcal{U}/\mathcal{U} \cap E^p \geqslant s + 1$. Taking into account (14), we obtain the assertion of the theorem.

§2. Congruence for Bernoulli numbers

Now we shall find a recurrent congruence for Bernoulli numbers using the properties of the Artin-Hasse function $E(x) = \exp(x + \frac{x^p}{p} + \ldots)$, more precisely the fact that the series $E(x)$ has p-integral coefficients.

By the definition of Bernoulli numbers, we have

$$t\frac{d}{dt}\log(\exp t - 1) = 1 + t + \sum_{m \geq 1} \frac{B_m}{m!} t^m.$$

Substituting the series $\lambda(x) = x + \frac{x^p}{p} + \ldots$ for t, we obtain

$$\lambda(x)\frac{d}{dx}\log(E(x) - 1) = \lambda(x)\frac{d}{d\lambda(x)}\log(E(x) - 1) \cdot \frac{d\lambda(x)}{dx}$$

$$= \frac{d}{dx}\lambda(x) + \frac{1}{2}\frac{d}{dx}\lambda(x)^2 + \sum_{m \geq 1}\frac{B_m}{(m+1)!}\frac{d}{dx}\lambda(x)^{m+1}$$

(17)
$$= x^{-1}\left\{(x + x^p + x^{p^2} + \ldots) + \sum_{i \geq 0}\frac{x^{2p^i}}{p^i} + \sum_{0 \leq i < j}\frac{p^i + p^j}{p^{i+j}}x^{p^i + p^j}\right.$$

$$\left. + \sum_{n \geq 2}nx^n \sum_{\substack{\alpha_0 + p\alpha_1 + \cdots = n \\ 2 \leq \alpha_0 + \alpha_1 + \cdots = m+1}}\frac{1}{\alpha_0!\,\alpha_1!\ldots} \cdot \frac{1}{p^{\alpha_1 + 2\alpha_2 + \ldots}} B_m\right\} \in \mathbb{Z}_p[\![x]\!].$$

The recurrent congruence for Bernoulli numbers can be obtained from (17) by considering coefficients at x^n for various n. For example, let us prove the following statement.

PROPOSITION. *Let $n \not\equiv -1 \mod p$ and $p - 1 < n \leq p^2 - 1$. Then*

$$B_n - \frac{1}{1!}B_{n-(p-1)} + \frac{1}{2!}B_{n-2(p-1)} - \cdots + \frac{(-1)^{[\frac{n+1}{2}]}}{[\frac{n+1}{2}]!}B_{n-[\frac{n+1}{2}](p-1)}$$

$$\equiv 0 \mod p^{[\frac{n+1}{p}]}.$$

PROOF. Calculating the coefficient of the x^{n+1} term in (17), we get

$$\frac{B_n}{(n+1)!} + \frac{B_{n-(p-1)}}{(n+1-p)!\,1!} \cdot \frac{1}{p} + \frac{B_{n-2(p-1)}}{(n+1-2p)!\,2!} \cdot \frac{1}{p^2} + \cdots$$

$$+ \frac{B_{n-[\frac{n+1}{p}](p-1)}}{(n+1-[\frac{n+1}{p}]p)!\,[\frac{n+1}{p}]!} \cdot \frac{1}{p^{[\frac{n+1}{p}]}} \in \mathbb{Z}_p.$$

This implies the congruence

$$\frac{p^{[\frac{n+1}{p}]}}{(n+1)!}B_n + \frac{p^{[\frac{n+1}{p}]-1}}{(n+1-p)!\,1!}B_{n-(p-1)} + \cdots$$

$$+ \frac{1}{(n+1-[\frac{n+1}{p}]p)!\,[\frac{n+1}{p}]!}B_{n-[\frac{n+1}{p}](p-1)} \equiv 0 \mod p^{[\frac{n+1}{p}]}.$$

By the Wilson theorem, we have

$$\frac{p^{[\frac{n}{p}]}}{n!} \equiv \frac{(-1)^{[\frac{n}{p}]}}{(n - [\frac{n}{p}]p)!} \mod p.$$

Hence we finally obtain the congruence

$$\frac{(-1)^{[\frac{n+1}{p}]}}{(n + 1 - [\frac{n+1}{p}]p)!} \left(B_n - \frac{1}{1!} B_{n-(p-1)} + \frac{1}{2!} B_{n-2(p-1)} + \cdots \right.$$
$$\left. + \frac{(-1)^{[\frac{n+1}{p}]}}{[\frac{n+1}{p}]!} B_{n-[\frac{n+1}{p}](p-1)} \right) \equiv 0 \mod p^{[\frac{n+1}{p}]}.$$

It only remains to take into account that under the condition on n the number $(n + 1 - [\frac{n+1}{p}]p)!$ is not divisible by p. The proposition is proved.

References

[La] S. Lang, *Cyclotomic fields*, Springer-Verlag, Heidelberg, 1978.
[LT] J. Lubin and J. Tate, *Formal complex multiplication in local fields*, Ann. of Math. (2) **81** (1965), 380–387.
[V1] S. V. Vostokov, *On the law of reciprocity in an algebraic number field*, Trudy Mat. Inst. Steklov. **148** (1978), 77–81; English transl. in Proc. Steklov Inst. Math. **1980**, no. 4.
[V2] _____, *Explicit form of the reciprocity law*, Izv. Akad. Nauk SSSR Ser. Mat. **42** (1978), 1288–1321; English transl. in Math. USSR-Izv. **13** (1979).
[Wa] L. C. Washington, *Introduction to cyclotomic fields*, Grad. Texts in Math., vol. 83, Springer-Verlag, New York, 1982.

Translated by THE AUTHOR

St. Petersburg University, Department of Mathematics and Mechanics, Bibliotechnaya pl. 2, St. Petersburg, 198904 Russia

Some Approaches to the Construction of Abelian Extensions for $\mathfrak{p}$-adic Fields

S. V. VOSTOKOV AND I. B. ZHUKOV

The paper concerns the problem of the explicit construction of abelian extensions for $\mathfrak{p}$-adic fields with arbitrary residue fields, and, in particular, for multidimensional local fields. By $\mathfrak{p}$-adic field we mean a complete field of characteristic 0 with residue field of prime characteristic.

One of the essential motivations of this study is related to the explicit description of the reciprocity map for higher local class field theory (see, e.g., [F1, F2]). Recall that this map is a homomorphism $\theta_K \colon K_n^{\text{top}} K \to \text{Gal}(K^{\text{ab}}/K)$, where K is an n-dimensional local field and $K_n^{\text{top}} K$ is the nth Parshin K-group of K.

A description of θ_K can be given in terms of a construction of abelian extensions of K. Say, for a field of characteristic p, abelian p-extensions can be constructed by means of Witt vectors. There is a duality between the Galois group of the maximal abelian extension of K of exponent p^m and the group of p-vectors of length m over K. Thus, the reciprocity map for p-extensions of multidimensional local fields of characteristic p can be constructed by using the pairing between $K_n^{\text{top}} K$ and $W_m(K)$ with values in $\mathbb{Z}/p^m\mathbb{Z}$. An explicit formula for this pairing was obtained in [P], see also [F2]. In the case $\text{char}\, K \neq p$, abelian extensions of exponent p^m are described by means of subgroups in $K^*/(K^*)^{p^m}$, provided K contains a primitive mth root of unity. The reciprocity map is described by the pairing of $K_n^{\text{top}} K$ and $K^*/(K^*)^{p^m}$ (the generalized Hilbert symbol). An explicit formula was established by the first author in [V]. Finally, in the case of one-dimensional local fields, the abelian extensions can be obtained by adjoining the roots of an isogeny of a Lubin-Tate formal group. Respectively, there exists a simple description of the reciprocity map in terms of endomorphisms of a formal group (see [LT]).

Thus, it is important to consider non-Kummer p-extensions of $\mathfrak{p}$-adic fields with imperfect residue field (p is the characteristic of the residue field).

Although we are far from the solution of the problem, we present three approaches to it. Each of these three topics gives an idea about the structure of

---

1991 *Mathematics Subject Classification.* Primary 11S15; Secondary 11S70, 11S31, 12F10, 12J10.

Key words and phrases. Local class field theory, multidimensional local fields, semiramified extensions, Witt theory.

Part of this work was completed in MPG-Arbeitsgruppe "Zahlentheorie" in Berlin.

© 1995, American Mathematical Society

abelian extensions in certain particular cases and, we hope, contains suggestions for further studies.

In §1 we construct semiramified extensions of any two-dimensional local field K (i.e., unramified extensions with respect to the first component of the valuation on K). For this kind of extensions one can generalize the Lubin-Tate construction and give an explicit description of the reciprocity homomorphism in terms of automorphisms of a formal group.

Section 2 contains some auxiliary results concerning non-Kummer abelian extensions of fields.

In §3 we construct cyclic extensions of degree p^2 with small depth of ramification by means of Witt vectors.

Finally, in §4 we deal with the case of p-adic, i.e., absolutely unramified field. We present a way of constructing cyclic p-extensions yielding in the compositum the maximal abelian p-extension of the given field. All necessary extensions are by subsequent adjoining zeroes of one series depending on a parameter. This series is constructed by means of a certain complicated algorithm.

§1. Semiramified abelian extensions of two-dimensional local fields

In this section we consider a two-dimensional local field F with first residue field of characteristic $p > 0$. Denote by $\overline{v} = (v^{(1)}, v^{(2)})$ a valuation of height 2 on F; $\mathcal{O}_F$ is the ring of valuation; $\mathfrak{M}_F$ is the maximal ideal; π, t are local parameters of F:

$$v^{(2)}(\pi) = 1, \qquad \overline{v}(t) = (0, 1).$$

(See [MZ].)

Let $F_{\mathrm{sr}}^{\mathrm{ab}}$ be the maximal abelian semiramified extension of F. Semiramified means unramified with respect to the valuation of height 1 on F. We know that $\mathrm{Gal}(K_{\mathrm{sr}}^{\mathrm{ab}}/K)$ is canonically isomorphic to $\mathrm{Gal}(\overline{F}^{\mathrm{ab}}/\overline{F})$, and we are going to determine constructively the action of the norm residue symbol on semi-ramified extensions.

Denote by g_t the polynomial $tX + X^q$, $q = p^f$ is the cardinality of the last residue field of F. For a power series Φ in one or several variables over $\mathcal{O}_F$ we denote by $\Phi^{(n)}$ the form of degree n in Φ. Put $\Phi_n = \Phi^{(1)} + \cdots + \Phi^{(n)}$.

As in the one-dimensional case ([LT, Lemma 1]; see also [Haz, Chapter I; FV, Chapter VIII, Lemma 1.2]), using the fact that $\mathcal{O}_F$ is a local ring with maximal ideal (t) and residue field isomorphic to $\mathbb{F}_q$, one can prove the following assertion.

LEMMA 1.1. *Let Q be a linear form in n variables over $\mathcal{O}_F$. Then there exists a unique series without the constant term $\Phi \in \mathcal{O}_F[[X_1, \ldots, X_n]]$ such that $\Phi^{(1)} = Q$, and*

$$\Phi(g_t(X_1), \ldots, g_t(X_n)) = g_t(\Phi(X_1, \ldots, X_n))$$

(it may be written as $\Phi \circ g_t = g_t \circ \Phi$). Furthermore,

$$\Phi^{(n+1)} \equiv \frac{\Phi_n \circ g_t - g_t \circ \Phi_n}{t(1 - t^n)} \mod \deg(n+2)$$

for $n = 1, 2, \ldots$.

Similarly to the one-dimensional case ([LT, Theorem 1; FV, Chapter VIII, Proposition 1.2]), this lemma has a number of consequences. First of all, there exists a unique formal group $G_t \in \mathcal{O}_F[\![X, Y]\!]$ such that $G_t \circ g_t = g_t \circ G_t$. Secondly, for any $a \in \mathcal{O}_F$ there exists a unique series $[a]_t \in \mathcal{O}_F[\![X]\!]$ such that

$$[a]_t \circ g_t = g_t \circ [a]_t, \qquad [a]_t \equiv aX \mod \deg 2.$$

In particular, $[t]_t = g_t$, $[1]_t = X$. As usual, we get $[a]_t \circ G_t = G_t \circ [a]_t$. The mapping $a \mapsto [a]_t$ is an embedding of $\mathcal{O}_F$ into the ring $\operatorname{End}(G_t)$.

Adjoining to F all the roots of the polynomial

$$[t^n]_t = \underbrace{g_t \circ \cdots \circ g_t}_{n},$$

we get a field we denote by $L_{t,n}$. Let F_{ur} be the maximal purely unramified extension of F. ($F_{\mathrm{ur}} = \varinjlim_{(m,p)=1} K(\mu_m)$).

PROPOSITION 1.1. *The field $L_t = \varinjlim L_{t,n}$ is a semiramified extension of F, linearly disjoint with F_{ur}/F, and $F_{\mathrm{sr}}^{\mathrm{ab}} = L_t F_{\mathrm{ur}}$.*

PROOF. It is clear that to get $L_{t,n}$ it suffices to adjoin the roots $\alpha_1, \ldots, \alpha_{q^n - q^{n-1}}$ of the polynomial

$$\frac{[t^n]_t}{[t^{n-1}]_t} = t + ([t^{n-1}]_t)^{q-1}.$$

This polynomial becomes an Eisenstein polynomial in the first residue fieldtop. Hence for any $i = 1, \ldots, q^n - q^{n-1}$:

$$(F(\alpha_i):F) = (\overline{F}(\overline{\alpha_i}):\overline{F}) = q^n - q^{n-1},$$

where $\overline{\alpha}_i$ is the image of α_i in the first residue field. It follows that $F(\alpha_i)/F$, hence $L_{t,n}/F$ is semiramified.

The Lubin-Tate theory for $\overline{F}$ asserts that $\overline{L}_t/\overline{F}$ is linearly disjoint with $(\overline{F})_{\mathrm{ur}}/\overline{F}$, and $\overline{F}^{\mathrm{ab}} = \overline{L}(\overline{F})_{\mathrm{ur}}$. Now it suffices to notice that $(\overline{F})_{\mathrm{ur}} = \overline{F}_{\mathrm{ur}}$, $\overline{F}^{\mathrm{ab}} = \overline{F_{\mathrm{sr}}^{\mathrm{ab}}}$. □

PROPOSITION 1.2. *Let K/F be a finite extension.*

1. *Let Φ be the series from Lemma 1.1. Then the substitution of elements of $\mathfrak{M}_K$ into Φ yields a series that converges in K, its sum lying in $\mathfrak{M}_K$.*

2. *Let $a, b \in \mathcal{O}_F$, $x, y \in \mathfrak{M}_K$. Then*

$$[a]_t(G_t(x, y)) = ([a]_t \circ G_t)(x, y);$$
$$G_t([a]_t(x), [a]_t(y)) = (G \circ [a]_t)(x, y);$$
$$[a]_t([b]_t(x)) = ([a]_t \circ [b]_t)(x).$$

(For the proof, see [VZ].)

Thus G_t determines an $\mathcal{O}_F$-module structure on $\mathfrak{M}_{L_{t,n}}$. Addition is given by substitution into G_t; multiplication by $a \in \mathcal{O}_F$ is given by the substitution into $[a]_t$.

The action of the norm residue homomorphism θ_F on $L_{t,n}$ is described by the following proposition.

PROPOSITION 1.3. *Let $x \in \mathfrak{M}_{L_{t,n}}$ be a root of $[t^n]_t/[t^{n-1}]_t$, $u \in 1 + \mathfrak{M}_F$. Then*

(i) $\theta_F(\{t, \pi\})\big|_{L_{t,n}} = 1$;

(ii) $\theta_F(\{t, u\})\big|_{L_{t,n}} = 1$;

(iii) $\theta_F(\{\pi, u\})(x) = [u]_t(x)$.

PROOF. The extension $L_{t,n}/F$ can be obtained by adjoining roots of irreducible polynomial $t + ([t^{n-1}]_t(X))^q$. This polynomial is of degree $q^n - q^{n-1} = [L_{t,n}:F]$, hence it is the characteristic polynomial for its roots. It follows that $t \in N_{L_{t,n}/F}(L_{t,n}^*)$, and we have proved (i) and (ii). To prove (iii), one needs to consider the residues of both parts and use the commutative diagram

$$\begin{array}{ccc} K_2^{\text{top}} F & \xrightarrow{\partial} & \overline{F}^* \\ \theta_F \downarrow & & \downarrow \theta_{\overline{F}} \\ \text{Gal}(F^{\text{ab}}/F) & \longrightarrow & \text{Gal}(\overline{F}^{\text{ab}}/\overline{F}) \end{array}$$

(for the case char $K = p$ it is established in [F2, Theorem 4.3]).

Here $\partial: K_2^{\text{top}} F \to \overline{F}^*$ is determined by the homomorphism $\partial: K_2 F \to \overline{F}^*$, which in turn is given by the property

$$\partial(\{x, y\}) = \overline{(-1)^{w(x)w(y)} x^{w(y)} y^{-w(x)}}$$

for any $x, y \in \overline{F}^*$ ($w = v^{(2)}: F^* \to \mathbb{Z}$ is the discrete valuation on F). For ∂ see [BT; FV, Chapter IX].

Thus, (iii) is equivalent to the well-known equality

$$\theta_{\overline{F}}(\overline{u}^{-1})(\overline{x}) = [\overline{u}]_{\overline{t}}(\overline{x}),$$

where $[\overline{u}]_{\overline{t}}(X) = \overline{u}X + \ldots$ is an endomorphism of the Lubin-Tate formal group $G_{\overline{t}}$, where $G_{\overline{t}}$ is the reduction of G_t.

§2. Cyclic non-Kummer extensions

1°. Here we introduce notations that will be used in the rest of the paper.

Let p be a fixed prime. For a field K, char $K \neq p$, denote by K_i the field $K(\zeta_{p^i})$. Let σ be a generator of $\text{Gal}(K_1/K)$, $\zeta_p^\sigma = \zeta_p^s$, $s \in \mathbb{Z}$. For $i \geq 1$ denote by $M_i = M_{i,K}$ the group $\{b \in (N_{K_i/K_1} K_i^*) \cdot (K_1^*)^p : b^{\sigma-s} \in (K_1^*)^p\}$. In particular, for $\zeta_p \in K$ we have

$$M_i = (N_{K_i/K} K_i^*) \cdot (K^*)^p.$$

Let us state some results from the Miki paper [M]. Fix prime p, a field K, char $K \neq p$, and $n \geq 1$.

PROPOSITION 2.1 ([M, §1, Proposition 2]). *Let σ be any generator of $\text{Gal}(K_n/K)$, $\zeta_{p^n}^\sigma = \zeta_{p^n}^s$, $s \in \mathbb{Z}$. Put $L = K_n(\alpha)$, $\alpha^{p^n} = u \in K_n$. Then L/K is abelian if and only if $u^{\sigma-s} \in (K_n^*)^{p^n}$.*

PROPOSITION 2.2 ([M, §1, Proposition 3]). *Denote* $N_1 = [K_n:K_1]$, $N_2 = [K_1:K]$, $N = [K_n:K]$. *Let* F_n/K *be the unique cyclic extension of degree* N_1 *in* K_n/K; σ_1 *be a generator of* $\mathrm{Gal}(K_n/K_1)$, σ_2 *be a generator of* $\mathrm{Gal}(K_n/F_n)$, $\zeta_{p^n}^{\sigma_i} = \zeta_{p^n}^{s_i}$, $i = 1, 2$. *Let* L/K_n *be a cyclic extension of degree* p^n. *Then the following conditions are equivalent.*

(i) L/K *is an abelian extension and its Galois group is the direct product of the group* $\mathrm{Gal}(L/K_n)$ *and a cyclic subgroup of order* N.

(ii) *There exists* $v \in K_n$ *such that* $L = K_n(\alpha)$, $\alpha^{p^n} = (v^{\Sigma_1})^{\Sigma_2}$ *for*

$$\Sigma_1 = \sigma_i^{N_i-1} + \sigma_i^{N_i-2}s_i + \cdots + \sigma_i s_i^{N_i-2} + s_i^{N_i-1}, \qquad i = 1, 2.$$

PROPOSITION 2.3 ([M, §1, Corollary to Proposition 3]). *Let* $K^{(1)}/K$ *be a cyclic extension of degree* p. *Then the following conditions are equivalent.*

(i) *There exists a cyclic extension* $K^{(n)}/K$ *of degree* p^n *such that* $K^{(1)} \subset K^{(n)}$.

(ii) $K^{(1)}(\zeta_p) = K_1(\sqrt[p]{a})$, $a \in M_{n,K}$.

2°. Let $\zeta_{p^l} \in K$ for some $l \geqslant 1$. Let us make Proposition 2.3 more precise.

PROPOSITION 2.4. *Let* $b \in K^*$, $b \notin \mu_{p^\infty} \cdot (K^*)^p$, $K^{(l)} = K(x)$, $x^{p^l} = b$, $n \geqslant 1$. *Then the following two conditions are equivalent.*

(i) *There exists a cyclic extension* $K^{(n)}/K$ *of degree* p^n *such that* $K^{(l)} \subset K^{(n)}$.

(ii) $b \in (N_{K_n/K} K_n^*)(K^*)^{p^l}$.

PROOF. Let (i) be valid. Condition $b \notin \mu_{p^\infty} \cdot (K^*)^p$ implies that $K^{(l)}/K$, and consequently $K^{(n)}/K$, is linearly disjoint with K_n/K. By Proposition 2.2, there exists $v \in K_n^*$ such that $L_v = K_n K^{(n)} = K_n(\alpha)$, $\alpha^{p^n} = v^{\Sigma_1}$. (Here $\Sigma_2 = 1$.) Let $\widetilde{K}^{(l)} = K(\widetilde{x})$, $(\widetilde{x})^{p^l} = N_{K_n/K} v$. Since

$$v^{\Sigma_1} \equiv N_{K_n/K}(v) \mod (K_n^*)^{p^l},$$

we have $K_n \widetilde{K}^{(l)} = K_n K^{(l)}$, i.e., $K(x, \zeta_{p^n}) = K(\widetilde{x}, \zeta_{p^n})$. By the Kummer theory, $b \in \langle N_{K_n/K} v \rangle \cdot \mu_{p^l} \cdot (K^*)^{p^l} \subset (N_{K_n/K} K_n^*) \cdot (K^*)^{p^l}$, i.e., we have proved (ii).

Conversely, let (ii) be valid. One may assume $b = N_{K_n/K}(v)$, $v \in K_n^*$. Put $L = K_n(\alpha)$, $\alpha^{p^n} = v^{\Sigma_1}$. Since $b \notin \mu_{p^\infty} \cdot (K^*)^p$, and $v^{\Sigma_1} \equiv b \mod (K_n^*)^{p^l}$, we conclude $v^{\Sigma_1} \notin (K_n^*)^p$, whence L/K_n is cyclic of degree p^n, and $K^{(l)} K_n \subset L$. By Proposition 2.2

$$\mathrm{Gal}(L/K) \approx \mathrm{Gal}(L/K_n) \oplus (\mathbb{Z}/N\mathbb{Z}) \approx (\mathbb{Z}/p^n\mathbb{Z}) \oplus (\mathbb{Z}/N\mathbb{Z}).$$

If $N = 1$, one may take $K^{(n)} = L$. Otherwise, we have $K^{(l)} \subset LK_{2n}$; $\mathrm{Gal}(LK_{2n}/K) \approx (\mathbb{Z}/p^n\mathbb{Z}) \oplus (\mathbb{Z}/p^n N\mathbb{Z})$, and by elementary group-theoretical reasons it is clear that $K^{(n)}$ exists. □

3°. Let now K be a p-adic field, i.e., a complete discretely valued field of characteristic 0 with residue field of characteristic p. Introduce some conventions for the rest of the paper.

• $v = v_K$ is the normalized valuation on K;

- $e = e_K = v_K(p)$ is the absolute ramification index;
- $\mathfrak{p}_K(i) = \{a \in K : v_K(a) \geq i\}$ for $i \geq 1$;
- $\mathfrak{M}_K = \mathfrak{p}_K(1)$;
- $U_{i,k} = 1 + \mathfrak{p}_K(i)$ for $i \geq 1$.

By **G** we denote the Lubin-Tate formal group over $\mathbb{Q}_p$ determined by its endomorphism of multiplication by p, namely, $[p]_{\mathbf{G}}(X) = pX + X^p$. Denote by $f_{\mathbf{G}}(X) = X + \ldots$ the series yielding an isomorphism $\mathbf{G}_m \to \mathbf{G}$ for the multiplicative group $\mathbf{G}_m$.

Let L/K be a cyclic extension of degree p, $d = d_k(L/K)$ be the depth of ramification (see [Hy], also [Z]). Then $0 \leq d \leq e_K$, and $d = 0$ corresponds to the case of an unramified extension. If $d > 0$, then for $G = \mathrm{Gal}(L/K)$ the ramification groups are $G_s = G$, $G_{s+1} = \{1\}$, $s = \frac{p}{p-1}d$.

It is known that cyclic extensions of degree p with $d < e_K$ can be described by means of Artin-Schreier polynomials (see, e.g., [FV, Chapter III, §2]). Namely, the following statement holds.

PROPOSITION 2.5.

1. Let $a \in K$ satisfy $-\frac{p}{p-1}e_K < v_K(a) \leq 0$, $v_K(a) \not\vdots p$ whenever $v_K(a) < 0$, and $\bar{a} \notin \wp(\overline{K})$ when $v_K(a) = 0$. If x is a root of the polynomial $\wp(X) - a = X^p - X - a$, then $K(x)/K$ is a cyclic extension of degree p, and

$$d_K(K(x)/K) = -\frac{p-1}{p}v_K(a).$$

2. Let L/K be a cyclic extension of degree p, $d = d_K(L/K) < e_K$. Then there exists an $a \in K$ satisfying $v_K(a) = -\frac{p}{p-1}d$, such that

$$L = K(x), \qquad x^p - x - a = 0.$$

REMARK. If $-\frac{p}{p-1}e_K < v_K(a) < 0$, and $v_K(a) \vdots p$, then we have either $K(x) = K$, or $K(x) = K(x_1)$, $x_1^p - x_1 = a_1 \in K$, $v_K(a_1) > v_K(a)$.

Notice also that for a cyclic extension L/K of degree p we have $d_K(L/K) = d_K(L(\zeta_p)/K(\zeta_p))$ and there exists a transparent correlation between Artin-Schreier extensions and those of Kummer.

PROPOSITION 2.6. *Let $\zeta_p \in K$, $a \in K$, $v_K(a) > -\frac{p}{p-1}e_K$, x be a root of the polynomial $X^p - X - a$. Then*

$$K(x) = K\left(\sqrt[p]{1 + f_{\mathbf{G}}^{-1}(\pi_0^p a)}\right) \quad \text{for } \pi_0 = \sqrt[p-1]{-p}.$$

PROOF. First, from the condition on a, one concludes that $\pi_0^p a \in \mathfrak{M}_K$, thus $f_{\mathbf{G}}^{-1}(\pi_0^p a)$ is well defined. ($\pi_0 \in K$ since $\pi_0 = f_{\mathbf{G}}(\zeta_p^i - 1)$ for some i.) Furthermore, $x^p - x = a$ implies $y^p + py = \pi_0^p a$ for $y = \pi_0 x$. Substituting both sides into $1 + f_{\mathbf{G}}^{-1}(x)$, one gets

$$(1 + f_{\mathbf{G}}^{-1}(y))^p = 1 + f_{\mathbf{G}}^{-1}(\pi_o^p a),$$

hence

$$K(x) = K(y) = K(1 + f_{\mathbf{G}}^{-1}(y)) = K\left(\sqrt[p]{1 + f_{\mathbf{G}}^{-1}(\pi_0^p a)}\right). \quad \square$$

§3. Extensions of degree p^2 with small depth of ramification

1°. We retain the conventions of the last subsection of §2. Let K be a p-adic field, L/K be a cyclic extension of degree p. We shall see that L/K can be extended to an extension with Galois group $\mathbb{Z}/p^2\mathbb{Z}$ provided that $d = d_K(L/K)$ is small.

PROPOSITION 3.1. *Let $d < e_K/p$. Then there exists a cyclic extension M/K of degree p^2 such that $L \subset M$.*

PROOF. By Proposition 2.3, one must prove that $L_1 = K_1(\sqrt[p]{a})$, $a \in M_{2,K}$. Since $d = d_K(L_1/K_1)$, we may assume $K = K_1$, i.e., $\zeta_p \in K$. By Propositions 2.5 and 2.6, it is sufficient to show that $U_{K,e_K+1} \subset M_{2,K}$.

Applying Propositions 2.5 and 2.6 again, we get

$$\lambda = d_{K_2}(K_2/K) \leqslant \frac{p-1}{p} e_{K_2}.$$

From the definition of d_{K_2} it follows that there exists a $b \in K_2$ such that $v_{K_2}(b) = -\lambda$, $\text{Tr}_{K_2/K} b = 1$. Then, for $\rho \in \mathfrak{p}_K(e_K+1)$, we have

$$N_{K_2/K}(1+\rho b) = 1 + \rho + \text{Tr}_{K_2/K}\varepsilon + N_{K_2/K}(\rho b)$$
$$\equiv 1 + \rho \mod \mathfrak{p}_K(v_K(\rho) + 1),$$

for $v_{K_2}(\varepsilon) \geqslant 2v_K(\rho b)$. Therefore, the elements $N_{K_2/K}(1+\rho b)$ generate $U_K(e_K+1)$. □

2°. Now we describe a construction of a cyclic extension of degree p^2 which contains a given extension L/K of degree p for $d_K(L/K) < \frac{1}{p+1}e_K$.

PROPOSITION 3.2. *Let $a \in K$, $-\frac{p}{p^2-1}e_K < v(a) \leqslant 0$. Put $M = K(x_0, x_1)$ for*

$$\wp(x_0, x_1) = (a, 0)$$

for $\wp = \text{Frob} - \text{id}: W_2(K) \to W_2(K)$. Then, if $x_0 \notin K$, M/K is a cyclic extension of degree p^2.

PROOF. Let $e_K = e$. We have

$$x_0^p - x_0 = a, \qquad x_1^p - x_1 = -p^{-1}((a+x_0)^p - a^p - x_0^p).$$

If $x_0 \notin K$, then by Proposition 2.5, $L = K(x_0)$ is a cyclic extension of degree p. It is also clear that either $M = L$ or M/L is cyclic of degree p. It follows that if M_1/K_1 is cyclic of degree p^2, then M/K is also cyclic of degree p^2. Thus one may assume $\zeta_p \in K$.

Put $y_1 = \pi_0 x_1$ (see Proposition 2.6). Then $M = L(y_1)$, and $y_1^p + py_1 = b$ for $b = \pi_0 \cdot (x_0^{p^2} - a^p - x_0^p)$.

Let σ be a generator of $\text{Gal}(L/K)$, $u = x_0^\sigma - x_0$. Proposition 2.6 implies $v(u) = 0$. Then,

$$b^\sigma = \pi_0 \cdot ((x_0^\sigma)^{p^2} - a^p - (x_0^\sigma)^p) = \pi_0 \cdot ((x_0+u)^{p^2} - a^p - (x_0+u)^p)$$

$$= \pi_0 \cdot \left(\sum_{i=0}^{p^2} C_{p^2}^i x_0^{p^2-i} u^i - a^p - \sum_{i=0}^{p} C_p^i x_0^{p-i} u^i \right).$$

Since $v(x_0^{p^2-i}) = \frac{p^2-i}{p}v(a) > -e$, for $i \not\equiv p$ we obviously have $v(C_{p^2}^i) = 2e$ and $C_{p^2}^{pi} \equiv C_p^i \mod \mathfrak{p}(2e)$, so that we get

$$b^\sigma \stackrel{\mathfrak{p}(\frac{p}{p-1}e+1)}{\equiv} \pi_0 \cdot \left(\sum_{i=0}^{p} C_p^i x_0^{p^2-pi} u^{pi} - a^p - \sum_{i=0}^{p} C_p^i x_0^{p-i} u^i\right)$$

$$= \pi_0 \cdot (x_0^{p^2} - a^p - x_0^p) - \sum_{i=1}^{p-1} \frac{C_p^i}{p}[p]_G(\pi_0 x_0^{p-i} u^i) + \pi_0(u^{p^2} - u^p)$$

$$= b - \sum_{i=1}^{p-1} \frac{C_p^i}{p}[p]_G(\pi_0 x_0^{p-i} u^i) + \pi_0(u^{p^2} - u^p).$$

The relation $(x_0 + u)^p - (x_0 + u) = a = x_0^p - x_0$ implies

$$\sum_{i=1}^{p-1} C_p^i x_0^{p-i} u^i + (u^p - u) = 0,$$

whence

$$u^p \equiv u \mod \mathfrak{p}(e - \frac{1}{p+1}e + 1), \qquad u^{p^2} \equiv u^p \mod \mathfrak{p}\left(2e - \frac{1}{p+1}e + 1\right).$$

Furthermore,

$$\sum_{i=1}^{p-1} \frac{C_p^i}{p}[p]_G(\pi_0 x_0^{p-i} u^i) \equiv [p]_G\left(\sum_{i=1}^{p-1} \frac{C_p^i}{p}\pi_0 x_0^{p-i} u^i\right) \mod \mathfrak{p}(\frac{p}{p-1}e+1)$$

(because $v(\pi_0 x_0^{p-i} u^i) > \frac{1}{p(p-1)}e$). Finally,

$$b + [p]_G c \equiv b +_G [p]_G c \mod \mathfrak{p}(\frac{p}{p-1}e+1)$$

for $c = -\sum_{i=1}^{p-1} \frac{C_p^i}{p}\pi_0 x_0^{p-i} u^i$, since

$$v(b) = \frac{p}{p-1}e + (p^2 - p + 1)v(x_0) > \frac{1}{p-1}e; \qquad v([p]_G c) > \frac{1}{p-1}e.$$

We conclude that $b^\sigma \equiv b +_G [p]_G c \mod \mathfrak{p}(\frac{p}{p-1}e+1)$. Passing to the multiplicative group and applying Lemma 1.2 of [Z], we deduce that M/K is abelian. It remains to show that

$$(*) \qquad N_{L/K}(1 + f_G^{-1}(c)) \neq 1.$$

Notice that

$$q = v_L(f_G^{-1}(c)) = \frac{1}{p-1}e_L - (p-1)v_L(x_0) = \frac{1}{p-1}e_L - d_L(L/K).$$

Since $d_L(L/K) = -\frac{p-1}{p}v_L(a) < \frac{1}{p}e_L$, we have $q > \frac{1}{p-1}d_L(L/K)$. For an appropriate σ, we have $u \equiv 1 \mod \mathfrak{M}_L$, and

$$f_{\mathbf{G}}^{-1}(c) \equiv c \equiv -\pi_0 x_0^{p-1} u \equiv -\pi_0 x_0^{p-1} \equiv -a\pi_0 x_0^{-1} \mod \mathfrak{p}_L(q+1).$$

It is clear that $\text{Tr}_{L/K} x_0^{-1} = -a^{-1}$, whence

$$N_{L/K}(1 + f_{\mathbf{G}}^{-1}(c)) \equiv 1 + \text{Tr}_{L/K} f_{\mathbf{G}}^{-1}(c) \equiv 1 + \pi_0 \mod \mathfrak{p}_L\left(\tfrac{1}{p-1}e_L + 1\right),$$

and we obtain (∗). From this it is easy to see that $M \neq L$, and, by Lemma 1.2 of [Z], M/K is cyclic of degree p^2. □

Let L/K be an arbitrary cyclic extension of degree p with $d_K(L/K) < \frac{1}{p+1}e_K$. Then

$$L = K(x), \qquad x^p - x = a \in K,$$
$$0 \geqslant v_K(a) = -\frac{p}{p-1}d_K(L/K) > -\frac{p}{p^2-1}e_K.$$

Proposition 3.2 enables us to construct a cyclic extension M/K of degree p^2, $L \subset M$.

REMARK. Instead of $(a, 0)$ one can take $(a, a_1) \in W_2(K)$, with $v_K(a_1) + v_K(a) > -\frac{p}{p-1}e_K$ and the condition on a as in Proposition. Thus the equation $\mathfrak{p}(x_0, x_1) = (a_0, a_1)$ may be used to determine all cyclic extensions M/K of degree p^2 with

$$d_K(M/K) < \frac{p^2+1}{p^2+p}e_K$$

(by the Hyodo inequality, see [Hy] or Proposition 1.4 of [Z]).

§4. Construction of abelian p-extensions of a p-adic field

1°. Fix a prime $p > 2$. In this section K denotes a p-adic field, i.e., an absolutely unramified p-adic field, whose residue field is of characteristic p. Now we stop using the convention $K_i = K(\zeta_{p^i})$.

PROPOSITION 4.1. L/K is cyclic of degree p if and only if $L = K(x)$, $x^p - x = a$, $a \in K$, $v(a) \geqslant -1$ and $x \notin K$. Such an L/K is unramified if and only if $v(a) = 0$.

REMARK. Condition $x \notin K$ clearly implies $v(a) \leqslant 0$.

PROOF. The proposition is a consequence of our Proposition 2.6 and Proposition 8 in [M] since

$$f_{\mathbf{G}}^{-1}(X) \equiv 1 + X + \frac{1}{2!}X^2 + \cdots + \frac{1}{(p-1)!}X^{p-1} \mod (X^p, pX). \quad \square$$

PROPOSITION 4.2. *Let* $K_1 = K(x)$, $x^p - x = -p^{-1}\alpha$, $\alpha \in \mathcal{O}_K$. *Then the following conditions are equivalent*:
  (i) *there exists a cyclic extension* K_j/K *of degree* p^j *such that* $K_1 \subset K_j$;
  (ii) $\overline{\alpha} \in (\overline{K})^{p^{j-1}}$.

PROOF. As the preceding one, this result clearly follows from Proposition 9 in [M]. □

PROPOSITION 4.3. *Denote by* K^{ab,p^n} (*respectively* $K_{\mathrm{ur}}^{\mathrm{ab},p^n}$) *the maximal abelian* (*respectively abelian unramified*) *extension of* K *of exponent* p^n. *Choose* $A_i \subset \mathcal{O}_K$, $1 \leqslant i \leqslant n$, *in such a way that* $\{\overline{\alpha} : \alpha \in A_i\}$ *is an* $\mathbb{F}_p$-*base of* $\overline{K}^{p^{i-1}}/\overline{K}^{p^i}$ *for* $i \leqslant n-1$ *and an* $\mathbb{F}_p$-*base of* $\overline{K}^{p^{n-1}}$ *for* $i = n$. *Let* $K_{i,\alpha}$ ($\alpha \in A_i$) *be any cyclic extension of degree* p^i *that contains* x *for* $x^p - x = -p^{-1}\alpha$. *Then* $K^{\mathrm{ab},p^n}/K$ *is the compositum of linearly disjoint extensions* $K_{i,\alpha}/K$ ($1 \leqslant i \leqslant n$; α *runs over* A_i) *and* $K_{\mathrm{ur}}^{\mathrm{ab},p^n}/K$.

PROOF. Denote by K_α/K the subextension of degree p in $K_{i,\alpha}/K$.

Suppose $K_{i,\alpha}/K$ (for all i, α) and $K_{\mathrm{ur}}^{\mathrm{ab},p^n}/K$ are not linearly disjoint. Then one finds $\alpha_1, \ldots, \alpha_k \in \bigcup_{i=1}^n A_i$ such that $[K_{\alpha_1} \ldots K_{\alpha_k} K_{\mathrm{ur}}^{\mathrm{ab},p} : K_{\mathrm{ur}}^{\mathrm{ab},p}] < p^k$. Hence also $[K_{\alpha_1} \ldots K_{\alpha_k} K_{\mathrm{ur}}^{\mathrm{ab},p}(\zeta_p) : K_{\mathrm{ur}}^{\mathrm{ab},p}(\zeta_p)] < p^k$. By Proposition 2.6

$$K_{\alpha_i}(\zeta_p) = K\left(\zeta_p, \sqrt[p]{1 + f_{\mathbf{G}}^{-1}(\pi_0^p(-p^{-1}\alpha))}\right) = K\left(\zeta_p, \sqrt[p]{1 + f_{\mathbf{G}}^{-1}(\pi_0 \alpha_i)}\right).$$

On the other hand, $K_{\mathrm{ur}}^{\mathrm{ab},p}(\zeta_p) = K(\zeta_p)_{\mathrm{ur}}^{\mathrm{ab},p} = K(\zeta_p, \sqrt[p]{U_{p,K(\zeta_p)}})$. By the Kummer theory we get

$$(\langle 1 + f_{\mathbf{G}}^{-1}(\pi_0 \alpha_i) \mid i = 1, \ldots, k\rangle \cdot U_{p,K(\zeta_p)} \cdot (K(\zeta_p)^*)^p : U_{p,K(\zeta_p)} \cdot (K(\zeta_p)^*)^p) < p^k,$$

which contradicts the linear independence of $\overline{\alpha}_1, \ldots, \overline{\alpha}_k$.

In particular, we obtain

$$\widetilde{X} \equiv \sum_{i,\alpha} X(K_{i,\alpha}/K) + X(K_{\mathrm{ur}}^{\mathrm{ab},p^n}/K) = \bigoplus_{i,\alpha} X(K_{i,\alpha}/K) \oplus X(K_{\mathrm{ur}}^{\mathrm{ab},p^n}/K).$$

It remains to prove that $\widetilde{X}$ coincides with $X(K^{\mathrm{ab},p^n}/K)$. Let $\chi_0 \in X(K^{\mathrm{ab},p^n}/K)$. Then for some j we have $p^{j-1}\chi_0 \neq 0$, $p^j \chi_0 = 0$. We proceed by induction on j.

Let $j = 1$. Then $\chi_0 \in X(L/K)$, L/K is unramified of degree p. By Propositions 4.1 and 2.6

$$L(\zeta_p) = K\left(\zeta_p, \sqrt[p]{1 + f_{\mathbf{G}}^{-1}(\pi_0^p a)}\right), \qquad a \in K, \quad v(a) \geqslant -1.$$

On the other hand,

$$1 + f_{\mathbf{G}}^{-1}(\pi_0^p a) \subset \langle 1 + f_{\mathbf{G}}^{-1}(\pi_0 \alpha_i) \mid i = 1, \ldots, k\rangle \cdot U_{p,K(\zeta_p)}$$

for some $\alpha_1, \ldots, \alpha_k \in \bigcup_{i=1}^n A_i$, whence by the Kummer theory

$$L(\zeta_p) \subset K_{\alpha_1} \ldots K_{\alpha_k} K_{\mathrm{ur}}^{\mathrm{ab},p}(\zeta_p),$$

and
$$\chi_0 \in X(L/K) \subset X(K_{\alpha_1}\ldots K_{\alpha_k}K_{\mathrm{ur}}^{\mathrm{ab},p}/K) \subset \widetilde{X}.$$

Let $j > 1$. Then
$$p^{j-1}\chi_0 = \sum_{i,\alpha}\chi_{i,\alpha} + \chi_{\mathrm{ur}}, \quad \chi_{i,\alpha} \in X(K_{i,\alpha}/K), \quad \chi_{\mathrm{ur}} \in X(K_{\mathrm{ur}}^{\mathrm{ab},p^n}/K).$$

Since $\chi_{\mathrm{ur}} \in X(K_{\mathrm{ur}}^{\mathrm{ab},p^n}/K)$ is p-divisible one may assume $\chi_{\mathrm{ur}} = 0$. Further, $p\chi_{i,\alpha} = 0$, hence
$$\chi_{i,\alpha} \in X(K_\alpha/K) = X(K_\alpha(\zeta_p)/K(\zeta_p)) = X\left(K\left(\zeta_p, \sqrt[p]{1 + f_{\mathbf{G}}^{-1}(\pi_0\alpha)}\right)/K(\zeta_p)\right).$$

On the other hand, by Proposition 4.2 we have
$$p^{j-1}\chi_0 \in X(K(x)/K), \quad x^p - x = -p^{-1}\beta, \quad \beta \in \mathcal{O}_K, \quad \overline{\beta} \in (\overline{K})^{p^{j-1}}.$$

By the Kummer theory
$$1 + f_{\mathbf{G}}^{-1}(\pi_0\beta) \equiv \prod_\alpha (1 + f_{\mathbf{G}}^{-1}(\pi_0\alpha))^{n_\alpha} \mod (K(\zeta_p)^*)^p,$$

where only a finite number of n_α do not vanish. By the choice of $A - i$, we get $n_\alpha = 0$ for $\alpha \in A_1 \cup \cdots \cup A_{j-1}$. Thus
$$p^{j-1}\chi_0 = \sum_{i=j}^n \sum_{\alpha \in A_i} \chi_{i,\alpha},$$

and $p\chi_{i,\alpha} = 0$. Hence
$$\chi_{i,\alpha} = p^{j-1}\chi'_{i,\alpha}, \quad \chi'_{i,\alpha} \in X(K_{i,\alpha}/K),$$
$$\chi_0 = \sum_{i=j}^n \sum_{\alpha \in A_i} \chi'_{i,\alpha} + \chi'_0, \quad p^{j-1}\chi'_0 = 0,$$

and it remains to apply the assumption of induction to χ'_0. □

2°. Now our purpose is to construct, for an arbitrary $\widetilde{\alpha} \in (\overline{K})^{p^{n-1}}$ ($n \geq 1$), a corresponding cyclic extension $K_{n,\alpha}/K$ of degree p^n, containing x, for $x^p - x = -p^{-1}\alpha$, $\alpha \in \mathcal{O}_k$, $\overline{\alpha} = \widetilde{\alpha}$.

The valuation v of $\mathbb{Q}_p$ is extended to the ring of polynomials $\mathbb{Q}_p[D]$:
$$v\left(\sum_{i=0}^n a_i D^i\right) = \min v(a_i),$$

and, in an obvious way, to $\mathbb{Q}_p(D)$. $\mathcal{O}_D$ denotes the ring of integers in the completion of this field. From now on we suppose $p > 3$.

PROPOSITION 4.4. *There exist $g_i = g_i(D) \in \mathcal{O}_D$, $i \in \mathbb{Z}$, and $R_i = R_i(D) \in \mathcal{O}_D$, $i \geqslant 0$, satisfying the following conditions*:
1. $g_0 \equiv 1 \mod p\mathcal{O}_D$, $g_i \equiv 0 \mod p\mathcal{O}_D$ for $i \neq 0$.
2. $R_0 = D$.
3. $v(g_i) \geqslant -i + 2 + \left[\frac{i}{p}\right] + \left[\frac{i-2}{p}\right]$ for $i \leqslant -1$.
4. *Let*

$$g(X) = g(X, D) = \sum_{i=-\infty}^{\infty} g_i X^{i(p-1)+1}, \qquad R(X, D) = \sum_{i=0}^{\infty} R_i(D) X^{i(p-1)+1}.$$

Then

(1) $\qquad g(X) +_G [p]_G R(g(X), D) = g(X +_G R([p]_G X, D^p))$.

Before passing to the proof, we introduce the following notation. Denote by $\mathcal{F}$ (respectively $\mathcal{F}_1$) the set of all series

$$\sum_{i=-\infty}^{\infty} a_i X^{i(p-1)+1}$$

with coefficients in $\mathcal{O}_D$ such that for $i \leqslant 0$ we have

$$v(a_i) \geqslant -i + 2 + \left[\frac{i}{p}\right] + \left[\frac{i-2}{p}\right] \quad \left(\text{respectively } v(a_i) \geqslant -i + 3 + \left[\frac{i}{p}\right] + \left[\frac{i-2}{p}\right]\right).$$

Notice that for such series the operations of multiplication and superposition are well defined. One proves

LEMMA 4.1. 1. *If $a(X) \in \mathcal{F}$, then $pa(X) \in \mathcal{F}_1$ and $(a(X))^p \in \mathcal{F}_1$.*
2. *If $a(X) \in \mathcal{F}$, $b(X) \in \mathcal{F}$ (respectively $b(X) \in \mathcal{F}_1$), $i \geqslant 1$, $1 \leqslant j \leqslant i(p-1)$, then $a(X)^{i(p-1)-j+1} b(X)^j \in \mathcal{F}$ (respectively $\mathcal{F}_1$).*
3. *If $a(X) \in \mathcal{F}$, $b(X) \in X\mathcal{O}_D[\![X^{p-1}]\!]$, $b(X) \equiv X \mod \mathcal{F}_1$, then $a(b(X)) \equiv a(X) \mod \mathcal{F}_1$.*

COROLLARY. *For arbitrary rows $\widetilde{g} \in \mathcal{F}$, $\widetilde{R} \in X\mathcal{O}_D[\![X^{p-1}]\!]$ let*

$$\Delta(\widetilde{g}, \widetilde{R}) = (\widetilde{g}(X) +_G [p]_G \widetilde{R}(\widetilde{g}(X), D)) - \widetilde{g}(X +_G \widetilde{R}([p]_G X, D^p)).$$

Then $\Delta(\widetilde{g}, \widetilde{R}) \in \mathcal{F}_1$.

PROOF OF THE COROLLARY. By part 2 of Lemma 4.1, $\widetilde{g} \in \mathcal{F}$ implies $\widetilde{R}(\widetilde{g}(X), D) \in \mathcal{F}$, and $\widetilde{R}([p]_G X, D^p) \in \mathcal{F}_1$. By part 1 of Lemma 4.1, $[p]_G \widetilde{R}(\widetilde{g}(X), D) \in \mathcal{F}_1$. Further, by part 2 again,

(2) $\qquad \widetilde{g}(X) +_G [p]_G \widetilde{R}(\widetilde{g}(X), D) \equiv \widetilde{g}(X) \mod \mathcal{F}_1$,

$$X +_G \widetilde{R}([p]_G X, D^p) \equiv X \mod \mathcal{F}_1.$$

By part 3 of Lemma 4.1 the last congruence implies

(3) $$\widetilde{g}(X +_G \widetilde{R}([p]_G X, D^p)) \equiv \widetilde{g}(X) \mod \mathcal{F}_1.$$

Bringing together (2) and (3) we complete the proof. □

Let us write
$$\Delta(\widetilde{g}, \widetilde{R}) = \sum_{i=-\infty}^{\infty} \Delta_i(\widetilde{g}, \widetilde{R}) X^{i(p-1)+1}.$$

PROOF OF PROPOSITION 4.4. Notice that

(4) $$G(X, Y) = X + Y + \sum_{i=1}^{\infty} \sum_{j=1}^{i(p-1)} f_{ij} X^{i(p-1)-j+1} Y^j,$$

and $f_{i1} \equiv (-1)^i \mod p$ for $i < p$.

Let $l \geq 1$. Denote by $\mathcal{P}_l$ the group of series
$$\sum_{i=-\infty}^{\infty} a_i X^{i(p-1)+1}$$

with coefficients in $\mathcal{O}_D$ such that $v(a_i) \geq l + \left[\frac{2-i}{p}\right]$ for any i.

Let us construct the series
$$g^{(\nu)}(X) = \sum_{i=-\infty}^{\infty} g_i^{(\nu)} X^{i(p-1)+1}, \quad R^{(\nu)}(X) = \sum_{i=0}^{\infty} R_i^{(\nu)} X^{i(p-1)+1},$$

$\nu = 0, 1, \cdots$, so that:
(a) $R_0^{(\nu)} = D$, $g_0^{(\nu)} \equiv 1 \mod p\mathcal{O}_D$, $g_i^{(\nu)} \equiv 0 \mod p\mathcal{O}_D$ for $i \neq 0$,
(b) $g^{(\nu)}(X) \in \mathcal{F}$;
(c) $\Delta(g^{(\nu)}, R^{(\nu)}) \in \mathcal{P}_{\nu+1}$;
(d) $g^{(\nu+1)}(X) \equiv g^{(\nu)}(X) \mod \mathcal{P}_{\nu+1}$, $R^{(\nu+1)}(X) \equiv R^{(\nu)}(X) \mod \mathcal{P}_{\nu+1}$.

This would complete the proof, since it is easy to see that $g(X) = \lim g^{(\nu)}(X)$ and $R(X) = \lim R^{(\nu)}(X)$ satisfy the conditions of Proposition 4.4.

Use recursion on ν. Put $g^{(0)}(X) = X$, $R^{(0)}(X) = DX$; then conditions (a)–(c) are obviously satisfied (for $\nu = 0$).

Now suppose that for $\nu = n - 1$ the series $g^{(\nu)}(X)$ and $R^{(\nu)}(X)$ with all required properties have been constructed; let us construct $g^{(n)}(X)$ and $R^{(n)}(X)$.

To do that we construct the series $g^{(n,m)}(X) \in \mathcal{F}$ and $R^{(n,m)}(X) \in X\mathcal{O}_D[\![X^{p-1}]\!]$ for any $m \in \mathbb{Z}$ in such a way that
(a) $v(\Delta_i(g^{(n,m)}, R^{(n,m)})) \geq n + 1 + \left[\frac{2-i}{p}\right]$, $i \leq m$;
(b) $v(\Delta_i(g^{(n,m)}, R^{(n,m)})) \geq n + \left[\frac{2-i}{p}\right]$, $i > m$;
(c) $g^{(n,m)}(X) \equiv g^{(n-1)} \mod \mathcal{P}_n$, $R^{(n,m)}(X) \equiv R^{(n-1)} \mod \mathcal{P}_n$.
Since $\Delta(g^{(n-1)}, R^{(n-1)}) \in \mathcal{F}_1$, we have

$$v(\Delta_i(g^{(n-1)}, R^{(n-1)})) \geq -i + 3 + \left[\frac{i}{p}\right] + \left[\frac{i-2}{p}\right].$$

Hence for sufficiently small i,

$$v(\Delta_i(g^{(n-1)}, R^{(n-1)})) \geqslant n + 1 + \left[\frac{2-i}{p}\right].$$

Therefore, for $m \ll 0$ one can take

$$g^{(n,m)}(X) = g^{(n-1)}(X), \qquad R^{(n,m)}(X) = R^{(n-1)}(X).$$

Now suppose that the series $g^{(n,m-1)}$ has been constructed; let us construct $g^{(n,m)}$.

Case 1. $-m + 2 + \left[\frac{m}{p}\right] + \left[\frac{m-2}{p}\right] \geqslant n + \left[\frac{2-m}{p}\right]$. Since $\Delta(g^{(n,m-1)}, R^{(n,m-1)}) \in \mathcal{F}_1$,

$$v(\Delta_m(g^{(n,m-1)}, R^{(n,m-1)})) \geqslant n + 1 + \left[\frac{2-m}{p}\right],$$

and one can take $g^{(n,m)}(X) = g^{(n,m-1)}(X)$, $R^{(n,m)}(X) = R^{(n,m-1)}(X)$.

Case 2. $-m + 2 + \left[\frac{m}{p}\right] + \left[\frac{m-2}{p}\right] < n + \left[\frac{2-m}{p}\right]$, $1 \leqslant n + \left[\frac{2-m}{p}\right]$, $m \not\equiv 2, 3 \mod p$. For $\alpha \in \mathcal{O}_D$ consider the series

$$g_\alpha(X) = g^{(n,m-1)}(X) + \alpha p^{n'} X^{(m-1)(p-1)+1},$$

where $n' = n + \left[\frac{-m+2}{p}\right]$. It is easy to see that

$$\Delta_m(g_\alpha, R^{(n,m-1)}) \equiv \Delta_m(g^{(n,m-1)}, R^{(n,m-1)})$$
$$+ (m-2)D^p \alpha p^{n'} \mod p^{n'+1} \mathcal{O}_D,$$

$$v(\Delta_i(g_\alpha, R^{(n,m-1)})) \geqslant n + 1 + \left[\frac{2-i}{p}\right] \quad \text{for } i < m.$$

Besides that, $n' > -m + 2 + \left[\frac{m}{p}\right] + \left[\frac{m-2}{p}\right]$, whence $n' \geqslant -(m-1) + 2 + \left[\frac{m-1}{p}\right] + \left[\frac{(m-1)-2}{p}\right]$. It follows that $\alpha p^{n'} X^{(m-1)(p-1)+1} \in \mathcal{F}$, and $g_\alpha(X) \in \mathcal{F}$. Therefore, one can take $g^{(n,m)} = g_\alpha$, $R^{(n,m)} = R^{(n,m-1)}$ for

$$\alpha = (m-2)^{-1} D^{-p} p^{-n'} \Delta_m(g^{(n,m-1)}, R^{(n,m-1)}).$$

Case 3. $-m + 2 + \left[\frac{m}{p}\right] + \left[\frac{m-2}{p}\right] < n + \left[\frac{2-m}{p}\right]$, $1 \leqslant n + \left[\frac{2-m}{p}\right]$, and $m = -qp + r$ for $r = 2$ or $r = 3$. When $r = 2$ suppose also $n + \left[\frac{2-m}{p}\right] \geqslant 2$.

Let $\alpha, \beta, \gamma \in \mathcal{O}_D$. Consider

$$g_{\alpha\beta\gamma}(X) = g^{(n,-qp+1)}(X) + \alpha p^{n'} X^{(-qp)(p-1)+1}$$
$$+ \beta p^{n'-1} X^{(-qp+1)(p-1)+1} + \gamma p^{n'-1} X^{(-qp+2)(p-1)+1}$$

for $n' = n + \left[\frac{2-(-qp+2)}{p}\right] = n + q$. Taking into account (4), it is easy to obtain by direct computation that

$$\Delta_{-qp+3}(g_{\alpha\beta\gamma}, R^{(n,m-1)}) \equiv \Delta_{-qp+3}(g^{(n,m-1)}, R^{(n,m-1)})$$
$$+ D^p(\beta + \gamma)p^{n'-1} \mod p^{n'}\mathcal{O}_D;$$

$$\Delta_{-qp+2}(g_{\alpha\beta\gamma}, R^{(n,m-1)}) \equiv \Delta_{-qp+2}(g^{(n,m-1)}, R^{(n,m-1)})$$
$$+ (2D^p\alpha + D^p\beta + D\gamma)p^{n'} \mod p^{n'+1}\mathcal{O}_D;$$

$$\Delta_{-qp+1}(g_{\alpha\beta\gamma}, R^{(n,m-1)}) \equiv \Delta_{-qp+1}(g^{(n,m-1)}, R^{(n,m-1)})$$
$$+ (-D^p\alpha + D\beta)p^{n'} \mod p^{n'+1}\mathcal{O}_D;$$

$$v(\Delta_i(g_{\alpha\beta\gamma}, R^{(n,m-1)})) \geq n + 1 + \left[\frac{2-i}{p}\right] \quad \text{for } i \leq -qp.$$

Besides that, one has (for both $r = 2$ and $r = 3$) $n' > qp - 2q$, so

$$n' \geq -(-qp) + 2 + \left[\frac{-qp}{p}\right] + \left[\frac{-qp-2}{p}\right],$$

$$n' - 1 \geq -(-qp+1) + 2 + \left[\frac{(-qp+1)}{p}\right] + \left[\frac{(-qp+1)-2}{p}\right]$$

$$\geq -(-qp+2) + 2 + \left[\frac{(-qp+2)}{p}\right] + \left[\frac{(-qp+2)-2}{p}\right].$$

This implies $g_{\alpha\beta\gamma}(X) \in \mathcal{F}$.

Notice that the matrix

$$\begin{pmatrix} 0 & D^p & D^p \\ 2D^p & D^p & D \\ -D^p & D & 0 \end{pmatrix}$$

is invertible over $\mathcal{O}_D$. Hence for an appropriate choice of α, β, γ we get

$$\Delta_{-qp+3}(g_{\alpha\beta\gamma}, R^{(n,m-1)}) \equiv 0 \mod p^{n'}\mathcal{O}_D;$$
$$\Delta_{-qp+2}(g_{\alpha\beta\gamma}, R^{(n,m-1)}) \equiv 0 \mod p^{n'+1}\mathcal{O}_D;$$
$$\Delta_{-qp+1}(g_{\alpha\beta\gamma}, R^{(n,m-1)}) \equiv 0 \mod p^{n'+1}\mathcal{O}_D,$$

and put

$$g^{(n,-qp+2)} = g^{(n,-qp+3)} = g_{\alpha\beta\gamma}, \qquad R^{(n,-qp+2)} = R^{(n,-qp+3)} = R^{(n,-qp+1)}.$$

Case 4. $m \equiv 2 \mod p$, $n + \left[\frac{2-m}{p}\right] = 1$ (i.e., $m = p(n-1)+2$). Let $\alpha \in \mathcal{O}_D$; put

$$R_\alpha(X) = R^{(n,m-1)}(X) + \alpha X^{m(p-1)+1}.$$

It is not difficult to verify that

$$\Delta_m(g^{(n,m-1)}, R_\alpha) \equiv \Delta_m(g^{(n,m-1)}, R^{(n,m-1)}) + \alpha p \mod p^2 \mathcal{O}_D;$$

$$v(\Delta_i(g^{(n,m-1)}, R_\alpha)) \geq n + 1 + \left[\frac{2-i}{p}\right] \quad \text{for } i < m.$$

Using the second part of Lemma 4.1, we see that one may take $g^{(n,m)} = g^{(n,m-1)}$, $R^{(n,m)} = R_\alpha$ for $\alpha = -p^{-1}\Delta_m(g^{(n,m-1)}, R^{(n,m-1)})$.

Case 5. $n + \left[\frac{2-m}{p}\right] \leq 0$. Then the second part of Lemma 4.1 implies that one may take $g^{(n,m)} = g^{(n,m-1)}$, $R^{(n,m)} = R^{(n,m-1)}$.

Now it is clear that one may put $g^{(n)} = g^{(n,m)}$, $R^{(n)} = R^{(n,m)}$ for sufficiently large m (e.g., for $m = pn$). $\square$

REMARK. Constructing $g(X)$ and $R(X)$ exactly as above, one obtains

$$g(X) \equiv p \cdot \frac{D^{1-p}-1}{2} \cdot X^{-p+2} + \left(1 + p \cdot \frac{D^{1-p}-1}{2}(1-D^p)\right) \cdot X$$
$$+ \ldots \mod p^2 \mathcal{O}_D\{\{X\}\},$$
$$R(X) \equiv DX \mod \deg(p^2+p-1).$$

Denote
$$S(X) = S(X, D) = \sum_{i=0}^{\infty} S_i(D) X^{i(p-1)+1}$$

the series which is inverse to $R(X)$ with respect to superposition. It is not difficult to show that $S(X) \equiv D^{-1}X \mod \deg(2p-1)$.

Now let $\widetilde{\alpha} \in (\overline{K})^{p^{n-1}}$, $n \geq 1$. If $\widetilde{\alpha} \neq 0$, one may assume $\widetilde{\alpha} = \overline{d}^{p^{n-1}}$ for $d \in U_K$. In K^{sep} choose elements $y_1, \ldots, y_n = y_{n,d}$ that satisfy the following equations:

$$y_1^p - y_1 = -p^{-1} \sum_{i=0}^{\infty} S_i(d^{p^{n-1}}) \cdot (-p)^i;$$

$$y_j^p - y_j = -p^{-1} \sum_{i=-\infty}^{\infty} g_i(d^{p^{n-j}}) \cdot (-p)^i \cdot y_{j-1}^{i(p-1)+1}, \quad j = 2, \ldots, n.$$

THEOREM 4.1. $K(y_{n,d})$ is a cyclic extension of degree p^n over K and it contains a zero of the polynomial $X^p - X + p^{-1}d^{p^{n-1}}$.

PROOF. Put $x_j = \pi_0 y_j$, $j = 1, \ldots, n$, for $\pi_0^{p-1} = -p$; $K_0 = K(\pi_0) = K(\zeta_p)$, $K_i = K_0(x_i)$. It is easy to see that

$$[p]_G x_1 = S(\pi_0, d^{p^{n-1}});$$
$$[p]_G x_j = \sum_{i=-\infty}^{\infty} g_i(d^{p^{n-j}}) \cdot x_{j-1}^{i(p-1)+1}, \quad j = 2, \ldots, n.$$

We shall prove by induction on j that K_j/K_0 is cyclic of degree p^j, K_j/K is abelian and $\text{Gal}(K_j/K)$ possesses a generator σ such that

$$x_j^\sigma = x_j +_G R([p]_G x_j, d^{p^{n-j}}).$$

Let us establish the base of induction ($j = 1$). K_1/K_0 is obviously cyclic of degree p. The fact that K_1/K is abelian follows, for example, from Proposition 4.1. Finally, for an appropriate $\sigma \in \text{Gal}(K_1/K_0)$ one has

$$x_1^\sigma = x_1 +_G \pi_0 = x_1 +_G R\big(S(\pi_0, d^{p^{n-1}}), d^{p^{n-1}}\big)$$
$$= x_1 +_G R\big([p]_G x_1, d^{p^{n-1}}\big).$$

Now assume K_{j-1} possesses the desired properties. We have $x_{j-1} = h([p]_G x_j)$, where $h(X)$ is the series inverse to $g(X, d^{p^{n-j}})$ in the sense of superposition. It follows that $K_{j-1} \subset K_j$. Further, for some generator $\sigma \in \text{Gal}(K_{j-1}/K_0)$, we have

$$x_{j-1}^\sigma = x_{j-1} +_G R\big([p]_G x_{j-1}, d^{p^{n-j+1}}\big),$$

hence it follows that

(5) $\quad g(x_{j-1}, d^{p^{n-j}})^\sigma = g\big(x_{j-1} +_G R([p]_G x_{j-1}, d^{p^{n-j+1}}), d^{p^{n-j}}\big)$
$\qquad\qquad = g(x_{j-1}, d^{p^{n-j}}) +_G [p]_G R\big(g(x_{j-1}, d^{p^{n-j}}), d^{p^{n-j}}\big)$

because of (1). Computing the values of the valuation, we obtain

$$N_{K_{j-1}/K_0}\big(1 + f_G^{-1}\big(R(g(x_{j-1}, d^{p^{n-j}}), d^{p^{n-j}})\big)\big) \neq 1,$$

whence by Lemma 1.1 of [Z] K_j/K_0 is cyclic of degree p^j. Further, one concludes from (5) that σ can be extended to a generator of $\text{Gal}(K_j/K_0)$ such that

$$x_j^\sigma = x_j +_G R\big([p]_G x_j, d^{p^{n-j}}\big).$$

Finally, since K_1/K is abelian and K_j/K is cyclic, it follows that K_j/K is abelian. In fact, the action of $\mathbb{Z}/(p-1)\mathbb{Z}$ on $\mathbb{Z}/p^j\mathbb{Z}$ is determined by its action on the quotient group isomorphic to $\mathbb{Z}/p\mathbb{Z}$.

Since $K_n = K(y_{n,d}, \pi_0)$ and $K(y_{n,d})/K$ is a p-extension, we conclude that $\text{Gal}(K(y_{n,d})/K) \approx \mathbb{Z}/p^n\mathbb{Z}$. It is clear that $y_1 \in K(y_{n,d})$. Notice that

$$-p^{-1} \sum_{i=0}^\infty S_i(d^{p^{n-1}}) \cdot p^i \equiv -p^{-1} d^{p^{n-1}} \mod p\mathcal{O}_D.$$

It easily follows that $K(y_{n,d})$ also contains a root of the polynomial $X^p - X + p^{-1} d^{p^{n-1}}$. $\square$

References

[BT] H. Bass and J. Tate, *The Milnor ring of a global field*, Lecture Notes in Math., vol. 342, Springer-Verlag, Berlin, Heidelberg, and New York, 1973, pp. 349–446.

[F1] I. B. Fesenko, *Class field theory of multidimensional local fields of characteristic 0 with the residue field of positive characteristic*, Algebra i Analiz 3 (1991), 165–196; English transl. in Leningrad Math. J. 3 (1992).

[F2] _____, *On class field theory of multidimensional local fields of positive characteristic*, Advances in Soviet Math., vol. 4, Amer. Math. Soc., Providence, RI, 1991, pp. 103–127.

[FV] I. B. Fesenko and S. V. Vostokov, *Local fields and their extensions: a constructive approach*, Amer. Math. Soc., Providence, RI, 1993.
[Haz] M. Hazewinkel, *Formal groups and applications*, Academic Press, New York, 1978.
[Hy] O. Hyodo, *Wild ramification in the imperfect residues case*, Adv. Stud. Pure Math. **12** (1987), 287–314.
[LT] J. Lubin and J. Tate, *Formal complex multiplication in local fields*, Ann. Math. **81** (1965), 380–387.
[M]   H. Miki, *On $\mathbb{Z}_p$-extensions of complete p-adic power series fields and function fields*, J. Fac. Sci. Univ. Tokyo, Sect. IA **21** (1974), 377–393.
[MZ] A. I. Madunts and I. B. Zhukov, *Multidimensional complete fields: topology and other basic constructions*, this volume.
[P]   A. N. Parshin, *Local class field theory*, Trudy Mat. Inst. Akad. Nauk SSSR **165** (1985), 143–170; English transl., Proc. Steklov Inst. Math. (1985), no. 3, 157–185.
[V]   S. V. Vostokov, *Explicit construction of class field theory for a multidimensional local field*, Izv. Akad Nauk SSSR. Ser. Mat. **49** (1985), 283–308; English transl., Math. USSR-Izv. **26** (1986), 263–287.
[VZ] S. V. Vostokov and I. B. Zhukov, *Abelian semiramified extensions of a 2-dimensional local field*, Rings and modules. Limit theorems of probability theory. No. 2, Izdat. Leningrad Univ., 1988. (Russian)
[Z]   I. B. Zhukov, *Structure theorems for complete fields*, this volume.

Translated by THE AUTHORS

St. Petersburg University, Department of Mathematics and Mechanics, Bibliotechnaya pl. 2, 198904 St. Petersburg, Russia

Structure Theorems for Complete Fields

I. B. ZHUKOV

The paper deals with certain structure theorems on $\mathfrak{p}$-adic fields, i.e., complete fields of characteristic 0 with arbitrary residue fields of prime characteristic p.

Section 1 contains preliminary results. Then we introduce "standard" fields which form the simplest class of complete fields with imperfect residue fields. They are the fields $k \otimes_{k_0} K_0$, where K_0 is an absolutely unramified field, k_0 is a subfield of K_0 with perfect residues, k is a finite extension of k_0. The central results are Theorems 2.1 and 2.2; each of them asserts that for any complete field K there is a finite extension of some simple type L/K such that L is a standard field. In §3 we use Theorem 2.1 to get a description of all Γ-extensions of arbitrary $\mathfrak{p}$-adic fields. This generalizes the corresponding result of H. Miki [M] for standard fields.

In the following section the results of §§2, 3 are applied to multidimensional complete fields of characteristic 0 with the first residue field of characteristic both p and 0. We give a new proof (and a generalization) of A. N. Parshin's classification theorem [P] and describe Γ-extensions of such fields.

I am very grateful to Professor S. V. Vostokov and to Dr. I. B. Fesenko for many valuable conversations on the subject.

This work was completed at Westfälische Wilhelms-Universität (Münster). I am very grateful to this University for its hospitality.

Conventions

- ζ_m: a primitive root of unity in the algebraic closure of the field under consideration;
- $\mathcal{O}_K$, $\mathfrak{M}_K$, $\overline{K}$, e_K: the ring, the ideal, the residue field, and the absolute ramification index of given discrete valuation on field K;
- $\mathfrak{p}_K(i) = \mathfrak{M}_K^i$ for $i \geq 1$;
- $U_K = \mathcal{O}_K^*$;
- $U_{i,k} = 1 + \mathfrak{p}_K(i)$ for $i \geq 1$;

1991 *Mathematics Subject Classification.* Primary 12J10; Secondary 11S15, 19F05.
Key words and phrases. Complete field, imperfect residue field, multidimensional local field.
This study was partially supported by a grant from PRO MATHEMATICA.

- $K_i = K(\zeta_{p^i})$ for fixed prime p, $i \geq 1$;
- $N_{j/i} = N_{K_j/K_i}$ for fixed K, $j \geq i \geq 1$.

§1. Depth of ramification

1. Let K be a complete field with respect to a discrete valuation $v = v_K$, the residue field $\overline{K}$ being of characteristic $p > 0$. We also denote by v_K the nonnormalized extension of the valuation to any finite extension of K.

Let L be a finite extension of K. For finite M/L, Hyodo [H] introduced the notion of depth of ramification:

$$d_K(M/L) \stackrel{\text{def}}{=} \inf\{v_K(\text{Tr}_{M/L}(y)/y) : y \in M^*\}.$$

Note that for any intermediate field N,

$$d_K(M/L) = d_K(M/N) + d_K(N/L)$$

[H, 2–4] and for tamely ramified M/L one has $d_K(M/L) = 0$ [H, 2–12].

Here we are concerned only with the case of a cyclic extension M/L of degree p. One immediately proves the following two propositions.

PROPOSITION 1.1. *Let* $\operatorname{char} K = p$, $L = K(x)$, $\wp(x) = x^p - x = a \in K$, $v_K(a) \leq 0$, *and* $v_K(a) \not\mid p$ *if* $v_K(a) < 0$. *Then* $d_K(L/K) = -\frac{p-1}{p}v_K(a)$. *Furthermore,* L/K *is unramified if and only if* $d_K(L/K) = 0$.

REMARK. Any nontrivial element of $K/\wp(K)$ possesses a representative $a \in K$ such that either $v(a) = 0$ or $v(a) < 0$ and $v(a) \not\mid p$. If $b \equiv a \mod \wp(K)$, then $v(b) \leq v(a)$.

PROPOSITION 1.2. *Let* $\operatorname{char} K = 0$, $\zeta_p \in K$, $a \in K^*$, $a \notin (K^*)^p$, $L = K(\sqrt[p]{a})$, $d = d_K(L/K)$, π *be a prime element of* K.

1. *Let* $v(a) \not\mid p$. *Then* L/K *is totally ramified and* $d = e_K$.
2. *Let* $v(a) = 0$, $\overline{a} \notin \overline{K}^p$. *Then* $e(L/K) = 1$, $d = e_K$.
3. *Let* $a = 1 + \pi^i u$, $v(u) = 0$, $i \not\mid p$. *Then* L/K *is totally ramified, and* $d = e_K - \frac{p-1}{p}i$.
4. *Let* $a = 1 + \pi^i u$, $v(u) = 0$, $i \vdots p$, $\overline{u} \notin \overline{K}^p$ *if* $i < \frac{p}{p-1}e_K$. *Then* $e(L/K) = 1$, $d = e_K - \frac{p-1}{p}i$. *In this case* $\overline{L}/\overline{K}$ *is separable if and only if* $i = \frac{p}{p-1}e_K$.

PROPOSITION 1.3. *Let* L/K *be a totally ramified cyclic extension of degree* p, π *be a prime of* L, $d = d_K(L/K)$. *Then*

1. $v_K(N_{L/K}\pi/\pi^p - 1) \geq d$.
2. *If* $v_K(u - 1) = i > 0$, *then* $v_K(N_{L/K}u - 1) \geq \min\{pi, i + d\}$.
3. *Let* $u \in U_{1,L}$, $\sigma \in \operatorname{Gal}(L/K)$. *Then*

$$v_K(u^{\sigma-1} - 1) \geq v_K(u - 1) + \frac{1}{p-1}d.$$

PROOF. To prove the first assertion denote
$$j(c) = v_K\left(\frac{N_{L/K}c}{c^p} - 1\right) \quad \text{for } c \in L^*.$$
Suppose $j(\pi) = \lambda < d$. This clearly implies that for any $c \in L^*$ we have $j(c) \geqslant \lambda$. Notice also that $a \in K$ implies $j(ac) = j(c)$ and that $j(c^i) = j(c)$ for $(i, p) = 1$. It follows that for any $c \in L^*$ if $v_L(c) \not\mid p$, then $j(c) = \lambda$.

Let char $K = 0$. We have
$$d = d_K(L_1/K_1) + d_K(K_1/K) - d_K(L_1/L)$$
$$= d_K(L_1/K_1) = e(K_1/K)^{-1}d_{K_1}(L_1/K_1).$$
Hence one can assume $\zeta_p \in K$. Then $L = K(x)$, $x^p \in K$. Obviously $j(x) = \infty$, whence $v_L(x) \vdots p$, and we can assume $x = 1 + b$, $b \in \mathfrak{M}_L$. Among all the possible values of x choose the one nearest to 1. Since L/K is totally ramified, by Proposition 1.1 we get $v_L(b) \not\mid p$. Then,
$$N_{L/K}b = N_{L/K}(x - 1) = N_{L/K}x + \text{Tr}_{L/K} A + (-1)^p = N_{L/K}x + (-1)^p$$
$$= (-1)^{p-1}x^p + (-1)^p = (-1)^{p-1}(1+b)^p + (-1)^p,$$
where A is a polynomial of degree $p - 1$ in x, $x^\sigma, \ldots, x^{\sigma^{p-1}}$, σ is a generator of $\text{Gal}(L/K)$. By Proposition 1.2, $v_K(b) = \frac{1}{p-1}(e_K - d)$; we calculate $j(b) = d$, which leads to a contradiction.

Finally, let char $K = p$. Then $L = K(x)$, $x^p - x = a \in K$. As L/K is totally ramified, one can assume $v_K(a) < 0$ and $v_K(a) \not\mid p$, i.e., $v_L(x) \not\mid p$. We have
$$N_{L/K}x = x(x+1)\ldots(x+p-1) = x^p + (p-1)!x = a,$$
whence $j(x) = -(p-1)v(x) = d$.

The second assertion of Proposition is evident in view of
$$N_{L/K}(1+b) = 1 + \text{Tr}_{L/K} b + \cdots + N_{L/K}b.$$
The third assertion is easily deduced from Proposition 1.2 and from the inequality
$$v_K(u^{\sigma-1} - 1) - v_K(u - 1) \geqslant v_K(\pi^{\sigma-1} - 1). \quad \square$$

2. Let M/K be a cyclic extension of degree p^2, L be the intermediate field. Hyodo established an inequality connecting $d_K(M/L)$ and $d_K(L/K)$ [H, 4-1]. We present a version of its proof not involving class field theory.

LEMMA 1.1. *Let K be an arbitrary field, char $K \neq p$, $\zeta_p \in K$, L/K be a cyclic extension of degree p^i, $M = L(\sqrt[p]{a})$, $a \in L$. Denote by σ a generator of $\text{Gal}(L/K)$.*
1. *M/K is abelian if and only if $a^{\sigma-1} \in (L^*)^p$.*
2. *Let $a^{\sigma-1} = b^p$, $b \in L^*$. Then M/K is cyclic if and only if $N_{L/K}b \neq 1$ (i.e., $N_{L/K}b$ is a primitive p th root of unity).*

PROOF. The first statement is evident. Denote by $\tilde{\sigma}$ any lifting of σ to the group $\text{Gal}(M/K)$, and let $x^p = a$. Then $x^{\tilde{\sigma}-1} = b \cdot \zeta_p^k$; $x^{\tilde{\sigma}^{p^i}-1} = N_{L/K}b$, and the second statement follows. $\square$

PROPOSITION 1.4. *Under the assumptions made at the beginning of the subsection, the following inequality holds*:

$$d_K(M/L) \geqslant \min\left(\left(p-1+\frac{1}{p}\right)d_K(L/K), \frac{p-1}{p}e_K + \frac{1}{p}d_K(L/K)\right).$$

PROOF. First, consider the case $\operatorname{char} K = p$. By the Artin–Schreier–Witt theory $L = K(z_1)$, $M = K(z_1, z_2)$, where $(z_1^p, z_2^p) = (z_1, z_2) + (b_1, b_2)$ in $W_2(K)$, i.e.,

$$z_1^p = z_1 + b_1, \qquad z_2^p = z_2 + b_2 - \sum_{k=1}^{p-1} \frac{C_p^k}{p} z_1^k b_1^{p-k},$$

where $b_1, b_2 \in K$. Since (b_1, b_2) determines the same extension as

$$(b_1, b_2) +_{W_2(K)} (b^p, 0) -_{W_2(K)} (b, 0) \quad \text{for any } b \in K,$$

one can assume $v_K(b_1) = -\frac{p}{p-1}d$, where $d = d_K(L/K)$. Similarly, one can assume $v_K(b_2) < 0$ and $v_K(b_2) \not\vdots p$ or else $v_K(b_2) = 0$ (see the Remark after Proposition 1.1).

If $d = 0$, then Proposition 1.4 is evident. Let $d > 0$, i.e., L/K is totally ramified. Assume

$$s = \sum_{k=1}^{p-1} \frac{C_p^k}{p} z_1^k b_1^{p-k}.$$

We then have

$$v_K(s) = v_K(z_1 b_1^{p-1}) = \left(p-1+\frac{1}{p}\right) v_K(b_1) = -\left(p-1+\frac{1}{p}\right) \cdot \frac{p}{p-1} d.$$

Let b_L be any element of L of kind $b_2 + b^p - b$, $b \in L$. We shall show that $v_K(b_L) \neq v_K(s)$. Further, if $v_L(b_L) \vdots p$, then this is so because $v_L(s) \not\vdots p$. If $v_K(b_L) \geqslant -\frac{p}{p-1}d$, we have $v_K(b_L) > v_K(s)$. Finally, from

$$v_K(b_2) \not\vdots p, \quad v_L(b_L) \not\vdots p, \quad b_L \equiv b_2 \mod \wp(L), \quad v_K(b_L) < -\frac{p}{p-1}d$$

one can deduce $v_K(b_L) = v_K(b_2) + d$. But then $v_K(b_L) = v_K(s)$ would mean

$$v_K(b_2) = -\frac{p^2}{p-1}d,$$

which is impossible since $v_K(b_2) \not\vdots p$.

Now let us calculate $d_K(M/L)$ using Proposition 1.1. We have

$$M = L(x), \quad x^p - x = a \in L, \quad v_K(a) = -\frac{p}{p-1}d_K(M/L).$$

Furthermore,

$$a = (b_2 - s) + (b^p - b) = (b_2 + b^p - b) - s = b_L - s,$$

$$d_K(M/L) = -\frac{p-1}{p}v_K(a) = -\frac{p-1}{p}\min(v_K(b_L), v_K(s))$$

$$\geq -\frac{p-1}{p}v_K(s) = \left(p - 1 + \frac{1}{p}\right)d.$$

We pass to the case $\operatorname{char} K = 0$. As in the proof of Proposition 1.3, one can assume $\zeta_p \in K$. If $d_K(M/L) = e_K$, then the required assertion is valid automatically. Otherwise, $M = L(\sqrt[p]{a})$, $v_K(a-1) = \frac{p}{p-1}(e_K - d_K(M/L))$. By Lemma 1.1 $a^{\sigma-1} = b^p$, $N_{L/K}b = \zeta_p^i$, $i \nmid p$, $\sigma \in \operatorname{Gal}(L/K)$. By assertion 2 of Proposition 1.3

$$v_K(b-1) \leq \max\left(\frac{1}{p(p-1)}e_K, \frac{1}{p-1}e_K - d\right),$$

where $d = d_K(L/K)$. Hence

$$v_K(a^{\sigma-1} - 1) \leq \max\left(\frac{1}{p-1}e_K, \frac{p}{p-1}e_K - pd\right).$$

On the other hand, by assertion 3 of Proposition 1.3,

$$v_K(a^{\sigma-1} - 1) \geq v_K(a-1) + \frac{1}{p-1}d.$$

Therefore,

$$\frac{p}{p-1}d_K(M/L) = \frac{p}{p-1}e_K - v_K(a-1) \geq \frac{p}{p-1}e_K - v_K(a^{\sigma-1} - 1) + \frac{1}{p-1}d$$

$$\geq \frac{p}{p-1}e_K - \max\left(\frac{1}{p-1}e_K, \frac{p}{p-1}e_K - pd\right) + \frac{1}{p-1}d$$

$$= \min\left(e_K + \frac{1}{p-1}d, \left(p + \frac{1}{p-1}\right)d\right). \quad \square$$

§2. Main theorems

1. In this section K denotes a p-adic field with residue field F of characteristic p. Introduce the following notations:
- $F_0 = \bigcap_{i=1}^{\infty} F^{p^i}$ (the maximal perfect subfield in F);
- k_0: the fraction field of $W(F_0)$, $k_0 \hookrightarrow K$;
- k: the algebraic closure of k_0 in K.

We call k the *constant subfield* of the field K. We say K is standard if $e_k = e_K$.

EXAMPLE. Let K be the completion of the field $k(X)$ under the discrete valuation

$$v\left(\frac{a_n X^n + \cdots + a_0}{b_m X^m + \cdots + b_0}\right) = \min v_k(a_i) - \min v_k(b_j),$$

where k is a number local field. Then k is the constant subfield of K, and K is standard. The field $K(\sqrt{\pi X})$, where π is a prime of k, is not standard.

Note the following obvious facts.

PROPOSITION 2.1. *Let K_0 be an absolutely unramified field, k_0 be its subfield with perfect residue field, k be a finite totally ramified extension of k_0. Then $k \otimes_{k_0} K_0$ is a standard field.*

Conversely, if K is standard, then $K = k \otimes_{k_0} K_0$, where K_0 is an absolutely unramified subfield of K with $\overline{K}_0 = \overline{K}$ (see [FV], Proposition 5-6, Chapter II); k_0 and k are defined above.

LEMMA 2.1. *If l is a finite extension of k, then $[lK:K] = [l:k]$.*

LEMMA 2.2. *K is standard if and only if it contains a prime element which is algebraic over k_0.*

COROLLARY. *Let L/K be a finite extension such that $e(L/K) = 1$. If K is standard, then L is standard.*

We establish one more auxiliary statement.

LEMMA 2.3. *Let the valuation of K be extended onto a field K', $e(K'/K) = 1$ and K'/K is not necessarily finite. If l is a finite extension of k then $e(lK'/lK) = 1$.*

PROOF. This lemma is evident for totally ramified l/k. For unramified l/k, first consider the maximal unramified algebraic subextension in K'/K. □

2. We say that an extension K'/K is *constant* if there exists a finite extension k'/k such that $K' = k'K$ (k' is obviously the constant subfield of K'). It is clear that a constant extension of a standard field is also standard.

THEOREM 2.1. *Let K be a $\mathfrak{p}$-adic field. Then there exists a solvable constant extension L/K such that L is standard.*

Almost as in the case of a local field (see [FV], Chapter II, §4) one proves

LEMMA 2.4. *Let L/K be a finite Galois extension. Then the inertia subgroup in $\mathrm{Gal}(L/K)$ is solvable.*

LEMMA 2.5. *Let K be standard and L/K be cyclic of prime degree. Then there exists a constant solvable extension K'/K such that $e(L'/K') = 1$, where $L' = LK'$.*

PROOF. Let $q = [L:K]$. Replacing K by its constant extension one can assume $\zeta_q \in K$ and $L = K(\sqrt[q]{a})$, $a \in \mathcal{O}_K$. Further, replace K by its totally ramified constant extension of degree divisible by q and assume that $v_K(a) = 0$. If $\overline{a} \notin (\overline{K}^*)^q$, then obviously $e(L/K) = 1$. In the opposite case one can suppose $a \in U_{1,K}$ and $q = p = \mathrm{char}\,\overline{K}$.

STRUCTURE THEOREMS FOR COMPLETE FIELDS

Fix $\alpha \in \mathbb{Q}$, $\alpha > 0$, and assume $v_K(a-1) > \alpha e_K$.

Note that for sufficiently large N, the field k_{N+1}/k_N is totally ramified of degree p. If π_{N+1} is a prime in k_{N+1} then by the first assertion of Proposition 1.3

$$v_k(N_{N+1/N}\pi_{N+1} - \pi_{N+1}^p) > d_k(k_{N+1}/k_N).$$

Since for large N we have $d_k(k_{N+1}/k_N) > 0$, Proposition 1.4 implies

$$d_k(k_{N+1}/k_N) \xrightarrow[N\to\infty]{} e_k.$$

Choose N so large that k_{N+1}/k_N is totally ramified of degree p and

(1) $\quad v_k(N_{N+1/N}\pi_{N+1} - \pi_{N+1}^p) > e_k - (p-1)\alpha e_k + p v_k(\pi_{N+1}).$

Replacing K by its constant extension, one can assume $k = k_N$. Denote by a_0 any element of $k^* \cdot (K^*)^p$ nearest to a and put $a = a_0 \cdot a_1$.

To begin with, let $v_K(a_1 - 1) \geq \frac{p}{p-1}e_k$. Let $K' = K(\sqrt[p]{a_0})$; then $L' = LK' = K'(\sqrt[p]{a_1})$. It is clear that K'/K is constant and $e(L'/K') = 1$ by item 4 of Proposition 1.1.

Now let $v_K(a_1 - 1) = j < \frac{p}{p-1}e_k$ and $a_1 = 1 + \pi^j b$, where π is a prime in K. It follows from the choice of a_0 that $\overline{b} \notin F_0$ and if $j \vdots p$, then we even have $\overline{b} \notin F^p$.

Let $s = \max\{i : \overline{b} \in F^{p^i}\}$. Proceed by induction on s.

Let $s = 0$. Assume $\widetilde{K} = K(\sqrt[p]{a_0})$. Let K' be any totally ramified constant extension of $\widetilde{K}$ of degree p. Then in K' we have $a \equiv \widetilde{a} \mod (K'^*)^p$,

$$\widetilde{a} \equiv 1 + \pi'^{pej}b \mod \mathfrak{M}_{K'}^{pej+1},$$

where π' is a prime in K', algebraic over k, $e = e(k(\sqrt[p]{a_0})/k)$. By item 4 of Proposition 1.2, K'/K is the desired extension.

Now let the required assertion be valid for $s < s_0$ and $\max\{i : \overline{b} \in F^{p^i}\} = s_0 > 0$. Then $j \not\vdots p$. Denote by a_1' any element of K nearest to a_1 that is of the form

$$\left(1 + \sum_{i \geq j} \pi^i d_i^p\right) u^p, \quad \text{where } d_i \in \mathcal{O}_K,\ \pi \text{ is a prime in } k,\ u \in 1 + \mathfrak{M}_K.$$

(It is not difficult to see that a_1' always exists.) Then $a_1 = a_1' \cdot a_1''$, where $a_1'' = 1$, or $a_1'' = 1 + \pi^r c$, $\overline{c} \notin F^p$.

It is clear that s does not depend on the choice of π, hence one can assume $\pi = N_{N+1/N}\pi_{N+1}$. Then $\pi = \pi_{N+1}^p + f$; $v_K(f) > e_K - ((p-1)j - p)v_K(\pi_{N+1})$ in view of (1). Thus

$$a_1 = a_1' \cdot a_1'' = \left(1 + \sum_{i \geq j}(\pi_{N+1}^p + f)^i d_i^p\right) \cdot a_1'' = \left(1 + \sum_{i \geq j}\pi_{N+1}^{pi} d_i^p\right) \cdot (1 + B) \cdot a_1'',$$

where $d_i \in \mathcal{O}_K$, $v_k(B) > e_k + j \cdot v_k(\pi_{N+1})$. Hence

$$a_1 = \left(1 + \sum_{i \geq j} \pi_{N+1}^i d_i\right)^p \cdot \tilde{a}_1,$$

where

$$\tilde{a}_1 = (1 - p\pi_{N+1}^j d_j + \ldots) \cdot (1 + B) \cdot a_1''.$$

Let $\tilde{a} = a_0 \cdot \tilde{a}_1 \in k_{N+1}K$. Then $\tilde{a}$ differs from a by a pth power and it is easy to see that the value of s computed for $\tilde{a}$, is less than s_0. Indeed,

$$\tilde{a} = a_0 \cdot (1 + \theta\pi_{N+1}^{e+j} d_j + \ldots) \cdot (1 + B) \cdot (1 + \pi_{N+1}^{pr} c + \ldots),$$

where $e = e_{k_{N+1}}$, $\theta \in \mathcal{O}_k^*$, $d_j, c \in \mathcal{O}_K$ (the last factor may be absent). If $e + j < pr \leq \infty$, then $s = \max\{i : \overline{d}_j \in F^{p^i}\} = s_0 - 1$. And if $e + j > pr$, then $s = \max\{i : \overline{c} \in F^{p^i}\} = 0$. The equality $e + j = pr$ is impossible because $e \vdots p$ and $j \not\vdots p$.

Finally, $v_K(\tilde{a} - 1) \geq v_K(a - 1) > \alpha e_K$ and, therefore, one can apply the assumption of induction. Since α is arbitrary, the proof is complete. □

COROLLARY. *Suppose that for a field K there exists a constant solvable extension which is a standard field. Let L/K be a finite extension. Then there exists a constant solvable extension K'/K such that K' is standard and $e(L'/K') = 1$ for $L' = LK'$.*

PROOF. Let K_2/K_0 be some finite extension of a p-adic field, K_1 be an intermediate field. It is clear that if the Corollary is valid for K_2/K_0, then it is valid for K_1/K_0. Besides, by Lemma 2.3, if the Corollary is valid for K_1/K_0 and for K_2/K_1, then it is valid for K_2/K_0. Hence one can assume that L/K is Galois and the inertia subfield coincides with K (since the Corollary is obvious for unramified extensions). Such an extension is solvable and it remains to use Lemma 2.5. □

PROOF OF THEOREM 2.1. Let K_0 be an arbitrary absolutely unramified subfield of K with $\overline{K}_0 = F$ (see [FV], Proposition 5–6, Chapter II). Then K_0 is standard, K/K_0 is finite and Theorem 2.1 follows immediately from the last Corollary. □

3. Now we can describe another way of embedding a p-adic field into a standard one.

THEOREM 2.2. *Let K be a p-adic field. Then there exists an unramified extension L/K and a natural number d such that $L(\zeta_d)$ is a standard field.*

PROOF. In view of Theorem 2.1, the assertion can be reduced to the following one. If K/K' is cyclic of prime degree, $L(\zeta_d)$ is standard, where L/K is unramified, $d \in \mathbb{N}$, then there exists an unramified extension L'/K' and $d' \in \mathbb{N}$ such that $L'(\zeta_{d'})$ is standard.

Let $M = L(\zeta_d)$, $M' = L_0(\zeta_d)$, where L_0/K' is the maximal unramified subextension in L/K'. If M' is standard, then there is nothing to prove. In the opposite case, M/M' is cyclic of prime degree q, $e(M/M') = 1$, $e(m/m') = q$, where m, m' are the constant subfields of M and M' respectively.

If M/M' is unramified, then L' can be taken to be the maximal unramified subextension in M/K', $d' = d$. Otherwise $q = p$ and the residue field extension for M/M' is inseparable. Increasing d, we can assume that $d \vdots p$.

By Proposition 1.1,

$$\text{either } M = M'(\sqrt[q]{a}), \ \overline{a} \notin (\overline{M}')^p, \text{ or } M = M'(\sqrt[p]{1 + \pi^i u}),$$

where π is a prime of M', $i \vdots p$, $i < \frac{p}{p-1}e'_M$, $\overline{u} \notin (\overline{M}')^p$. We also have $M = M'(\sqrt[q]{c})$, $c \in m'$, so the first case is impossible. It then follows that $c \equiv 1 + \pi^i u \mod (M'^*)^p$. Having taken N sufficiently large, we get $c \equiv c_N$ mod $(m'^*_N)^p$, where $v_{M'}(c_N - 1) > i$. Then $1 + \pi^i u \equiv c_N \mod (M'^*_N)^p$, which is possible only if $\overline{u} \in (\overline{M'_N})^p$. Thus, $\overline{M'_N} \neq \overline{M'}$, $\overline{M'_N} = \overline{M}_N$ (since $\overline{M'} \subset \overline{M'_N} \subset \overline{M}_N = \overline{M}$, $[\overline{M}:\overline{M'}] = p$). Therefore, M'_N is standard, and one takes L_0 for L' and l.c.m.(d, p^N) for d'. $\square$

§3. Γ-extensions and cyclic p-extensions

1. We start with some general remarks on Γ-extensions, i.e., extensions of fields with Galois group $\mathbb{Z}_p$, where p is a fixed prime integer.

Let K be an arbitrary field, p be prime, char $K \neq p$. For a Galois extension L/K denote by $X^{(p)}(L/K)$ the group of all p-characters of its Galois group, i.e., $\text{Hom}_{\text{cont}}(\text{Gal}(L/K), \mathbb{Q}_p/\mathbb{Z}_p)$, where $\mathbb{Q}_p/\mathbb{Z}_p$ is discrete. There exists a one-to-one correspondence between subgroups of $X^{(p)}(K) = X^{(p)}(K^{\text{sep}}/K) = X^{(p)}(K^{\text{ab}}/K)$ and abelian p-extensions of K. Γ-extensions correspond to subgroups isomorphic to $\mathbb{Q}_p/\mathbb{Z}_p$. For a Γ-extension L/K, denote by $\chi_i(L/K)$ an arbitrary element of $X^{(p)}(L/K)$ of order p^i, assuming $p\chi_{i+1}(L/K) = \chi_i(L/K)$, $i = 1, 2, \ldots$.

The restriction $\text{Gal}(L) \to \text{Gal}(K)$ determines a mapping $i_{L/K}: X^{(p)}(K) \to X^{(p)}(L)$.

When $\zeta_p \in K$, we denote by $M_{\infty,K}$ the set of all $a \in K^*$ such that $a \in (K_\Gamma^*)^p$ for some Γ-extension K_Γ/K.

LEMMA 3.1. $M_{\infty,K}$ is a subgroup of K^*.

PROOF. The Kummer theory establishes an isomorphism

$$K^*/(K^*)^p \xrightarrow{\sim} {}_pX^{(p)}(K^{\text{sep}}/K).$$

The image of $M_{\infty,K}/(K^*)^p$ is a subgroup, namely, the intersection of the subgroup ${}_pX^{(p)}(K^{\text{sep}}/K)$ and the maximal divisible subgroup in $X^{(p)}(K^{\text{sep}}/K)$. $\square$

Let L/K be a finite solvable extension, $\zeta_p \in L$. We shall use the homomorphism $\eta = \eta_{L/K}: \text{Gal}(L/K) \to (\mathbb{Z}/p\mathbb{Z})^*$ determined by the condition $\zeta_p^g = \zeta_p^{\eta(g)}$ for all $g \in \text{Gal}(L/K)$.

LEMMA 3.2. Let L_Γ/L be a Γ-extension; $a \in L$, $a \in (L_\Gamma^*)^p$, $a \notin (L^*)^p$. Then the following conditions are equivalent:
  (i) $L_\Gamma = K_\Gamma L$ for some Γ-extension K_Γ/K;

(ii) $a^{g-\eta(g)} \in (L^*)^p$ for all $g \in \mathrm{Gal}(L/K)$;
(iii) $\chi(L_\Gamma/L) \in i_{L/K}(X^{(p)}(K))$.

PROOF. Let (i) be satisfied. Then L_Γ/K is a Galois extension and the subgroup $\mathrm{Gal}(L_\Gamma/L)$ lies in the center of $\mathrm{Gal}(L_\Gamma/K)$. This means that for a suitable $h \in \mathrm{Gal}(L_\Gamma/L)$ and any $g \in \mathrm{Gal}(L_\Gamma/K)$, $x \in L_\Gamma$, $x^p = a$, we have

$$x^g \zeta_p^{\eta(g)} = (x\zeta_p)^g = x^{hg} = x^{gh},$$

whence $x^{g-\eta(g)} \in L^*$ and $a^{g-\eta(g)} \in (L^*)^p$.

Exactly in the same way, from the fact that $a^{g-\eta(g)} \in (L^*)^p$ for all $g \in \mathrm{Gal}(L/K)$ it follows that L_Γ/K is a Galois extension and $\mathrm{Gal}(L_\Gamma/L)$ lies in the center of $\mathrm{Gal}(L_\Gamma/K)$. Proceeding by induction on the length of the composition row for $\mathrm{Gal}(L/K)$, one sees that in L_Γ/K there exists a subextension K_Γ/K with Galois group $\mathbb{Z}_p$. Then $K_\Gamma L/L$ is a Γ-extension lying in L_Γ/L. Hence $L_\Gamma = K_\Gamma L$.

Now the equivalence (i) $\iff$ (iii) can be easily deduced. $\square$

Let K be a $\mathfrak{p}$-adic field with residue field of characteristic p; $e = e_K$; $\zeta_p = K$.

LEMMA 3.3. *If* $u \in 1 + \mathfrak{p}_K(\frac{p}{p-1}e)$, *then there exists an unramified Γ-extension* K_Γ/K *such that* $u \in (K_\Gamma^*)^p$. *In particular,* $1 + \mathfrak{p}_K(\frac{p}{p-1}e) \subset M_{\infty,K}$.

PROOF. $K(\sqrt[p]{u})/K$ is unramified. It follows from the Artin–Schreier–Witt theory that the residue field extension $\overline{K(\sqrt[p]{u})}/\overline{K}$ can be extended to a Γ-extension $F/\overline{K}$. Let K_Γ/K be the corresponding unramified extension of K. Then $K(\sqrt[p]{u}) \subset K_\Gamma$ and $u \in M_{\infty,K}$. $\square$

2. Let K and k have the same meaning as in §2. Using Theorem 2.1, we shall prove a theorem generalizing the result of Miki in [M].

LEMMA 3.4. *Let* l/k *be a finite tamely ramified Galois extension and* $G = \mathrm{Gal}(l/k)$. *Let* $L = lK$. *Then the homomorphism*

$$H^1(G, \overline{l}/\wp(\overline{l})) \to H^1(G, \overline{L}/\wp(\overline{L}))$$

induced by the embedding $\overline{l}/\wp(\overline{l}) \to \overline{L}/\wp(\overline{L})$ *is an isomorphism.* (*Here* $\wp$ *is the Artin–Schreier homomorphism.*)

PROOF. It is clear that G can be identified with $\mathrm{Gal}(L/K)$. Let G_0 (G_0') be the inertia subgroup of $\mathrm{Gal}(l/k)$ (respectively of $\mathrm{Gal}(L/K)$). Since Card G_0 is prime to p, $H^i(G_0, \overline{l}) = 0$ and $H^i(G_0', \overline{L}) = 0$, $i \geq 1$. The G/G_0-module $\overline{l}$, just as the G/G_0'-module $\overline{L}$, is cohomologically trivial. Hence $H^i(G, \overline{l}) = H^i(G, \overline{L}) = 0$, $i \geq 1$. The assertion of Lemma 3.4 is now deduced from the following diagrams with exact rows:

$$\begin{array}{ccccccccc} 0 & \longrightarrow & \wp(\overline{l}) & \longrightarrow & \overline{l} & \longrightarrow & \overline{l}/\wp(\overline{l}) & \longrightarrow & 0 \\ & & \downarrow & & \downarrow & & \downarrow & & \\ 0 & \longrightarrow & \wp(\overline{L}) & \longrightarrow & \overline{L} & \longrightarrow & \overline{L}/\wp(\overline{L}) & \longrightarrow & 0 \end{array}$$

$$
\begin{array}{ccccccccc}
0 & \longrightarrow & \mathbb{Z}/p\mathbb{Z} & \longrightarrow & \bar{l} & \longrightarrow & \wp(\bar{l}) & \longrightarrow & 0 \\
& & {\scriptstyle \text{id}}\downarrow & & \downarrow & & \downarrow & & \\
0 & \longrightarrow & \mathbb{Z}/p\mathbb{Z} & \longrightarrow & \bar{L} & \longrightarrow & \wp(\bar{L}) & \longrightarrow & 0 \quad \square
\end{array}
$$

THEOREM 3.1. *Let K_Γ/K be a Γ-extension. Then there exists a Γ-extension k_c/k and an unramified Γ-extension K_u/K such that $K_\Gamma \subset k_c K_u$.*

PROOF. A. Suppose at first that K is standard. This means that the pair (K, k) satisfies conditions (i)–(iii) from Miki's paper [M], and thus the Theorem is valid for K.

B. Now let there exist a finite solvable *tamely ramified* extension l/k such that for $L = lK$ the assertion of the Theorem is valid, and $\zeta_p \in L$. Denote $G = \text{Gal}(l/k) = \text{Gal}(L/K)$.

Let $L_\Gamma = lK_\Gamma$. Then $\text{Gal}(L_\Gamma/L) \hookrightarrow \text{Gal}(K_\Gamma/K)$, hence L_Γ/L is a Γ-extension. By our assumption, $L_\Gamma = l_c L_u$, where l_c/l is a Γ-extension, L_u/L is an unramified Γ-extension. Denote by $L_\Gamma^{(1)}$, $l_c^{(1)}$, $L_u^{(1)}$ the subextensions of degree p in L_Γ/L, l_c/l, L_u/L respectively; then

$$l_c^{(1)} = l(\sqrt[p]{c}), \quad c \in l, \quad L_u^{(1)} = L(\sqrt[p]{u}), \quad u \in 1 + \mathfrak{p}_L(\frac{p}{p-1}e_L).$$

By the Kummer theory, we can also assume

$$L_\Gamma^{(1)} = L\left(\sqrt[p]{c^\alpha u^{-\beta}}\right),$$

where $\chi_1(L_\Gamma/L) = \alpha\chi_1(l_c L/L) - \beta\chi_1(L_u/L)$, $0 \leq \alpha, \beta \leq p-1$. Furthermore, we can assume that $L_\Gamma^{(1)}/L$ is not unramified. (Otherwise increase L.) By Lemma 3.2 $(c^\alpha u^{-\beta})^g \equiv (c^\alpha u^{-\beta})^{\eta(g)} \mod (L^*)^p$ for all $g \in G$. One can see that

(1) $\qquad c^{\alpha(g-\eta(g))} \equiv c_g \mod (l^*)^p, \qquad u^{\beta(g-\eta(g))} \equiv c_g \mod (L^*)^p,$

where $c_g \in l \cap 1 + \mathfrak{p}_L(\frac{p}{p-1}e_L)$.

Let $u^\beta = 1 + (\zeta_p - 1)^p b$. Then

$$u^{\beta(g-\eta(g))} \equiv 1 + \eta(g)(\zeta_p - 1)^p (b^g - b) \mod \mathfrak{p}_L(\frac{p}{p-1}e_L + 1).$$

It follows that $\overline{\eta(g)^{-1}(\zeta_p-1)^{-p}(c_g-1)} = \bar{b}^g - \bar{b} + x^p - x$ for some $x \in \bar{L}$. Therefore, the mapping

$$G \to \bar{l}/\mathfrak{p}(\bar{l}), \qquad g \mapsto \overline{\eta(g)^{-1}(\zeta_p-1)^{-p}(c_g-1)}$$

turns out to be a crossed homomorphism, whose image in $H^1(G, \bar{L}/\mathfrak{p}(\bar{L}))$ is 0. Then, by Lemma 3.4, its image in $H^1(G, \bar{l}/\mathfrak{p}(\bar{l}))$ also vanishes. This in fact means $c_g \equiv u_c^{g-\eta(g)} \mod (l^*)^p$, where $u_c \in 1 + \mathfrak{p}_l(\frac{p}{p-1}e_L)$. Let $u_c \notin (l^*)^p$.

By Lemma 3.3 there exists an unramified Γ-extension l_u/l such that $u_c \in (l_u^*)^p$. Notice that for an appropriate choice of $\hat{\alpha}, \hat{\beta} \in \mathbb{Z}_p$, $\hat{\alpha} \equiv \alpha \mod p$, $\hat{\beta} \equiv \beta \mod p$ we have

$$\chi_i(L_\Gamma/L) = \hat{\alpha}\chi_i(l_c/l) - \hat{\beta}\chi_i(L_u/L)$$
$$= (\hat{\alpha}\chi_i(l_c/l) - \chi_i(l_u/l)) - (\hat{\beta}\chi_i(L_u/L) - \chi_i(l_u/l)) = \chi_{i,c} - \chi_{i,u},$$

and $\chi_{i,c} = p\chi_{i+1,c}$, $\chi_{i,u} = p\chi_{i+1,u}$, $i = 1, 2, \ldots$. Choose the minimal m such that $\chi_{m,u} \neq 0$. Then $\chi_{i+m-1,u} = \chi_i(L_u'/L)$, where L_u'/L is an unramified Γ-extension, since $l_u' \subset l_u L_u$. Exactly in the same way we get $\chi_{i,c} = \chi_i(l_c'/l)$, where l_c'/l is some Γ-extension. In fact, $\chi_{1,c} \neq 0$ since $L_\Gamma^{(1)}/L$ is not unramified. By Lemma 3.2, there exists a Γ-extension k_c/k such that $k_c l = l_c'$. Therefore, $\chi_1(L_u'/L) = \chi_{m,c} - \chi_m(L_\Gamma/L) \in i_{L/K}(X(K))$, and there exists a Γ-extension K_u/K with $K_u L = L_u'$. The inertia subfield in K_u/K cannot be of finite degree over K, hence K_u/K is unramified. Finally, $K_\Gamma \subset l_c' L_u' = k_c K_u L$ easily implies $K_\Gamma \subset k_c K_u$.

The case $u_c \in (l^*)^p$ can be treated similarly with $\chi_i(l_u/l)$ formally made equal to zero.

C. Now we pass to the general case. In view of Theorem 2.1, one finds a finite solvable extension l/k such that for $L = lK$ the assertion of the Theorem is valid. Now we can assume that $\zeta_p \in k$, and that $g = \text{Gal}(L/K)$ coincides with the ramification subgroup.

As well as in the previous case, we get the congruences (1), but now $\eta(g) = 1$ for all $g \in G$ (since G is a p-group), and

$$u^g \equiv u \mod \mathfrak{p}_L\left(\frac{p}{p-1}e_L + 1\right).$$

This implies $c_g \in (l^*)^p$, and we repeat the preceding argument with $u_c = 1$. □

3. Now we present some auxiliary results to be used in §4.

For a field F, $\zeta_p \in F$, and a positive integer j, we shall denote by $M_{j,F}$ the group $(N_{F_j/F} F_j^*)(F^*)^p$. Miki [M] showed that the p-extensions of F can be described by means of this group. Namely, the following result holds.

PROPOSITION 3.1. *Let* $\zeta_p \in F$, $j \geq 1$, $K = F(x) \neq K$, $x^p = a \in F$. *The following two assertions are equivalent*:
(i) *there exists an extension* L/F, $\text{Gal}(L/F) \approx \mathbb{Z}/p^j\mathbb{Z}$, $K \subset L$;
(ii) $a \in M_{j,F}$.

Denote $M_{(\infty),F} = \bigcap_{j=1}^\infty M_{j,F}$.

COROLLARY. $M_{\infty,F} \subset M_{(\infty),F}$.

PROPOSITION 3.2. *Let* K *be a* $\mathfrak{p}$-*adic field*, $\text{char}\,\overline{K} = p$, $\zeta_p \in K$, k *be the constant subfield. Then* $M_{(\infty),K} \subset (1 + \widetilde{\mathfrak{p}}_K) \cdot k^* \cdot (K^*)^p$, *where* $\widetilde{\mathfrak{p}}_K = \mathfrak{p}_K(\frac{p}{p-1}e_K)$.

PROOF. Suppose at first that K is standard. Let π be a prime in k. Denote by Λ_i the set of all elements

$$1 + \sum_{\beta=1}^\infty a_\beta \pi^\beta,$$

where

$$a_\beta \in \mathcal{O}_K^{p^{i-\alpha}} \text{ for } \beta < (1-p^{-\alpha-1})\frac{p}{p-1}e_K, \alpha = 0, 1, \ldots, i-1;$$

$$a_\beta \in \mathcal{O}_K \quad \text{for } \beta \geq (1-p^{-i})\frac{p}{p-1}e_K.$$

It is easy to show that Λ_i is a subgroup in K^* independent of the choice of π.

Proposition 4 in [M] implies that for some h $N_{K_j/K}U_{1,K_j} \subset \Lambda_{j-h}$ for any $j \geq h$. This assertion can also be proved immediately by using Propositions 1.3 and 1.4.

Suppose there exists an $a \in M_{(\infty),K}$, $a \notin (1+\tilde{\mathfrak{p}}_K) \cdot k^* \cdot (K^*)^p$. Choose $c \in k^*$, $d \subset K^*$ in such a way that $v_K(acd^p - 1)$ has its maximal possible value. This might be done since we always have $v_K(acd^p - 1) < \frac{p}{p-1}e_K$ by the choice of a. Let

$$acd^p = 1 + \gamma\pi^r, \qquad \gamma \in U_K.$$

By the choice of c, we have $\overline{\gamma} \notin \bigcap_{i=1}^\infty \overline{K}^{p^i}$, whence $\overline{\gamma} \notin \overline{K}^{p^l}$ for some l.

Furthermore, let g denote the entier of $\log_p(\frac{p}{p-1}e_K)$. One has $a \in M_{(\infty),K}$, hence $a \in M_{g+h+l,K}$. Let

$$a \equiv N_{K_{g+h+l}/K}b \mod (K^*)^p.$$

Since K is standard, $K_{g+h+l}^* = K^*k_{g+h+l}^*U_{1,K_{g+h+l}}$ and one can assume without loss of generality that $b \in U_{1,K_{g+h+l}}$. We get $a_0 = N_{K_{g+h+l}/K}b \in \Lambda_{g+l}$. By the definitions of g and Λ_{g+l},

$$a_0 \equiv 1 + \sum_{i=1}^{\frac{p}{p-1}e_K - 1} u_i^p \pi^i \mod \tilde{\mathfrak{p}}_K,$$

$u_i \in \mathcal{O}_K^{p^{g+l-\alpha-1}}$ for $i < (1-p^{-\alpha-1})\frac{p}{p-1}e_K$, $\alpha = 0, 1, \ldots, g+l-1$ (notice $(1-p^{-g-l}) \cdot \frac{p}{p-1}e_K > \frac{p}{p-1}e_K - 1$ by the definition of g). One can assume that $u_i \in U_k$ or $u_i = 0$. If $v(a_0 - 1) \vdots p$. Put

$$a_1 = a_0 \cdot \left(1 + \sum_{i=1}^{\frac{1}{p-1}e_K - 1} u_{pi}\pi^i\right)^{-p}.$$

It is easy to see that $a_1 \in \Lambda_{g+l}$, and $v(a_1 - 1) > v(a_0 - 1)$. Then, if

$$j = v(a_0 - 1) \not\vdots p, \quad a_0 \equiv c_0 \mod \mathfrak{p}_K(j+1), \quad c_0 \in k,$$

consider $a_1 = c_0^{-1}a_0 \in \Lambda_{g+l}$, $v(a_1 - 1) > v(a_0 - 1)$. Continuing in a similar way and taking into account the inequalities

$$v(a_0 - 1) < v(a_1 - 1) < \ldots \leq r = v(acd^p - 1),$$

we conclude that

$$a_s = ac'd'^p \in \Lambda_{g+l}, \qquad c' \in k^*, \quad d' \in K^*,$$

$$r' = v(a_s - 1) \not{\mid} p, \qquad \overline{\pi^{-r'}(a_s - 1)} \notin \overline{k} = \bigcap_{i=1}^{\infty} \overline{K}^{p^i}.$$

Clearly for $\widetilde{c} \in k^*$, $\widetilde{d} \in K^*$ we have $v(a_s\widetilde{c}\widetilde{d}^p - 1) \leqslant v(a_s - 1)$, whence

$$r = v(acd^p - 1) \leqslant v(a_s - 1), \qquad v(acd^p - 1) = v(ac'd'^p - 1).$$

This is possible only when

$$a_s = ac'd'^p \equiv acd^p = 1 + \gamma\pi^r \mod \mathfrak{p}_K(r+1).$$

The last congruence contradicts the conditions $a_s \in \Lambda_{g+l}$ and $\overline{\gamma} \notin \overline{K}^{p^l}$.

Now suppose the field K is not standard. By Theorem 2.1, there exists a finite k'/k such that $k'K$ is standard. Let $a \in M_{(\infty),K} \subset M_{(\infty),k'K}$ (the embedding is a consequence of Proposition 3.1). Then we clearly have $a \in (1 + \widetilde{\mathfrak{p}}_{k'K}) \cdot (k')^* \cdot ((k'K)^*)^p$. Therefore, the extension $k'K(\sqrt[p]{a})/k'K$ can be embedded into the compositum of a constant extension and an unramified one; the same is true of $k'K(\sqrt[p]{a})/K$ and of $K(\sqrt[p]{a})/K$. By the Kummer theory, $a = a_c a_u$, $K(\sqrt[p]{a_c})/K$ being constant, $K(\sqrt[p]{a_u})/K$ being unramified. Then $a_c \in k^* \cdot (K^*)^p$, $a_u \in (1 + \widetilde{\mathfrak{p}}_K) \cdot (K^*)^p$, which completes the proof. □

§4. Multidimensional complete fields

1. In this section K denotes an n-dimensional complete field of characteristic 0, the last residue field of which is perfect of characteristic p (see [MZ]). Then for some m, $1 \leqslant m \leqslant n$, the $(m-1)$th residue field of K is of characteristic 0 and the mth residue field is of characteristic p. The letter m will always have this meaning in the sequel.

PROPOSITION 4.1. *Let k be a complete field of characteristic 0 with perfect residue field of characteristic p. Then $k\{\{T_1\}\}\ldots\{\{T_{n-1}\}\}$ is a standard field with constant subfield k.*

This is evident.

PROPOSITION 4.2. *Let $m = 1$, K be standard, k be the constant subfield of K, and $t_1, \ldots, t_n$ be local parameters of K. Then there exists an isomorphism of n-dimensional complete fields $\varphi\colon k\{\{T_1\}\}\ldots\{\{T_{n-1}\}\} \xrightarrow{\sim} K$ such that $\varphi(T_i) = t_i$, $i = 1, \ldots, n-1$.*

PROOF. Let $a \in K_0 = k\{\{T_1\}\}\ldots\{\{T_{n-1}\}\}$,

$$(1) \qquad a = \sum_{\overline{r} \in I} \theta_{\overline{r}} T_1^{r^{(1)}} \ldots T_{n-1}^{r^{(n-1)}} \pi^{r^{(n)}},$$

where $I \in \mathbb{Z}^n$ is an admissible set, $\theta_{\bar{r}} \in \mathcal{R}_k$, π is a prime in k. Assume

(2) $$f(a) = \sum_{\bar{r} \in I} \theta_{\bar{r}} t_1^{r^{(1)}} \ldots t_{n-1}^{r^{(n-1)}} \pi^{r^{(n)}};$$

here the sum is well defined. (In fact, this homomorphism arises when we define the Parshin topology on K; see [MZ, Prop. 1.1].)

It is immediately seen that $f|_k = \text{id}$. It follows that if a_1 and a_2 are expanded into finite sums of the form (1), then

$$f(a_1 + a_2) = f(a_1) + f(a_2), \qquad f(a_1 a_2) = f(a_1) f(a_2).$$

The sequential continuity of addition and multiplication implies that f is a ring homomorphism. Since any element of K is expandable into a series in $t_1, \ldots, t_{n-1}, \pi$ with coefficients from $\mathcal{R}_k \cup \{0\}$, f is surjective. Finally, it follows immediately from (1) and (2) that f preserves the valuation of rank n. □

COROLLARY. K is a finite extension of its subfield of the form

$$k\{\{t_1\}\} \ldots \{\{t_{n-m}\}\}((t_{n-m+2})) \ldots ((t_n)),$$

where $t_1, \ldots, t_{n-m}, t_{n-m+2}, \ldots, t_n$ are the corresponding parameters of K and k is a field with perfect residue field. Besides that, K possesses a finite solvable extension of the form K'/K, where

$$K' = k'\{\{t_1'\}\} \ldots \{\{t_{n-m}'\}\}((t_{n-m+2}')) \ldots ((t_n')),$$

k' being a finite extension of k, and $K' = k'K$.

PROOF. If F is a complete field under a discrete valuation, π is a prime, $\text{char } F = \text{char } \overline{F}$, then there exists an isomorphism $\overline{F}((X)) \to F$ of valuated fields which maps X to π. This is known and can be easily verified.

Therefore, one can assume $m = 1$. Denote by K_0 any absolutely unramified subfield of K with $\overline{K}_0 = \overline{K}$. Then by Proposition 4.2 one can assume that $K_0 = k\{\{t_1\}\} \ldots \{\{t_{n-1}\}\}$, where k is the constant subfield of K_0, $t_1, \ldots, t_{n-1}$ are the first $n-1$ local parameters of K_0 (and K). The second part of the Corollary follows from Theorem 2.1 and Proposition 4.2. □

2. Now K is an n-dimensional local field of characteristic 0, i.e., we suppose in addition that the last residue field of K is finite. We shall give a complete description of the Γ-extensions of K.

It is known that for a discretely valuated field F there exists a unique homomorphism

$$\partial_F \colon K_n^M F \to K_{n-1}^M \overline{F}$$

such that $\partial_F(\{a_1, \ldots, a_{n-1}, \pi\}) = \{\overline{a_1}, \ldots, \overline{a_{n-1}}\}$ for any $a_1, \ldots, a_{n-1} \in F$, $\pi \in F$, $v(\pi) = 1$ (see [FV, Ch. IX]).

For a multidimensional field K, it can be shown that ∂_F is continuous and thus determines the homomorphism

$$\partial_K \colon K_n^{\text{top}} K \to K_{n-1}^{\text{top}} \overline{K}.$$

It is known (see [F] for char $K = p$) that the diagram

$$\begin{array}{ccc} K_n^{\text{top}} & \xrightarrow{\partial_K} & K_{n-1}^{\text{top}}\overline{K} \\ {\scriptstyle \theta_K}\downarrow & & \downarrow{\scriptstyle \theta_{\overline{K}}} \\ \text{Gal}(K^{\text{ab}}/K) & \longrightarrow & \text{Gal}(\overline{K}^{\text{ab}}/\overline{K}) \end{array}$$

commutes, where θ_K and $\theta_{\overline{K}}$ are the reciprocity homomorphisms for K and $\overline{K}$ respectively.

By k we shall denote the algebraic closure of $\mathbb{Q}_p$ in K. In particular, for $m = 1$, k is the constant subfield of K. Fix local parameters $t_1, \ldots, t_n$ of K and consider the homomorphism $\alpha\colon k^* \to K_n^{\text{top}}K$, given by $\alpha(a) = \{t_1 \ldots, t_{n-m}, a, t_{n-m+2}, \ldots, t_n\}$. Let

$$\theta_k\colon k^* \to \text{Gal}(k^{\text{ab}}/k), \qquad \theta_K\colon K_n^{\text{top}}K \to \text{Gal}(K^{\text{ab}}/K)$$

be the reciprocity homomorphisms for k and K respectively.

PROPOSITION 4.3. *Let* $\varkappa\colon \text{Gal}(K^{\text{ab}}/K) \to \text{Gal}(k^{\text{ab}}K/K) \approx \text{Gal}(k^{\text{ab}}/k)$ *be the natural projection. Then* $\theta_k = \varkappa \circ \theta_K \circ \alpha$.

PROOF. Applying $n - 1$ times the above diagram, we conclude that the element

$$\theta_K(\{t_1, \ldots, t_{n-m}, a, t_{n-m+2}, \ldots, t_n\})$$

acts on $k^{\text{ur}}K/K$ as the $v_k(a)$th degree of Frobenius automorphism. Besides that, if l/k is a finite abelian extension and $a \in N_{L/K}l^*$, then

$$\{t_1, \ldots, t_{n-m}, a, t_{n-m+2}, \ldots, t_n\} \in N_{lK/K}K_n^{\text{top}}(lK),$$

whence $\varkappa \circ \theta_K \circ \alpha(a)\,|_{l/k} = \text{id}$.

Thus, the homomorphism $\varkappa \circ \theta_K \circ \alpha\colon k^* \to \text{Gal}(k^{\text{ab}}/k)$ coincides with θ_k on unramified extensions and is maps $N_{l/k}l^*$ into $\text{Gal}(k^{\text{ab}}/l)$ for any l. Therefore, $\varkappa \circ \theta_K \circ \alpha = \theta_k$. □

COROLLARY. *Let* $\zeta_p \in K$. *Then* $k^* \subset M_{j,K}$ *if and only if* $K = K_j$ (*i.e.,* $\zeta_{p^j} \in K$).

PROOF. Let $\zeta_{p^j} \notin K$. We can assume $K = K_{j-1}$ (decrease j otherwise). Then $\zeta_{p^j} \notin k$ and one finds $a \in k$ such that $\theta_{k_j/k}(a) \neq 1$. Then, by Proposition 4.3,

$$\theta_{K_j/K}(\{t_1 \ldots, t_{n-m}, a, t_{n-m+2}, \ldots, t_n\}) \neq 1.$$

Hence $a \notin (N_{K_j/K}K_j^*) \cdot (K^*)^p$. The inverse assertion is evident. □

THEOREM 4.1. *Let K, k, m be as above. Denote*

$$\widetilde{\mathfrak{p}}_K = \mathfrak{p}_K \left(\frac{p}{p-1} e_k^{(n-m+1)}, \underbrace{0, \ldots, 0}_{m-1} \right).$$

(*For the last notation see* [MZ]).

1. *Let K_Γ/K be a Γ-extension. Then there exists a Γ-extension k_c/k and a semiramified Γ-extension K_u/K such that $K_\Gamma \subset k_c K_u$.*
2. *Let $\zeta_p \in K$. Then $M_{\infty,K} = M_{(\infty),K} = (1 + \widetilde{\mathfrak{p}}_K) \cdot M_{\infty,k} \cdot (K^*)^p$.*
3. *Let $\zeta_p \in K$. Then $M_{\infty,K}$ is a subgroup of index p in $(1 + \widetilde{\mathfrak{p}}_K) \cdot k^* \cdot (K^*)^p$.*

PROOF. Use induction on m and begin with the case $m = 1$. Then the first assertion is a particular case of Theorem 3.1. Let $\zeta_p \in K$. By the Kummer theory, we conclude $M_{\infty,K} \subset (1 + \widetilde{\mathfrak{p}}_K) \cdot M_{\infty,K} \cdot (K^*)^p$. The inverse inclusion follows from Lemma 3.3.

If K is one-dimensional, then the relation $M_{\infty,K} = M_{(\infty),K}$ follows from class field theory. In the general case, $M_{(\infty),K} \subset (1 + \widetilde{\mathfrak{p}}_K) \cdot k^* \cdot (K^*)^p$ by Proposition 3.2; hence

$$(1 + \widetilde{\mathfrak{p}}_K) \cdot M_{\infty,K} \cdot (K^*)^p = M_{\infty,K} \subset M_{(\infty),K} \subset (1 + \widetilde{\mathfrak{p}}_K) \cdot k^* \cdot (K^*)^p,$$

and, since $M_{\infty,k}$ is of index p in k^*, the assumption $M_{\infty,K} \neq M_{(\infty),K}$ would imply $M_{(\infty),K} = (1 + \widetilde{\mathfrak{p}}_K) \cdot k^* \cdot (K^*)^p$. This is impossible since $k^* \not\subset M_{(\infty),K}$ by the Corollary to Proposition 4.3.

Now let $m > 1$. Then $K = E((X))$ for an $(n-1)$-dimensional local field E. Let $\zeta_p \in E$. It is easy to see that $1 + \mathfrak{p}_K(1) \subset (K^*)^p \subset M_{(\infty),K}$, and for $v_K^{(n)} \not| \, p$ one has $a \notin M_{(\infty),K}$. Thus $M_{(\infty),K} = M_{(\infty),E} \cdot (K^*)^p$. On the other hand, by the assumption of induction,

$$M_{(\infty),E} \cdot (K^*)^p = M_{\infty,E} \cdot (K^*)^p \subset M_{\infty,K} \subset M_{(\infty),K},$$

whence

$$M_{\infty,K} = M_{(\infty),K} = (1 + \widetilde{\mathfrak{p}}_E) \cdot M_{\infty,k} \cdot (K^*)^p = (1 + \widetilde{\mathfrak{p}}_K) \cdot M_{\infty,k} \cdot (K^*)^p.$$

Assertions 1 and 3 are now obvious for K.

Finally, let $\zeta_p \notin E$ and K_Γ/K be a Γ-extension. Then $K_\Gamma(\zeta_p) = \widetilde{E}_\Gamma K$, where $\widetilde{E}_\Gamma/E(\zeta_p)$ is a Γ-extension. Taking into account Lemma 3.2, we see that $\widetilde{E}_\Gamma = E_\Gamma(\zeta_p)$, E_Γ/E is a Γ-extension, and we can apply the assumption of induction to E_Γ/E. □

COROLLARY. *Let K be a multidimensional local field of characteristic 0.*

1. *Let L/K be an extension with Galois group $\mathbb{Z}/p\mathbb{Z}$; suppose L/K can be extended to an extension with Galois group $\mathbb{Z}/p^i\mathbb{Z}$ for any i. Then L/K can be extended to a Γ-extension.*
2. *The subgroup of all divisible elements in $X^{(p)}(K^{\mathrm{sep}}/K)$ is a divisible group.*

PROOF. In the case $\zeta_p \in K$ the first assertion means exactly that $M_{\infty,K} = M_{(\infty),K}$. Let $\zeta_p \notin K$. Then $L(\zeta_p)/K(\zeta_p)$ can be extended to a Γ-extension $\widetilde{K}_\Gamma/K(\zeta_p)$, and by Lemma 3.2 $\widetilde{K}_\Gamma = K_\Gamma(\zeta_p)$ for a Γ-extension K_Γ/K.

The second assertion is a reformulation of the first one in the language of characters. □

References

[MZ] A. I. Madunts, I. B. Zhukov, *Multidimensional complete fields: topology and other basic constructions*, this volume.

[F] I. B. Fesenko, *On class field theory of multidimensional local fields of positive characteristic*, Advances in Soviet Math., vol. 4, Amer. Math. Soc., Providence, RI, 1991, pp. 103–127.

[FV] I. B. Fesenko and S. V. Vostokov, *Local fields and their extensions. A constructive approach*, Amer. Math. Soc., Providence, RI, 1993.

[M] H. Miki, *On $\mathbb{Z}_p$-extensions of complete p-adic power series fields and function fields*, J. Fac. Sci. Univ. Tokyo, Sect. 1A **21** (1974), no. 3, 377–393.

[H] O. Hyodo, *Wild ramification in the imperfect residue field case*, Adv. Studies in Pure Math. **12** (1987), 287–314.

[P] A. N. Parshin, *Abelian coverings of arithmetic schemes*, Dokl. Akad. Nauk SSSR **243** (1978), no. 4, 855–858; English transl., Soviet Math. Dokl. **19** (1978), no. 6, 1438–1442.

Translated by THE AUTHOR

DEPARTMENT OF MATHEMATICS AND MECHANICS, ST. PETERSBURG UNIVERSITY, BIBLIOTECHNAYA PL. 2, 198904 ST. PETERSBURG, RUSSIA

Unit Fractions

O. IZHBOLDIN AND L. KURLIANDCHIK

One of the most ancient problems in mathematics is the representation of the unit in the form

(1) $$1 = \frac{1}{a_1} + \frac{1}{a_2} + \cdots + \frac{1}{a_n}$$

with $a_1 < a_2 < \cdots < a_n$, $a_i \in \mathbb{N}$.

In our opinion, the following three theorems are rather interesting.

THEOREM 1. *Let the sequence $r_1, r_2, r_3, \ldots$ be determined by the conditions $r_1 = 2$, $r_{n+1} = r_1 \cdot r_2 \cdot \ldots \cdot r_n + 1$, so that $r_2 = 3$, $r_3 = 7$, $r_4 = 43, \ldots$. If among all the sequences $(a_1, a_2, \ldots, a_n)$ of natural numbers satisfying conditions (1) we choose the one for which the value of a_n is maximal, then this maximal value equals $r_n - 1$.*

THEOREM 2. *Let the sequence $(a_1, a_2, \ldots, a_n)$ satisfy condition (1). Then this sequence is not an arithmetic progression.*

THEOREM 3. *From any increasing infinite arithmetic progression it is possible to choose some numbers satisfying conditions (1).*

PROOF OF THEOREM 1. This result is due to Curtiss [1], but his proof is not that simple. In this paper we give a much simpler proof.

LEMMA 1. *Let $x_1, x_2, \ldots, x_n$ be n real numbers satisfying the conditions*

(2)
$$x_1 \geq x_2 \geq \cdots \geq x_n \geq 0$$
$$x_1 + x_2 + \cdots + x_n = 1$$
$$x_1 \cdot x_2 \cdot \ldots \cdot x_k \leq x_{k+1} + x_{k+2} + \cdots + x_n \quad (k = 1, 2, \ldots n-1).$$

Then $x_n \geq \dfrac{1}{r_n - 1}$.

1991 *Mathematics Subject Classification.* Primary 11B25.
Key words and phrases. Unit fraction.

© 1995, American Mathematical Society

Theorem 1 can be easily obtained from this lemma. Indeed, it is obvious that $x_i = 1/a_i$ satisfy the conditions (2). Hence $a_n \leq r_n - 1$, and the maximal value of a_n equals $r_n - 1$ since

$$\frac{1}{r_1} + \frac{1}{r_2} + \cdots + \frac{1}{r_{n-1}} + \frac{1}{r_n - 1} = 1.$$

PROOF OF LEMMA 1. Let $y_1, y_2, \ldots, y_n$ be n real numbers satisfying conditions (2) with minimal product $y_1 \cdot y_2 \cdot \ldots \cdot y_n$. It is not difficult to see that for this sequence we have

(3)
$$y_1 > y_2 > \cdots > y_n > 0, \qquad y_1 + y_2 + \cdots + y_n = 1$$
$$y_1 \cdot y_2 \cdot \ldots \cdot y_k = y_{k+1} + y_{k+2} + \cdots + y_n \quad (k = 1, 2, \ldots n-1).$$

Indeed, if some of the conditions (3) are not true, then we can choose $\varepsilon > 0$ so that the numbers

$$y_1, y_2, \ldots, y_{i-1}, y_i + \varepsilon, y_{i+1}, \ldots, y_{k-1}, y_k - \varepsilon, y_{k+1}, \ldots, y_n$$

satisfy conditions (2). On the other hand

$$y_1 \cdot y_2 \cdot \ldots \cdot y_{i-1} \cdot (y_i + \varepsilon) \cdot y_{i+1} \cdot \ldots \cdot y_{k-1} \cdot (y_k - \varepsilon) \cdot y_{k+1} \cdot \ldots \cdot y_n < y_1 \cdot y_2 \cdot \ldots \cdot y_n$$

and this contradicts the minimality of the product $y_1 \cdot y_2 \cdot \ldots \cdot y_n$.

From conditions (3) we get

$$y_1 = \frac{1}{2}, \quad y_2 = \frac{1}{3}, \ldots, y_{n-1} = \frac{1}{r_{n-1}}, \quad y_n = \frac{1}{r_n - 1}.$$

Finally we conclude that

$$x_n^2 \geq x_1 \cdot x_2 \cdot \ldots \cdot x_n \geq y_1 \cdot y_2 \cdot \ldots \cdot y_n = y_n^2.$$

Hence, $x_n \geq y_n = 1/r_n - 1$. This completes the proof of the lemma and, as stated above, also that of Theorem 1.

REMARK. In this proof we do not use the fact that a_n is an integer and so we may conclude that the sum

$$\frac{1}{r_1} + \frac{1}{r_2} + \cdots + \frac{1}{r_{n-1}}$$

is the closest strict approximation of 1 from below by a sum of $n-1$ unit fractions.

PROOF OF THEOREM 2. It is well known that the sum

$$\frac{1}{n+1} + \frac{1}{n+2} + \cdots + \frac{1}{n+k}$$

is not an integer for any natural numbers n, k. This simple problem is special case of Theorem 2. In this paper we shall prove Theorem 2 in the general case.

Our proof is by *reductio ad absurdum*. Let the natural numbers a, d, n satisfy

(4)
$$\sum_{i=0}^{n} \frac{1}{a + id} = 1.$$

Lemma 2. *The number d is even.*

Proof. Suppose that d is odd. From the denominators of (4) choose the one divisible by the maximal power of 2. There are at least two such denominators. Let $p \cdot 2^k$ and $q \cdot 2^k$ two of them with $p < q$, where p, q are odd. Since $q \cdot 2^k - p \cdot 2^k$ is divisible by d, we see that $q - p$ is also divisible by d and hence $(p + d) \cdot 2^k$ is one of the denominators. But it divisible by a greater power of 2, namely 2^{k+1}, contradiction.

Now we may assume that $a \geq 3$. Indeed, if $a = 2$, then equation (4) can be rewritten in the form

$$\sum_{i=0}^{n-1} \frac{1}{(1 + \frac{d}{2}) + \frac{d}{2} \cdot i} = 1$$

and $1 + d/2 \geq 3$, since for $a = 2, d = 2$ the sum is not equal to 1.

Lemma 3. $\dfrac{a + nd}{n} \leq \dfrac{3}{2} \sum_{i=1}^{n} \dfrac{1}{i}.$

Proof. We have

$$1 = \sum_{i=0}^{n} \frac{1}{a + id} \leq \frac{1}{3} + \sum_{i=1}^{n} \frac{1}{a + id} \leq \frac{1}{3} + \sum_{i=1}^{n} \frac{n}{(a + nd)i} = \frac{1}{3} + \frac{n}{a + nd} \sum_{i=1}^{n} \frac{1}{i}.$$

Corollary. $d < \dfrac{3}{2} \sum_{i=1}^{n} \dfrac{1}{i}.$

Denote by p the greatest prime less then n. Using Bertrand's postulate, we see that $n/2 < p \leq n$.

Lemma 4. $d < p$.

Proof. We consider three cases.
Case 1. $n \geq 9$. We have

$$\frac{d}{p} < \frac{2}{n} d < \frac{2}{n} \cdot \frac{3}{2} \cdot \sum_{i=1}^{n} \frac{1}{i} = 3 \cdot \frac{1}{n} \cdot \sum_{i=1}^{n} \frac{1}{i} \leq 3 \cdot \frac{1}{9} \cdot \sum_{i=1}^{9} \frac{1}{i} = 0.94 \cdots < 1.$$

Case 2. $5 \leq n \leq 8$. In this case $p \geq 5$, and we have

$$d < \frac{3}{2} \sum_{i=1}^{n} \frac{1}{i} \leq \frac{3}{2} \sum_{i=1}^{8} \frac{1}{i} = 4.07 \cdots < p.$$

Case 3. $n \leq 4$. We have

$$1 = \sum_{i=0}^{n} \frac{1}{a + id} \leq \sum_{i=0}^{4} \frac{1}{3 + 2i} = 0.87 \cdots < 1.$$

This is a contradiction.

LEMMA 5. *Among the denominators in equation* (4) *there are exactly two divisible by* p.

PROOF. Since $p \leq n$ and $(d, p) = 1$, we conclude that there are at least two denominators divisible by p, because otherwise after the sum will be reduced to a common denominator we shall obtain a fraction with the denominator divisible by p and the numerator nondivisible by p. Let the denominators mentioned above be $a + id$, $a + jd$ ($i < j$). It is clear that $j - i$ is divisible by p, and since $j - i \leq n < 2p$ we get $j - i = p$. Thus there are exactly two denominators divisible by p: $a + (j - p)d$, and $a + jd$.

LEMMA 6. $2a + 2nd - pd \geq 2p^2$.

PROOF. Since $(a + (j - p)d)(a + jd)$ is divisible by p^2 and

$$\frac{1}{a + (j - p)d} + \frac{1}{a + jd} = \frac{2a + 2jd - pd}{(a + (j - p)d)(a + jd)},$$

we see that $2a + 2jd - pd$ is also divisible by p^2. Since d is even and $p \neq 2$, we have $2a + 2jd - pd \:\vdots\: 2p^2$. Therefore $2a + 2nd - pd \geq 2p^2$.

LEMMA 7. $a < n$.

PROOF. If $a \geq n$, then

$$1 = \sum_{i=0}^{n} \frac{1}{a + id} \leq \sum_{i=0}^{n} \frac{1}{n + 2i} = \frac{1}{2} \sum_{i=0}^{n} \left(\frac{1}{n + 2i} + \frac{1}{n + 2(n - i)} \right)$$

$$= 2n \cdot \sum_{i=0}^{n} \frac{1}{(n + 2i)(n + 2(n - i))} \leq 2n \cdot (n + 1) \cdot \frac{1}{n} \cdot \frac{1}{3n} = \frac{2(n + 1)}{3n} < 1,$$

a contradiction.

LEMMA 8. $p \leq \frac{2}{3}n$.

PROOF. Since $2a + 2nd - pd \geq 2p^2$, $a \leq n - 1$, $d \leq p - 1$, we get $2(n - 1) + (2n - p)(p - 1) \geq 2p^2$. Hence $2n + 1 \geq 3p + \frac{2}{p}$. Therefore $3p \leq 2n$.

LEMMA 9. $n \leq 22$.

PROOF. We have

$$\frac{3}{2} \sum_{i=1}^{n} \frac{1}{i} \geq \frac{a + nd}{n} = \frac{(2a + 2nd - pd) + pd}{2n} \geq \frac{2p^2 + pd}{2n} > \frac{p^2}{n} = \frac{(\frac{n}{2})^2}{n} = \frac{n}{4}.$$

Hence

$$\frac{1}{n} \sum_{i=1}^{n} \frac{1}{i} > \frac{1}{6}.$$

If $n \geq 23$, then

$$\frac{1}{n} \sum_{i=1}^{n} \frac{1}{i} \leq \frac{1}{23} \sum_{i=1}^{23} \frac{1}{i} < \frac{1}{6}.$$

This is a contradiction.

If $n \leq 22$, it is easy to verify that $p > \frac{2}{3}n$. (Reminder: p is the largest prime less than n). This completes proof of Theorem 2.

PROOF OF THEOREM 3. A no less ancient problem in mathematics is the representation of rationals in the form

$$\sum_{i=1}^{n} \frac{1}{a_i} \quad \text{with } a_1 < a_2 < \cdots < a_n, \ a_i \in \mathbb{N}.$$

Perhaps the first result on the subject was due to Leonardo Pisano (Fibonacci) in 1202. He proved that any positive rational can be expressed as a finite sum of distinct unit fractions. Stewart and Breusch in 1954 proved that it is always possible to represent $a/(2b+1)$ as a sum of distinct unit fractions of the form $1/(2m+1)$. Graham in 1963 proved that any simplified fraction s/t can be expressed as a sum of distinct unit fractions of the form $1/(an+b)$ if and only if

$$\left(\frac{t}{(t,(a,b))}, \frac{a}{(a,b)} \right) = 1.$$

Graham's result and Theorem 3 are both corollaries of the following fact.

PROPOSITION. *Let a be a natural number. Any natural number b can be expressed as a sum of distinct unit fractions from the set*

$$\left\{ \frac{1}{a+1}, \frac{1}{2a+1}, \frac{1}{3a+1}, \ldots \right\}.$$

Graham's proof [2] is very complicated. The proof given in this paper is substantially simpler.

We put $p = a+1$, $q = 2a+1$. We shall use the following notation:

$$A_n = \left\{ \sum_{0 \leq \alpha, \beta \leq n} \varepsilon_{\alpha,\beta} p^\alpha q^\beta \ \bigg| \ \varepsilon_{\alpha,\beta} \in \{0, 1\} \right\},$$

$$B_n = \{ x \mid x \in A_n, \ x \equiv 1 \pmod{apq} \},$$

$$B_\infty = \bigcup_{n=0}^{\infty} B_n,$$

$$n(x) = \min\{ n \mid x \in B_n \}, \quad \text{where } x \in B_\infty.$$

Using this notation, we see that every element x of B_∞ can be written as

$$x = \sum_i p^{\alpha_i} q^{\beta_i}, \quad 0 \leq \alpha_i, \beta_i \leq n(x), \quad (\alpha_i, \beta_i) \neq (\alpha_j, \beta_j) \text{ for } i \neq j.$$

Hence

$$\frac{1}{(pq)^{n(x)}} = \sum_i \frac{1}{x p^{n(x)-\alpha_i} q^{n(x)-\beta_i}}.$$

It is easy to verify that all denominators of the sum

$$\sum_{x \in B_\infty} \sum_i \frac{1}{x p^{n(x)-\alpha_i} q^{n(x)-\beta_i}}$$

are distinct integers $\equiv 1 \pmod{a}$. Consequently it is sufficient to prove that any natural b can be expressed as the sum

$$b = \sum_{i=1}^{k} \frac{1}{(pq)^{n(x_i)}}, \quad \text{with } x_1 < x_2 < \cdots < x_k, \ x_i \in B_\infty.$$

Hence it is sufficient to show that

$$\sum_{x \in B_\infty} \frac{1}{(pq)^{n(x)}} = \infty.$$

Since

$$\sum_{x \in B_\infty} \frac{1}{(pq)^{n(x)}} = \sum_{x \in B_0} \frac{1}{(pq)^{n(x)}} + \sum_{x \in B_1 \setminus B_0} \frac{1}{(pq)^{n(x)}} + \cdots + \sum_{x \in B_n \setminus B_{n-1}} \frac{1}{(pq)^{n(x)}} + \cdots$$

$$= \frac{|B_0|}{(pq)^0} + \frac{|B_1| - |B_0|}{(pq)^1} + \cdots + \frac{|B_n| - |B_{n-1}|}{(pq)^n} + \cdots = \sum_{i=0}^{\infty} \frac{|B_n|}{(pq)^n} \left(1 - \frac{1}{pq}\right),$$

it suffices it to prove the following statement.

STATEMENT. $\varliminf \dfrac{|B_n|}{(pq)^n} > 0.$

The proof is not too difficult, but complicated. We shall deduce this statement from several lemmas.

LEMMA 10. $|B_n| \geq |A_{n-2a}|$.

PROOF. Consider the function

$$f(x) = x \cdot p^{2a} \cdot q^{2a} + p^a \cdot q^a \cdot (1 + p + p^2 + \cdots + p^t) + 1,$$

with $0 \leq t < a$, $t \equiv -1 - x \pmod{a}$. This function is injective because

$$x = \left[\frac{f(x)}{p^{2a} \cdot q^{2a}}\right].$$

If $x \in A_{n-2a}$, then $f(x) \in A_n$, and since $f(x) \equiv x + (t+1) + 1 \pmod{a}$ and $f(x) \equiv 1 \pmod{pq}$, we conclude that $f(x) \in B_n$. Hence $f(A_{n-2a}) \subset B_n$. Finally, have $|B_n| \geq |A_{n-2a}|$.

COROLLARY. $\varliminf \dfrac{|B_n|}{(pq)^n} \geq \dfrac{1}{(pq)^{2a}} \varliminf \dfrac{|A_n|}{(pq)^n}$.

Let $M = \{m_1, m_2, \ldots, m_n\}$ be an arbitrary set of natural numbers. Denote by $S(M)$ the following set

$$S(M) = \left\{ \sum_{i=1}^n \varepsilon_i m_i \ \Big| \ \varepsilon_i \in \{0, 1\} \right\}.$$

It is clear that $A_n = S(\{p^\alpha q^\beta \mid 0 \leq \alpha, \beta \leq n\})$.

LEMMA 11. *Let M be a set of n natural numbers $m_1 < m_2 < \cdots < m_n$, such that $m_{i+1} < 2m_i$ ($i = 1, 2, \ldots, n-1$). Then*

$$|S(M)| \geq \frac{2m_n}{m_1}.$$

PROOF. It is easy to verify by induction over n that

$$[0, 2m_n) \subset [0, m_1) + S(M) \quad (\text{where } A + B = \{x + y \mid x \in A, \ y \in B\}).$$

Now evidently

$$|S(M)| \geq \frac{2m_n}{m_1}.$$

LEMMA 12. *Let u, v be natural numbers such that $q^u < p^v < 2q^u$, $u > a$, $n > \max(u, v)$.*
Let $M_n = \{p^\alpha q^\beta \mid 0 \leq \alpha, \beta \leq n\} \cap [q^u, p^{n-v}q^u]$. Then

$$|S(M_n)| \geq 2p^{n-v}q^{n-u}.$$

PROOF. It is sufficient to realize the evident fact that the set M_n is as in Lemma 2.

Finally we have

$$\varliminf \frac{|B_n|}{(pq)^n} \geq \frac{1}{(pq)^{2a}} \varliminf \frac{|A_n|}{(pq)^n} \geq \frac{1}{(pq)^{2a}} \varliminf \frac{|S(M_n)|}{(pq)^n}$$
$$\geq \frac{1}{(pq)^{2a}} \varliminf \frac{2p^{n-v}q^{n-u}}{(pq)^n} = \frac{1}{(pq)^{2a}} \cdot \frac{2}{p^v \cdot q^u} > 0.$$

This completes proof of Theorem 3.

Appendix

We have considered three problems connected with sums of unit fractions. However there are many other problems devoted to the subject. The reader can find some of them in [3]. In conclusion we would like to pose two problems which are, in our opinion, of considerable interest.

PROBLEM. *Let $a_1 < a_2 < \cdots < a_n$ be natural numbers such that*

$$\frac{1}{a_1} + \frac{1}{a_2} + \cdots + \frac{1}{a_n} = 1.$$

Is it true that $a_{k+1} - a_k \geq 3$ for some k?

PROBLEM. *Let A be any set of natural numbers. Does A or $\mathbb{N} \setminus A$ always contain some integers such that the sum of their reciprocals equals 1?*

References

1. D. R. Curtiss, *On Kellog's diophante problem*, Amer. Math. Monthly **29** (1922), 380–387.
2. R. L. Graham, *On finite sums of unit fractions*, Proc. London Math. Soc **14** (1964), 193–207.
3. P. Erdős and R. L. Graham, *Old and new problems and results in combinatorial number theory*, Univ. de Geneve, Geneva, 1980.

Translated by THE AUTHORS

St. Petersburg State University, Department of Mathematics and Mechanics, 198904, St. Petersburg, Russia

Collections of Multiple Sums

D. V. FOMIN AND O. T. IZHBOLDIN

ABSTRACT. Is it possible to recover a collection of n numbers given the collection of its k-sums? This question is investigated from the algebraic and combinatorial point of view. The problem is almost completely solved if $k \leq 4$.

§1. Introduction

This article is devoted to one specific arithmetic/combinatorial question, which appears to be a very interesting and deep problem in algebra. As far as we know, it was originally posed as contest question in the following form:

PROBLEM A. *Let A be a collection of n real numbers $a_1, a_2, \ldots, a_n$ (repetitions allowed). Consider the collection $A^{(2)}$ of $\binom{n}{2}$ 2-sums of collection A, i.e., the collection of all sums of the form*

$$a_{i_1} + a_{i_2}, \quad \text{where } i_1 < i_2.$$

Is it possible to restore the collection A given the collection $A^{(2)}$?

This problem was solved completely in 1987 and some years later our attention was attracted to its evident generalization:

PROBLEM B. *Consider any positive integers n and k such that $n \geq k$ and any n-collection A (that is, $|A| = n$). We shall denote by $A^{(k)}$ the collection of k-sums of collection A, i.e., the collection of all sums of the form*

$$a_{i_1} + a_{i_2} + \cdots + a_{i_k}, \quad \text{where } 1 \leq i_1 < i_2 < \cdots < i_k \leq n.$$

Do there exist two distinct n-collections A and B such that collections $A^{(k)}$ and $B^{(k)}$ coincide?

EXAMPLE. If $A = \{0, 1, 2, 5\}$, $k = 2$, then $A^{(2)} = \{1, 2, 3, 5, 6, 7\}$, $A^{(3)} = \{3, 6, 7, 8\}$.

1991 *Mathematics Subject Classification.* Primary 05E05; Secondary 11D72.
Key words and phrases. Multiple sums, symmetric polynomials.

© 1995, American Mathematical Society

DEFINITION. We say that collections A and B are "k-conjugate" if $A^{(k)} = B^{(k)}$, while $A \neq B$. We shall denote this relation by the sign "$\sim$". Furthermore, we shall also call A and B "*singular n-collections*". Also we shall call a pair (n, k) "*singular*" if there exist k-conjugate n-collections.

Structure of the paper. In §2 a complete investigation of the case $k = 2$ will be presented. Thus Problem A is fully solved. It turns out that the initial collection can be always recovered if and only if n is not a power of two.

In §3 a certain strong necessary condition for a singularity is found. To check this condition one has to solve a certain Diophantine equation.

Section 4 is devoted to the resolution of Problem B in cases $k = 3$ and $k = 4$. Here we use the results of §3 and solve the corresponding Diophantine equations.

In §§5, 6 we discuss several interesting questions connected with the basic problems of the subject.

Acknowledgements. We would like to thank Professor Alexander Merkuriev for his contribution to the problem investigated here.

§2. First steps: $k = 2$ and the series $(2k, k)$

The description of singular pairs in the case $k = 2$ is given by the following theorem (proved in 1987).

THEOREM 2.1. *The pair $(n, 2)$ is singular if and only if n is the power of 2.*

PROOF. First we must present an example of 2-conjugate n-collections if $n = 2^m$. We do this by induction on m.

Base. $m = 1$, $n = 2$. $A = \{0, 3\}$, $B = \{1, 2\}$.

Induction step. Assume that $|A_m| = |B_m| = 2^m$ and $A_m \sim B_m$. Choose number d such that collections

$$A_{m+1} = A_m \cup (B_m + d), \qquad B_{m+1} = B_m \cup (A_m + d)$$

do not coincide (here $B_m + d$ or $A_m + d$ denote the collections obtained by translation). Then these new 2^{m+1}-collections A_{m+1} and B_{m+1} are the desired ones. Indeed, for each pair (α_1, α_2) from A_{m+1} we can find the corresponding pair (β_1, β_2) from B_{m+1} with the same sum and this correspondence is one-to-one. In fact,

Case 1. $\alpha_1, \alpha_2 \in A_m$. Then we can take the pair (β_1, β_2) such that $\beta_1, \beta_2 \in B_m$.

Case 2. $\alpha_1, \alpha_2 \in B_m + d$. Then take the pair (β_1, β_2) such that $\beta_1, \beta_2 \in A_m + d$.

Case 3. $\alpha_1 \in A_m$, $\alpha_2 \in B_m + d$. Then we can take the pair $(\alpha_1 + d, \alpha_2 - d)$; $\alpha_1 + d \in A_m + d$, $\alpha_2 - d \in B_m$.

Case 4. $\alpha_1 \in B_m + d$, $\alpha_2 \in A_m$. The situation is absolutely similar to that of Case 3.

EXAMPLE. The 2-collections $\{0, 3\}$ and $\{1, 2\}$ generate the 2-conjugate 4-collections $\{1, 1, 2, 4\}$ and $\{0, 2, 3, 3\}$. The next step gives 8-collections $\{1, 1, 1, 2, 3, 4, 4, 4\}$ and $\{0, 2, 2, 2, 3, 3, 3, 5\}$.

Thus, it remains left to show that if n is not an exact power of 2 then $A^{(2)} = B^{(2)}$ implies $A = B$ for all two n-collections A and B. To prove this, we must verify that standard symmetric polynomials (power sums)

$$s_p(x_1, \ldots, x_n) = x_1^p + \cdots + x_n^p$$

give the same results when calculated for the given collections A and B. In order to check this fact, it suffices to show that $s_p(A)$ can be uniquely determined by the values of power sums of the collection $A^{(2)}$.

First, $s_1(A^{(2)}) = \binom{n-1}{1} s_1(A)$ and $s_1(A) = \binom{n-1}{1}^{-1} s_1(A^{(2)})$. Further, it would be sufficient to prove that for each p

$$s_p(A^{(2)}) = \lambda_p s_p(A) + \langle s_1(A), \ldots, s_{p-1}(A) \rangle,$$

where λ_p is a nonzero constant and $\langle \ldots \rangle$ is an expression depending only on the polynomials $s_1(A), \ldots, s_{p-1}(A)$, whose values have been already determined at the time of the current calculation.

Hence, our goal is to evaluate λ_p. We have

$$s_p(A^{(2)}) = \sum_{i<j}(a_i + a_j)^p = \frac{1}{2}\sum_{i \neq j}(a_i + a_j)^p = \frac{1}{2}\sum_{i \neq j}\sum_{r=0}^{p}\binom{p}{r}a_i^r a_j^{p-r}$$

$$= \frac{1}{2}\sum_{r=0}^{p}\binom{p}{r}\sum_{i \neq j}a_i^r a_j^{p-r} = \frac{1}{2}\sum_{r=0}^{p}\binom{p}{r}[s_r(A)s_{p-r}(A) - s_p(A)]$$

$$= \frac{1}{2}\sum_{r=1}^{p-1}\binom{p}{r}s_r(A)s_{p-r}(A) + (n - 2^{p-1})s_p(A),$$

since $s_0(A) = n$.

Thus, $\lambda_p = n - 2^{p-1} \neq 0$ if n is not an exact power of 2. This completes the proof. □

Two years after the solution of Problem A, in 1989, it was noticed that the phenomenon of conjugation is more interesting than it seemed before. One of the authors of the present article, while considering some properties of collections A_m, B_m constructed above, unexpectedly discovered the series of conjugate collections. More precisely, it turned out that each pair ($2k$, k) is singular and it is quite easy to give an example of two k-conjugate $2k$-collections. To do this it is convenient to introduce the following definition.

DEFINITION. A collection $A = \{a_1, a_2, \ldots, a_n\}$ is said to be *self-conjugate* (for fixed k) if $A \sim A'$, where A' is the collection symmetric to A with respect to its arithmetic mean, i.e.,

$$A' = (2\overline{a} - a_1, 2\overline{a} - a_2, \ldots, 2\overline{a} - a_n),$$

where $\bar{a} = (a_1 + a_2 + \cdots + a_n)/n$. We also say that A is *nontrivial* self-conjugate collection if A is self-conjugate and does not coincide with A'. In other words, such a collection A is not symmetric, while the collection $A^{(k)}$ is symmetric with respect to some point.

Consider now any $2k$-collection A which is not symmetric with respect to any point on the real axis. We claim that A is a nontrivial self-conjugate collection. Indeed, for any k-subcollection

$$\{a_{i_1}, \ldots, a_{i_k}\} \subset A$$

one can point out the k-subcollection

$$\{b_{j_1}, \ldots, b_{j_k}\} \subset A',$$

where the indices $j_1, \ldots, j_k$ form the complement to the set of indices $\{i_1, \ldots, i_k\}$ in $\{1, 2, \ldots, 2k\}$. Obviously

$$b_{j_1} + \cdots + b_{j_k} = (2\bar{a} - a_{j_1}) + \cdots + (2\bar{a} - a_{j_k})$$
$$= 2k\bar{a} - (a_{j_1} + \cdots + a_{j_k}) = a_{i_1} + \cdots + a_{i_k}.$$

So if $n = 2k$, then almost all n-collections are singular.

This remarkable fact encouraged us to fight the generalized Problem B once again, and our first intention was to proceed in the same algebraic way, using the properties of symmetric functions for the cases $k = 3, 4, \ldots$. Some other properties of self-conjugacy are explained in the next section.

§3. Merkuriev–Izhboldin Theorem

Now we consider the quantity

$$s_p(A^{(k)}) = \sum_{i_1 < \cdots < i_k} (a_{i_1} + \cdots + a_{i_k})^p$$

and its expression as a polynomial $R(s_1, \ldots, s_n)$, where $s_1, \ldots, s_n$ are power sums for the initial collection $\{a_1, \ldots, a_n\}$. Since R is of degree p when expressed in terms of a_i and so is s_p, we have the unique representation

$$s_p(A^{(k)}) = \lambda_p s_p + \langle s_1, \ldots, s_{p-1}\rangle \quad (p \leq n),$$

where $\langle \ldots \rangle$ is some polynomial that does not concern us at the moment and λ_p is a number that depends only on k, p and n.

Thus, as in the proof of Theorem 2.1, we must calculate the coefficient λ_p and investigate whether or not it vanishes for some p. The explicit formula for λ_p was found in April, 1991, by Alexander Merkur′ev and proved in April, 1992, by Oleg Izhboldin.

THEOREM 3.1. *We have*

$$\lambda_p = \sum_{i=1}^{k} (-1)^{i-1} i^{p-1} \binom{n}{k-i}.$$

This theorem can be immediately derived from the proof of the following result, which is one we really need for our further investigation.

DEFINITION. Set

$$F_{k,p}(n) = \sum_{i=1}^{k} (-1)^{i-1} i^{p-1} \binom{n}{k-i}.$$

THEOREM 3.2. *If the pair (n, k) is singular, then $F_{k,p}(n) = 0$ for some $p \leq n$.*

PROOF. Let $a_1, a_2, \ldots, a_n$ be the initial collection of numbers. Let us consider the functions

$$f(t) = \sum_{i=1}^{n} e^{a_i t}, \qquad g(x, t) = \prod_{i=1}^{n} \left(1 + xe^{a_i t}\right).$$

Then we can represent these functions as formal series in the variables t and x. First of all,

$$f(t) = \sum_{i=1}^{n} e^{a_i t} = \sum_{i=1}^{n} \sum_{l=0}^{\infty} \frac{1}{l!} (a_i t)^l = \sum_{l=0}^{\infty} \frac{1}{l!} \sum_{i=1}^{n} a_i^l t^l.$$

Setting $\mathfrak{s}_l = \mathfrak{s}_l(a_1, a_2, \ldots, a_n) = \frac{1}{l!} \sum_{i=1}^{n} a_i^l$, we get

$$f(t) = \sum_{l=0}^{\infty} \mathfrak{s}_l t^l.$$

Similar calculations for $g(x, t)$ yield

$$g(x, t) = \prod_{i=1}^{n} \left(1 + xe^{a_i t}\right) = \sum_{k=0}^{n} \sum_{i_1 < \cdots < i_k} e^{(a_{i_1} + \cdots + a_{i_k})t} x^k$$

$$= \sum_{k=0}^{n} \sum_{i_1 < \cdots < i_k} \sum_{l=0}^{\infty} \frac{1}{l!} \left((a_{i_1} + \cdots + a_{i_k})t\right)^l x^k$$

$$= \sum_{k=0}^{n} \sum_{l=0}^{\infty} \frac{1}{l!} \sum_{i_1 < \cdots < i_k} (a_{i_1} + \cdots + a_{i_k})^l x^k t^l;$$

and setting

$$\mathfrak{S}_{k,l} = \mathfrak{S}_{k,l}(a_1, a_2, \ldots, a_n) = \frac{1}{l!} \sum_{i_1 < \cdots < i_k} (a_{i_1} + \cdots + a_{i_k})^l,$$

we obtain

$$g(x, t) = \sum_{k, l \geq 0} \mathfrak{S}_{k,l} x^k t^l.$$

Now we set $\mathfrak{S}_{k,l} = 0$ if $k > n$.

LEMMA 3.3. *We have*

(A) $$g(x, t) = \exp\left(\sum_{k=1}^{\infty}(-1)^{k-1}f(kt)\frac{x^k}{k}\right).$$

PROOF. Indeed,

$$\frac{d}{dx}\ln g(x, t) = \frac{d}{dx}\ln\prod_{i=1}^{n}(1 + xe^{a_i t}) = \sum_{i=1}^{n}\frac{1}{x}\frac{xe^{a_i t}}{1 + xe^{a_i t}}$$

$$= \sum_{i=1}^{n}\frac{1}{x}\sum_{k=1}^{\infty}(-1)^{k-1}(xe^{a_i t})^k = \sum_{k=1}^{\infty}(-x)^{k-1}\sum_{i=1}^{n}e^{ka_i t}$$

$$= \sum_{k=1}^{\infty}(-x)^{k-1}f(kt).$$

Thus we have

$$g(x, t) = \exp\left(\int_0^x \sum_{k=1}^{\infty}(-x)^{k-1}f(kt)\,dx\right) = \exp\left(\sum_{k=1}^{\infty}(-1)^{k-1}f(kt)\frac{x^k}{k}\right).\quad\square$$

COROLLARY 3.4. *We have*

(B) $$\sum_{k,l\geq 0}\mathfrak{S}_{k,l}x^k t^l = (1 + x)^n \exp\left(\sum_{k,l\geq 1}(-1)^{k-1}\mathfrak{s}_l k^{l-1}x^k t^l\right).$$

PROOF. Substituting the values of the functions

$$g(x, t) = \sum_{k,l\geq 0}\mathfrak{S}_{k,l}x^k t^l; \qquad f(t) = \sum_{l=0}^{\infty}\mathfrak{s}_l t^l = n + \sum_{l=1}^{\infty}\mathfrak{s}_l t^l,$$

into formula (A), we obtain

$$\sum_{k,l\geq 0}\mathfrak{S}_{k,l}x^k t^l = \exp\left(\sum_{k=1}^{\infty}(-1)^{k-1}\left(n + \sum_{l=0}^{\infty}\mathfrak{s}_l(kt)^l\right)\frac{x^k}{k}\right)$$

$$= \exp\left(\sum_{k=1}^{\infty}(-1)^{k-1}n\frac{x^k}{k}\right)\cdot\exp\left(\sum_{k=1}^{\infty}(-1)^{k-1}\sum_{l=1}^{\infty}\mathfrak{s}_l(kt)^l\frac{x^k}{k}\right)$$

$$= \left(\exp\left(\sum_{k=1}^{\infty}(-1)^{k-1}\frac{x^k}{k}\right)\right)^n\cdot\exp\left(\sum_{k,l\geq 1}(-1)^{k-1}\mathfrak{s}_l k^{l-1}x^k t^l\right).$$

It remains to note that

$$\exp\left(\sum_{k=1}^{\infty}(-1)^{k-1}\frac{x^k}{k}\right) = \exp(\ln(1 + x)) = 1 + x.\quad\square$$

DEFINITION. Let $\varphi_{k,l}(\mathfrak{s}_1, \mathfrak{s}_2, \ldots)$ denote the coefficient of the term $x^k t^l$ in the expansion of the function
$$(1+x)^n \exp\left(\sum_{k,l \geq 1} (-1)^{k-1} \mathfrak{s}_l k^{l-1} x^k t^l\right).$$

It is quite obvious that $\varphi_{k,l}$ depends only on the variables $\mathfrak{s}_1, \ldots, \mathfrak{s}_l$ and that $\varphi_{k,l}$ can be written in the following way:
$$\varphi_{k,l}(\mathfrak{s}_1, \ldots, \mathfrak{s}_l) = \tilde{\varphi}_{k,l}(\mathfrak{s}_1, \ldots, \mathfrak{s}_{l-1}) + \lambda_{k,l} \cdot \mathfrak{s}_l,$$
where $\lambda_{k,l}$ is some constant depending only on n, k and l. Using this notation, we rewrite formula (B) as
$$(C) \qquad \mathfrak{S}_{k,l} = \varphi_{k,l}(\mathfrak{s}_1, \ldots, \mathfrak{s}_l) = \tilde{\varphi}_{k,l}(\mathfrak{s}_1, \ldots, \mathfrak{s}_{l-1}) + \lambda_{k,l} \cdot \mathfrak{s}_l.$$

LEMMA 3.5. *We have*
$$\lambda_{k,p} = \sum_{i=1}^k (-1)^{i-1} i^{p-1} \binom{n}{k-i} = F_{k,p}(n).$$

PROOF. Taking (C) into account, we get $\lambda_{k,p} = \frac{\partial \varphi_{k,p}}{\partial \mathfrak{s}_p}$, or, informally, $\lambda_{k,p} = \frac{\partial \mathfrak{S}_{k,p}}{\partial \mathfrak{s}_p}$. Evaluating the partial derivative $\frac{\partial}{\partial \mathfrak{s}_p}$ in both sides of (B), we get
$$\sum_{k,l \geq 0} \frac{\partial \mathfrak{S}_{k,l}}{\partial \mathfrak{s}_p} x^k t^l$$
$$= (1+x)^n \left(\sum_{k=1}^\infty (-1)^{k-1} k^{p-1} x^k t^p\right) \cdot \exp\left(\sum_{k,l \geq 1} (-1)^{k-1} \mathfrak{s}_l k^{l-1} x^k t^l\right).$$

Dividing by t^p and substituting $t = 0$, we obtain
$$\sum_{k \geq 0} \frac{\partial \mathfrak{S}_{k,p}}{\partial \mathfrak{s}_p} x^k = (1+x)^n \sum_{k \geq 1} (-1)^{k-1} k^{p-1} x^k.$$

So, using the equality $\lambda_{k,p} = \frac{\partial \mathfrak{S}_{k,p}}{\partial \mathfrak{s}_p}$, we get
$$\sum_{k \geq 0} \lambda_{k,p} x^k = (1+x)^n \sum_{k \geq 1} (-1)^{k-1} k^{p-1} x^k = \sum_{i \geq 1} (-1)^{i-1} i^{p-1} x^i (1+x)^n$$
$$= \sum_{i \geq 1} (-1)^{i-1} i^{p-1} x^i \cdot \sum_{j=0}^n \binom{n}{j} x^j = \sum_{i,j} (-1)^{i-1} i^{p-1} \binom{n}{j} x^{i+j}$$
$$= \sum_{k=0}^n \left(\sum_{i=1}^n (-1)^{i-1} i^{p-1} \binom{n}{k-i}\right) x^k = \sum_{k=0}^n F_{k,p}(n) x^k.$$

Thus we have $\lambda_{k,p} = F_{k,p}(n)$, and the proof of the lemma is complete. □

Now let us fix k and simplify the notation as follows:
$$\mathfrak{S}_p = \mathfrak{S}_{k,p}; \quad \varphi_p = \varphi_{k,p}; \quad \tilde{\varphi}_p = \tilde{\varphi}_{k,p}; \quad \lambda_p = \lambda_{k,p}.$$

LEMMA 3.6. *If all numbers $\lambda_1, \ldots, \lambda_n$ are nonzero, then the collection $\{\mathfrak{S}_1, \ldots, \mathfrak{S}_n\}$ uniquely determines the collection $\{\mathfrak{s}_1, \ldots, \mathfrak{s}_n\}$.*

PROOF. Using the equality $\mathfrak{S}_p = \varphi_p(\mathfrak{s}_1, \ldots, \mathfrak{s}_p) = \varphi_p(\mathfrak{s}_1, \ldots, \mathfrak{s}_{p-1}) + \lambda_p \cdot \mathfrak{s}_p$, we can calculate the numbers $\mathfrak{s}_1, \ldots, \mathfrak{s}_n$ by using formulas

$$\mathfrak{s}_1 = \frac{1}{\lambda_1}\mathfrak{S}_1, \quad \mathfrak{s}_2 = \frac{1}{\lambda_2}\left(\mathfrak{S}_2 - \tilde{\varphi}_2(\mathfrak{s}_1)\right), \quad \mathfrak{s}_3 = \frac{1}{\lambda_3}\left(\mathfrak{S}_3 - \tilde{\varphi}_3(\mathfrak{s}_1, \mathfrak{s}_2)\right), \ldots$$

$$\mathfrak{s}_n = \frac{1}{\lambda_n}\left(\mathfrak{S}_n - \tilde{\varphi}_n(\mathfrak{s}_1, \ldots, \mathfrak{s}_{n-1})\right). \quad \square$$

Now we can easily complete the proof of Theorem 3.2. Suppose that the numbers $\lambda_p = F_{k,p}(n)$ are nonzero. Since, by assumption, the set of k-sums is known, we can calculate the numbers

$$\mathfrak{S}_p = \frac{1}{p!}\sum_{i_1 < \cdots < i_k}(a_{i_1} + \cdots + a_{i_k})^p, \quad p = 1, 2, \ldots, n,$$

and, consequently, we can find the numbers

$$\mathfrak{s}_p = \frac{1}{p!}\sum_{i=1}^n a_i^p, \quad p = 1, 2, \ldots, n,$$

that enable us to determine the collection $\{a_1, a_2, \ldots, a_n\}$. $\quad\square$

§4. Cases $k = 3$ and $k = 4$

Our next goal is to clear up the situation for $k = 3$ and $k = 4$. Let us consider the corresponding polynomials $F_{k,p}(n)$.

$$F_{3,p}(n) = \binom{n}{2} - 2^{p-1}\binom{n}{1} + 3^p = \frac{1}{2}(n^2 - (2^p + 1)n + 2 \cdot 3^{p-1}),$$

$$F_{4,p}(n) = \binom{n}{3} - 2^{p-1}\binom{n}{2} + 3^{p-1}\binom{n}{1} - 4^{p-1}$$

$$= \frac{1}{6}(n^3 - (3 \cdot 2^{p-1} + 3)n^2 + (2 + 3 \cdot 2^{p-1} + 2 \cdot 3^p)n - 6 \cdot 4^{p-1}).$$

PROPOSITION 4.1. *The set of positive integer roots of the polynomials $F_{3,p}(n)$ is $\{1, 2, 3, 6, 27, 486\}$.*

PROOF. First, we check that the roots of polynomials $F_{3,p}(n)$ belong to the set described above. To solve the equation

(D) $$F_{3,p}(n) = \frac{1}{2}(n^2 - (2^p + 1)n + 2 \cdot 3^{p-1}) = 0,$$

we need two lemmas.

LEMMA 4.2. *If $4^p - 1$ is divisible by 3^r, then $3p$ is divisible by 3^r.*

PROOF. Easy induction on r. □

LEMMA 4.3. *If p is a positive integer greater than 12, then*

$$2 + \log_3 p \leq (p-1) \log_3 \frac{3}{2}.$$

PROOF. This is another easy exercise. To verify the inequality, it suffices to check that the function $f(x) = 2 + \log_3 x - (x-1)\log_3(3/2)$ satisfies the conditions $f(13) < 0$ and $f'(x) < 0$ if $x \geq 13$. □

Let us return to equation (D). Let $m = (2^p + 1) - n$. Then we have

$$\begin{cases} m + n = 2^p + 1, \\ mn = 2 \cdot 3^{p-1}. \end{cases}$$

Hence,

$$m = \frac{2 \cdot 3^{p-1}}{n} \geq \frac{2 \cdot 3^{p-1}}{n + m - 1} = \left(\frac{3}{2}\right)^{p-1} \geq 3^{[(p-1)\log_3 \frac{3}{2}]}.$$

Set $r = [(p-1)\log_3(3/2)]$. Then $m \geq 3^r$. Since m is a divisor of $2 \cdot 3^{p-1}$, we that $m = 2 \cdot 3^i$ or $m = 3^i$. In any case, the inequality $m \geq 3^r$ implies that $i \geq r$ and consequently, m is divisible by 3^r. Similarly, n is divisible by 3^r. Therefore,

$$4^p - 1 = (2^p - 1)(2^p + 1) = (m+n)(2^p - 1) \text{ is divisible by } 3^r.$$

Now it follows from Lemma 4.2 that $3p$ is divisible by 3^r, hence $3p \geq 3^r$. Thus,

$$1 + \log_3 p \geq r = \left[(p-1)\log_3 \frac{3}{2}\right] \geq (p-1)\log_3 \frac{3}{2} - 1,$$

and Lemma 4.3 implies that $p \leq 12$. Further, investigating all the values of p from 1 through 12, one can easily verify that the quadratic equation $n^2 - (2^p + 1)n + 3 \cdot 2^{p-1} = 0$ has integer roots if and only if $p = 1, 3, 5, 9$. Indeed,

$$F_{3,1} = \frac{1}{2}(n-1)(n-2), \quad F_{3,2} = \frac{1}{2}(n-2)(n-3),$$

$$F_{3,3} = \frac{1}{2}(n-3)(n-6), \quad F_{3,5} = \frac{1}{2}(n-6)(n-27),$$

$$F_{3,9} = \frac{1}{2}(n-27)(n-486),$$

and we get the required result. □

THEOREM 4.4. *If $k = 3$, then the singular pairs are exactly the following ones*: $(3, 3)$, $(3, 6)$, $(3, 27)$, $(3, 486)$.

PROOF. The only thing that is left to prove is that the pairs $(3, 27)$ and $(3, 486)$ are singular. It is sufficient to present examples of nontrivial self-conjugate collections with 27 and 486 elements respectively. Here they are:

$$n = 27 \quad A_{27} = \{\underbrace{0, \ldots, 0}_{5}, \underbrace{1, \ldots, 1}_{10}, \underbrace{2, \ldots, 2}_{10}, \underbrace{3, 3}_{2}\}$$

$$n = 486 \quad A_{486} = \{\underbrace{0, \ldots, 0}_{22}, \underbrace{1, \ldots, 1}_{176}, \underbrace{2, \ldots, 2}_{231}, \underbrace{3, \ldots, 3}_{56}, \underbrace{4}_{1}\}. \quad \square$$

In the conclusion we prove the following fact.

PROPOSITION 4.5. *The set of positive integer roots of the polynomials $F_{4,p}(n)$ is $\{1, 2, 3, 4, 8, 12\}$.*

PROOF. First, factorize our polynomials for $p \leq 7$

$$F_{4,1} = \frac{1}{6}(n-1)(n-2)(n-3)$$

$$F_{4,2} = \frac{1}{6}(n-2)(n-3)(n-4)$$

$$F_{4,3} = \frac{1}{6}(n-3)(n-4)(n-8)$$

$$F_{4,4} = \frac{1}{6}(n-4)(n^2 - 23n + 96)$$

$$F_{4,5} = \frac{1}{6}(n-8)(n^2 - 43n + 192)$$

$$F_{4,6} = \frac{1}{6}(n-12)(n^2 - 87n + 512)$$

$$F_{4,7} = \frac{1}{6}(n-8)(n^2 - 187n + 3072).$$

Therefore, if $p \leq 7$, the relation $F_{4,p}(n) = 0$ is possible only if $n = 1, 2, 3, 4, 8, 12$. So we can consider only values of p greater than or equal to 8.

LEMMA 4.6. $2 + 3 \cdot 2^{p-1} + 2 \cdot 3^p$ *is not divisible by* 16.

PROOF. One more easy exercise in elementary number theory. $\square$

Rewrite the equality $F_{4,p}(n) = 0$ in the following way

(E) $$n(n - (3 \cdot 2^{p-1} + 3)) + (2 + 3 \cdot 2^{p-1} + 2 \cdot 3^p) = \frac{6 \cdot 4^{p-1}}{n}.$$

Since $2 + 3 \cdot 2^{p-1} + 2 \cdot 3^p$ is not divisible by 16, we see that either n is not divisible by 16 or $6 \cdot 4^{p-1}/n$ is not divisible by 16.

Case 1. n is not divisible by 16. In this case, since $6 \cdot 4^{p-1}$ is divisible by n, n is equal to 2^i or to $3 \cdot 2^i$. Because n is not a multiple of 16, we have $i \leq 3$ and therefore n belongs to the set $\{1, 2, 4, 8, 3, 6, 12, 24\}$. If $n = 6$ or $n = 24$, then (E) cannot be true since

$$n(n - (3 \cdot 2^{p-1} + 3)) + (2 + 3 \cdot 2^{p-1} + 2 \cdot 3^p) \equiv 2 \pmod 3$$

$$\frac{6 \cdot 4^{p-1}}{n} = \frac{6 \cdot 4^{p-1}}{6 \cdot 4^i} = 4^{p-1-i} \equiv 1 \pmod 3.$$

Thus, n can be equal only to $1, 2, 3, 8, 12$.

Case 2. $6 \cdot 4^{p-1}/n$ is not divisible by 16. The number $6 \cdot 4^{p-1}/n$ is a divisor of $6 \cdot 4^{p-1}$ and is therefore equal to 2^i or to $3 \cdot 2^i$. In addition, since $6 \cdot 4^{p-1}/n$ is not divisible by 16, $i \leq 3$ and hence $6 \cdot 4^{p-1}/n \leq 3 \cdot 2^3 = 24$. Thus,

$$n \geq \frac{6 \cdot 4^{p-1}}{24} > \frac{6 \cdot 4^{p-1}}{2^{p-1}} = 6 \cdot 2^{p-1}, \text{ since } p \geq 8.$$

So we have

$$6 \cdot 4^{p-1} = 6 \cdot 2^{p-1}(6 \cdot 2^{p-1} - 5 \cdot 2^{p-1}) < n(n - 5 \cdot 2^{p-1})$$

$$< n(n - (3 \cdot 2^{p-1} + 3)) = \frac{6 \cdot 4^{p-1}}{n} - (3 + 3 \cdot 2^{p-1} + 2 \cdot 3^p) < 6 \cdot 4^{p-1}.$$

Finally, we come to a contradiction, and the proof is complete. □

REMARK. Moreover, it can be easily verified that the polynomials $F_{4,p}(n)$ have integer roots only if $p \leq 7$.

THEOREM 4.7. *The pairs* $(4, 4)$, $(4, 8)$ *are singular. There are no other singular pairs* $(4, n)$ *except, possibly, the pair* $(4, 12)$.

REMARK. Unfortunately, we still do not know whether or not the pair $(4, 12)$ is singular. The authors recently proved that there exist no nontrivial self-conjugate collections containing 12 elements.

§5. Almost all collections are not singular if $n \neq k, 2k$

It seems rather obvious that conjugate collections consist of numbers that are connected by plenty of arithmetic relations; that is, such collections are not of the "general kind". This is confirmed by the following theorem (proved in 1989).

THEOREM 5.1. *Assume that the n-collections A and B are k-conjugate. Suppose, moreover, that the elements of at least one of these collections are linearly independent over $\mathbb{Q}$. Then either $n = k$ or $n = 2k$ and in the latter case A and B are symmetric with respect to some point on the real axis.*

PROOF.

1. Without loss of generality we can suppose that the numbers of the collection B are linearly independent over $\mathbb{Q}$. Let $x, y \in A$. Then there are $\binom{n-2}{k-1}$ pairs (X, Y), such that $X, Y \subset A$, $|X| = |Y| = k$ and $X \triangle Y = \{x, y\}$, $x \in X$, $y \in Y$. (Here $X \triangle Y$ denotes the "symmetric difference" $X \setminus Y \cup Y \setminus X$).

Let us denote by $\sigma(S)$ the sum of numbers in the subset S. Then for any of the mentioned pairs (x, y) we have

$$x - y = \sigma(X) - \sigma(Y) = \sigma(X \setminus Y) - \sigma(Y \setminus X).$$

Now we need one evident auxiliary proposition that will come in handy a little later in this proof.

LEMMA 5.2. *For any pair of subsets $\mathcal{A}$ and $\mathcal{B}$ in the set of numbers that are linearly independent over $\mathbb{Q}$, we can uniquely recover the differences of sets $\mathcal{A}\setminus\mathcal{B}$ and $\mathcal{B}\setminus\mathcal{A}$, known the difference $\sigma(\mathcal{A}) - \sigma(\mathcal{B})$.*

For any X, Y we have the corresponding subsets X', Y' in B with the same sum of numbers, i.e.,

$$x - y = \sigma(X) - \sigma(Y) = \sigma(X') - \sigma(Y') = \sigma(X' \setminus Y') - \sigma(Y' \setminus X').$$

Thus, for every pair (U, V) of k-subsets with $U \setminus V = \{x\}$, $V \setminus U = \{y\}$ (there are $\binom{n-2}{k-1}$ of them) we have

$$x - y = \sigma(U \setminus V) - \sigma(V \setminus U) = \sigma(U' \setminus V') - \sigma(V' \setminus U').$$

Hence, for any such pair

$$U' \setminus V' = \text{const}, \quad V' \setminus U' = \text{const} \quad \text{(see Lemma 5.2).}$$

Suppose that each of these "const"-subsets contains p elements ($p \geq 1$). Thus, in all, there are $\binom{n-2p}{k-p}$ pairs with the indicated differences of the sets. Taking into account the inequality

$$\binom{n-2p}{k-p} < \binom{n-2}{k-1}, \quad \text{if } p > 1,$$

we get $p = 1$.

2. So, we have obtained a correspondence which maps each pair (x, y) in A to the pair (p, q) in B such that

$$x - y = p - q.$$

If we take the pairs (x, y), (y, z), and the dual pairs (p, q), (r, s), then we get

$$x - y = p - q, \quad y - z = r - s, \quad x - z = p + r - q - s.$$

But there exists some pair (u, v) in B that corresponds to the pair (x, z) and we have $x - z = u - v$. Therefore, the pairs (p, q) and (r, s) must have a nonempty intersection.

3. Consider all pairs (x, y_i) $(i = 1, 2, \ldots, n-1)$ and the corresponding dual pairs. The case $n \leq 4$ is obvious and we leave it to the reader. So, if $n > 4$, then we can apply the following simple fact: if $n - 1$ pairs are chosen within a set in such a way that any two of them intersect, then they all must have a common point. Therefore, all dual pairs also have a common element $\bar{x}$ in B. Moreover, this element actually occupies the same place in all those pairs.

Thus, the dual pairs are of the form a) $(\bar{x}, y_i')$ or b) $(y_i', \bar{x})$.

In case a):

$$\sum_i (x - y_i) = \sum_i (\bar{x} - y_i') \implies nx - \sigma(A) = n\bar{x} - \sigma(B) \implies x = \bar{x}$$

(since $\sigma(A) = \sigma(B)$), and therefore $y_i = y'_i$, i.e., $A = B$.

In case b):

$$\sum_i (x - y_i) = \sum_i (y'_i - \bar{x}) \implies nx - \sigma(A) = \sigma(B) - n\bar{x} \implies x + \bar{x} = \frac{2\sigma}{n},$$

$$y_i + y'_i = x + \bar{x} = \frac{2\sigma}{n},$$

(where $\sigma = \sigma(A) = \sigma(B)$) i.e., the collections A and B are symmetric with respect to the point σ/n.

Thus, for $n > k$ the only remaining possibility is that our collections are symmetric to each other. But if $n \neq 2k$, then for each subcollection X in A we have the subcollection $B \setminus \bar{X}$ with the same sum of numbers ($\bar{X}$ here denotes the subcollection of B symmetric to X). This subcollection contains $n - k$ numbers, but the dual subcollection $Y \subset B$ contains k numbers and has the same sum of numbers. Thus, because of the linear independence over $\mathbb{Q}$, $n - k = k$, $n = 2k$. □

COROLLARY 5.3. *In the situation described above (i.e., when at least one of n-collections consists of numbers that are linearly independent over $\mathbb{Q}$) there cannot exist three distinct k-conjugate n-collections.*

§6. Conclusion

Summing up, we would like to point out some unsolved problems that naturally arose during our investigation.

PROBLEM 6.1. *Determine all the solutions of equation $F_{k,p}(n) = 0$. This problem seems to be extremely difficult, so it is rather relevant to formulate two "subproblems":*

(6.1.1) *For fixed $k > 2$, is it true that the equation $F_{k,p}(n) = 0$ has a finite number of solutions (that is, does there exist only finite number of pairs (p, n), such that $F_{k,p}(n) = 0$)?*

(6.1.2) *Find at least one solution of equation $F_{k,p}(n) = 0$ such that $n > 2k$ and $k \geq 5$ (we consider the case $n > 2k$ since the pairs (n, k) and $(n, n - k)$ are singular simultaneously).*

PROBLEM 6.2. *Let the triple (n, k, p) be a root of the equation $F_{k,p}(n) = 0$ and $n \geq k$. Is it true that the pair (n, k) is singular? In particular, is the pair $(4, 12)$ singular?*

PROBLEM 6.3. *Do there exist three n-collections that are pairwise k-conjugated for some n and k?*

PROBLEM 6.4. *Does there exist an example of conjugate collections that are not self-conjugate for some $k > 2$?*

Translated by THE AUTHORS

ST. PETERSBURG STATE UNIVERSITY, DEPARTMENT OF MATHEMATICS AND MECHANICS, 198904, ST. PETERSBURG, RUSSIA

E-mail address: dvfom@hq.math.lgu.spb.su

On Convergence of Series over Local Fields

A. I. MADUNTS

The problem of the convergence of series over local fields is important in many parts of the arithmetic of local fields. G. Henniart investigated this topic (see [4]) but there are defects in his paper.

Our paper is devoted to this problem. We prove sufficient conditions for the convergence of superpositions of formal sums of series over local fields in §1 and apply these conditions to series determining π_0^n-primary elements (see [1, 2]) in §2.

§1. Sufficient conditions for convergence of formal sums of series

Let k be a local field (a finite extension of $\mathbb{Q}_p$), $\mathfrak{o}_0$ be the ring of integers of this field, $F(X, Y)$ be a formal group over $\mathfrak{o}_0$. We shall write $X +_F Y$ instead of $F(X, Y)$.

Let K be a finite extension of k, $K[[X]]$ be the field of formal power series over K, v be a valuation of K.

THEOREM 1.1. *Let $x \in K$, $a(X), b(X), d(X) \in K[[X]]$, where*

$$a(X) = \sum_{r \geq 1} a_r X^r, \qquad b(X) = \sum_{r \geq 1} b_r X^r,$$

$$d(X) = a(X) +_F b(X) = \sum_{r \geq 1} d_r X^r.$$

If

$$\inf_{\substack{1 \leq t \leq r \\ r_1 + \cdots + r_t = r}} v(a_{r_1} \ldots a_{r_t} x^r) \xrightarrow[r \to +\infty]{} +\infty$$

and

$$\inf_{\substack{1 \leq t \leq r \\ r_1 + \cdots + r_t = r}} v(b_{r_1} \ldots b_{r_t} x^r) \xrightarrow[r \to +\infty]{} +\infty,$$

then $d(x)$ is finite, and $d(x) = a(x) +_F b(x)$.

1991 *Mathematics Subject Classification.* Primary 11S31; Secondary 13J05.
Key words and phrases. Local field, series, formal group, convergence, π_0^n-primary unit.

© 1995, American Mathematical Society

PROOF.

$$v(a_r x^r) \xrightarrow[r \to +\infty]{} +\infty \quad \text{and} \quad v(b_r x^r) \xrightarrow[r \to +\infty]{} +\infty$$

from the conditions of our theorem for $t = 1$. Therefore $a(x)$, $b(x)$ are finite. If $v(a_r x^r) > 0$ and $v(b_r x^r) > 0$ for all r, then $a(x) +_F b(x)$ is finite too. But for all r

$$tv(a_r x^r) = v(a_r \ldots a_r x^{tr}) \xrightarrow[t \to +\infty]{} +\infty$$

by our conditions. Hence $v(a_r x^r) > 0$ and similarly $b(x) > 0$.

Therefore $a(x) +_F b(x)$ is finite. Now we consider the series $d(X)$.

$$d_r = a_r + b_r + \sum_{i,j \geq 1} \sum_{\substack{\lambda_1 + \cdots + \lambda_i + \\ +\mu_1 + \cdots + \mu_j = r}} a_{\lambda_1} \ldots a_{\lambda_i} b_{\mu_1} \ldots b_{\mu_j}$$

(we do not write the integer multiplicands determined by our formal group).

The series $d(x)$ is finite if $v(d_r x^r) \xrightarrow[r \to +\infty]{} +\infty$. But

$$v(d_r x^r) \geq \inf_{\substack{1 \leq i+j \leq r \\ \lambda_1 + \cdots + \lambda_i + \mu_1 + \cdots + \mu_j = r}} v(a_{\lambda_1} \ldots a_{\lambda_i} b_{\mu_1} \ldots b_{\mu_j} x^r)$$

$$= \inf_{\substack{1 \leq i+j \leq r \\ \lambda_1 + \cdots + \lambda_i = r_1 \\ \mu_1 + \cdots + \mu_j = r_2 \\ r_1 + r_2 = r}} \left(v(a_{\lambda_1} \ldots a_{\lambda_i} x^{r_1}) + v(b_{\mu_1} \ldots b_{\mu_j} x^{r_2}) \right)$$

$$\geq \inf_{r_1 + r_2 = r} \left(\inf_{\substack{1 \leq i \leq r_1 \\ \lambda_1 + \cdots + \lambda_i = r_1}} v(a_{\lambda_1} \ldots a_{\lambda_i} x^{r_1}) + \inf_{\substack{1 \leq j \leq r_2 \\ \mu_1 + \cdots + \mu_j = r_2}} v(b_{\mu_1} \ldots b_{\mu_j} x^{r_2}) \right).$$

We have $r_1 > r/2$ or $r_2 > r/2$. Therefore the corresponding summand is greater than any given number if r is large. The other summand is a member of a sequence tending to infinity and this summand is greater than some number depending only on the series but not on r. Hence

$$v(d_r x^r) \xrightarrow[r \to +\infty]{} +\infty,$$

and $d(x)$ is finite.

We must prove that $a(x) +_F b(x)$ and $d(x)$ are equal.

Consider $(a(x) +_F b(x)) - \sum_{r=1}^m d_r x^r$. We have

$$v\left((a(x) +_F b(x)) - \sum_{r=1}^m d_r x^r \right)$$

$$\geq v\left(\sum_{\substack{r = \lambda_1 + \cdots + \lambda_i + \mu_1 + \cdots + \mu_j \geq m+1 \\ 1 \leq i+j \leq r}} a_{\lambda_1} \ldots a_{\lambda_i} b_{\mu_1} \ldots b_{\mu_j} x^r \right)$$

$$\geq \inf_{\substack{r = \lambda_1 + \cdots + \lambda_i + \mu_1 + \cdots + \mu_j \geq m+1 \\ 1 \leq i+j \leq r}} v(a_{\lambda_1} \ldots a_{\lambda_i} b_{\mu_1} \ldots b_{\mu_j} x^r) \xrightarrow[m \to +\infty]{} +\infty,$$

because we proved before that

$$\inf_{\substack{1 \leq i+j \leq r \\ r = \lambda_1 + \cdots + \lambda_i + \mu_1 + \cdots + \mu_j}} v(a_{\lambda_1} \ldots a_{\lambda_i} b_{\mu_1} \ldots b_{\mu_j} x^r) \xrightarrow[r \to +\infty]{} +\infty.$$

The theorem is proved.

COROLLARY. *Suppose $x \in K$, I is fixed natural number, and for any $i \leqslant I$*

$$a^{(i)}(X) = \sum_{r \geqslant 1} a_r^{(i)} X^r \in K[[X]], \qquad d(X) = \sum_{r=1}^{I} {}_F a^{(i)}(X).$$

If for any $i \geqslant I$ we have

$$\inf_{\substack{1 \leqslant t \leqslant r \\ r_1 + \cdots + r_t = r}} v(a_{r_1}^{(i)} \ldots a_{r_t}^{(i)} x^r) \xrightarrow[r \to +\infty]{} +\infty,$$

then $d(x)$ is finite, and $d(x) = \sum_{i=1}^{I} {}_F a^{(i)}(x)$.

This is an easy generalization.

REMARK. If for all natural r we have $v(a_r x^r) \geqslant Mr$, where $M > 0$ and M does not depend on r, then the condition of Theorem 1.1 holds for the series $a(X) = \sum_{r \geqslant 1} a_r X^r$ and the element x.

THEOREM 1.2. *Suppose that $x \in K$ and for all natural i we have*

$$a^{(i)}(X) = \sum_{r \geqslant 1} a_r^{(i)} X^r \in K[[X]].$$

Let

$$\lambda_i(r, x) = \inf_{\substack{1 \leqslant t \leqslant r \\ r_1 + \cdots + r_t = r}} v(a_{r_1}^{(i)} \ldots a_{r_t}^{(i)} x^r),$$

$$\lambda_i(x) = \inf_{r \geqslant 1} \lambda_i(r, x).$$

If
1. *for all natural i we have $\lambda_i(r, x) \xrightarrow[r \to +\infty]{} +\infty$,*
2. *$\lambda_i(x) \xrightarrow[i \to +\infty]{} +\infty$,*

then the series $d(X) = \sum_{i \geqslant 1} {}_F a^{(i)}(X)$ exists, $d(x)$ is finite and

$$d(x) = \sum_{i \geqslant 1} {}_F a^{(i)}(x).$$

PROOF. We consider the series

$$d^{(n)}(X) = \sum_{i=1}^{n} {}_F a^{(i)}(X) = \sum_{r \geqslant 1} d_r^{(n)} X^r.$$

We have $d(X) = \lim_{n \to +\infty} d^{(n)}(X)$, hence $d_r = \lim_{n \to +\infty} d_r^{(n)}$ for all r. When does $d(X)$ exist, i.e., when are all the d_r finite? We have

$$d^{(n+1)}(X) = d^{(n)}(X) + a^{(n+1)}(X) + \sum_{i,j \geqslant 1} d^{(n)}(X)^i a^{(n+1)}(X)^j$$

(we do not write integral multiplicands determined by our formal group).
Therefore,

$$d_r^{(n+1)} = d_r^{(n)} + a_r^{(n+1)}$$
$$+ \sum_{i,j \geq 1} \sum_{\lambda_1 + \cdots + \lambda_i + \mu_1 + \cdots + \mu_j = r} d_{\lambda_1}^{(n)} \ldots d_{\lambda_i}^{(n)} a_{\mu_1}^{(n+1)} \ldots a_{\mu_j}^{(n+1)};$$

(1)
$$v(d_r^{(n+1)} - d_r^{(n)})$$
$$\geq \inf_{\substack{2 \leq i+j \leq r \\ \lambda_1 + \cdots + \lambda_i + \mu_1 + \cdots + \mu_j = r \\ i,j \geq 1}} (v(a_r^{(n+1)}), v(d_{\lambda_1}^{(n)} \ldots d_{\lambda_i}^{(n)} a_{\mu_1}^{(n+1)} \ldots a_{\mu_j}^{(n+1)})).$$

If for all r we have $v(a_r^{(n)}) \xrightarrow[n \to +\infty]{} +\infty$, then for all r we also have

$$v(d_r^{(n+1)} - d_r^{(n)}) \xrightarrow[n \to +\infty]{} +\infty$$

and the series $d(X)$ exists.

We shall prove this proposition by induction on r.

Let $v(a_r^{(n)}) \xrightarrow[n \to +\infty]{} +\infty$ for all r. Then

$$r = 1 \implies d_1^{(n)} = a_1^{(1)} + \cdots + a_1^{(n)},$$

hence $d_1 = \sum_{n \geq 1} a_1^{(n)}$ is finite if $v(a_1^{(n)}) \xrightarrow[n \to +\infty]{} +\infty$.

Let d_l exist for all $l < r$. Then $v(d_l^{(n)}) \geq M_l$ for these l and for any n, because $v(d_l - d_l^{(n)}) \xrightarrow[n \to +\infty]{} +\infty$, and

$$v(d_l^{(n)}) \geq \inf(v(d_l), v(d_l - d_l^{(n)})).$$

Let us consider the inequality (1).

For $\alpha \leq i$ we have $\lambda_\alpha < r$, therefore $v(d_{\lambda_\alpha}^{(n)}) \geq M_{\lambda_\alpha}$. If r is fixed, then there is finite number of sets $\lambda_1, \ldots, \lambda_i$, where $\lambda_1 + \cdots + \lambda_i < r$, $i < r$. Hence all $v(d_{\lambda_1}^{(n)} \ldots d_{\lambda_i}^{(n)})$ are greater than some constant not depending on n. But $j \geq 1$ in (1), and we have the factor $a_{\mu_1}^{(n+1)}$ in every $d_{\lambda_1}^{(n)} \ldots d_{\lambda_i}^{(n)} a_{\mu_1}^{(n+1)} \ldots a_{\mu_j}^{(n+1)}$.

Furthermore, $v(a_{\mu_1}^{(n+1)}) \xrightarrow[n \to +\infty]{} +\infty$ by our conditions.

If r is fixed, then there is finite number of sets $\mu_1, \ldots, \mu_j$. Therefore,

$$\inf_{\mu_1 < r} v(a_{\mu_1}^{(n)}) \xrightarrow[n \to +\infty]{} +\infty,$$

and the other summands in the sum

$$v(d_{\lambda_1}^{(n)}) + \cdots + v(d_{\lambda_i}^{(n)}) + v(a_{\mu_1}^{(n+1)}) + \cdots + v(a_{\mu_j}^{(n+1)})$$

are greater than some constant.

Hence $v(d_r^{(n+1)} - d_r^{(n)}) \xrightarrow[n \to +\infty]{} +\infty$. The induction is complete.

We have
$$\inf_r v(a_r^{(i)} x^r) \xrightarrow[i \to +\infty]{} +\infty$$
by condition 2 of our theorem.

Let r be fixed. For all N, there is an I such that $v(a_r^{(i)} x^r) > N + rv(x)$ for $i > I$, i.e., $v(a_r^{(i)}) > N$. Therefore $v(a_r^{(i)}) \xrightarrow[r \to +\infty]{} +\infty$ for all r. We have proved that d_r exists for all r and that $d(X)$ exists.

We can prove by an easy induction that

$$d_r^{(n)} = \sum_{\substack{1 \leq i_1 < i_2 < \cdots < i_t \leq n \\ 1 \leq t \leq n \\ 1 \leq l_1 + \cdots + l_t \leq r \\ \alpha_1^{(1)} + \cdots + \alpha_{l_1}^{(1)} + \cdots + \alpha_1^{(t)} + \cdots + \alpha_{l_t}^{(t)} = r}} a_{\alpha_1^{(1)}}^{(i_1)} \cdots a_{\alpha_{l_1}^{(1)}}^{(i_1)} \cdots a_{\alpha_1^{(t)}}^{(i_t)} \cdots a_{\alpha_{l_t}^{(t)}}^{(i_t)},$$

(2)
$$d_r^{(n+1)} - d_r^{(n)} = \sum_{\substack{1 \leq i_1 < i_2 < \cdots < i_t = n+1 \\ 1 \leq t \leq n+1 \\ 1 \leq l_1 + \cdots + l_t \leq r \\ l_t \geq 1 \\ \alpha_1^{(1)} + \cdots + \alpha_{l_1}^{(1)} + \cdots + \alpha_1^{(t)} + \cdots \alpha_{l_t}^{(t)} = r}} a_{\alpha_1^{(1)}}^{(i_1)} \cdots a_{\alpha_{l_1}^{(1)}}^{(i_1)} \cdots a_{\alpha_1^{(t)}}^{(i_t)} \cdots a_{\alpha_{l_t}^{(t)}}^{(i_t)}$$

(we omit the integer factors).

We have multiplication by $a_\alpha^{(n+1)}$ in every summand of the last sum. But

$$d_r = \sum_{n \geq 1}(d_r^{(n+1)} - d_r^{(n)}) + d_r^{(1)},$$

$$d_r - d_r^{(n)} = \sum_{m \geq n} \sum_{\substack{1 \leq t \leq m \\ 1 \leq i_1 < \cdots < i_t = m \\ 1 \leq l_1 + \cdots + l_t \leq r \\ l_t \geq 1 \\ \alpha_1^{(1)} + \cdots + \alpha_{l_1}^{(1)} + \cdots + \\ + \alpha_1^{(t)} + \cdots + \alpha_{l_t}^{(t)} = r}} a_{\alpha_1^{(1)}}^{(i_1)} \cdots a_{\alpha_{l_1}^{(1)}}^{(i_1)} \cdots a_{\alpha_1^{(t)}}^{(i_t)} \cdots a_{\alpha_{l_t}^{(t)}}^{(i_t)},$$

and

$$v((d_r - d_r^{(n)})x^r) \geq \inf_{m \geq n} \inf_{\substack{1 \leq t \leq m \\ 1 \leq i_1 < \cdots < i_t = m \\ 1 \leq l_1 + \cdots + l_t \leq r \\ l_t \geq 1 \\ \alpha_1^{(1)} + \cdots + \alpha_{l_1}^{(1)} = r_1 \\ \cdots \\ \alpha_1^{(t)} + \cdots + \alpha_{l_t}^{(t)} = r_t \\ r_1 + \cdots + r_t = r}} v(a_{\alpha_1^{(1)}}^{(i_1)} \cdots a_{\alpha_{l_1}^{(1)}}^{(i_1)} \cdots a_{\alpha_1^{(t)}}^{(i_t)} \cdots a_{\alpha_{l_t}^{(t)}}^{(i_t)} x^r)$$

$$\geq \inf_{m \geq n} \left(\inf_{\substack{1 \leq t \leq m-1 \\ 1 \leq i_1 < \cdots < i_t \leq m-1}} (\lambda_{i_1}(x) + \cdots + \lambda_{i_n}(x)) + \lambda_m(x) \right)$$

$$\geq \inf_{m \geq n} \left(\inf_{\substack{1 \leq t \leq m-1 \\ 1 \leq i_1 < \cdots < i_t \leq m-1}} \sum_{\alpha \leq t} \inf(0, \lambda_{i_\alpha}(x)) + \lambda_m(x) \right)$$

$$\geqslant \inf_{m \geqslant n} \left(\sum_{t \geqslant 1} \inf(0, \lambda_t(x)) + \lambda_m(x) \right).$$

But $\lambda_i(x) \xrightarrow[i \to +\infty]{} +\infty$ by condition 2 of our theorem. Therefore $\lambda_i(x) < 0$ for a finite set of i and $\sum_{t \geqslant 1} \inf(0, \lambda_t(x)) = M > -\infty$. Hence

$$v((d_r - d_r^{(n)})x^r) \geqslant \inf_{m \geqslant n} \lambda_m(x) + M \xrightarrow[n \to +\infty]{} +\infty$$

by condition 2 of our theorem and for all A there exists N such that for $n > N$

$$v((d_r - d_r^{(n)})x^r) > A \quad \text{for all } r.$$

Furthermore, $d_r^{(n)}(x)$ is finite for every n, and $d_r^{(n)}(x) = \sum_{i \leqslant n} {}_F a^{(i)}(x)$. But

$$v(d_r x^r) \geqslant \inf(v(d_r^{(n)} x^r), v((d_r - d_r^{(n)})x^r))).$$

If $n > N$, then for all r we have $v((d_r - d_r^{(n)})x^r) > A$.
Now, $v(d_r^{(n)} x^r) \xrightarrow[r \to +\infty]{} +\infty$, therefore for $r > r_0$ we have $v(d_r^{(n)} x^r) > A$.
Finally for all A we have found r_0 such that $v(d_r x^r) > A$ whenever $r > r_0$.
Hence $d(x)$ is finite.
We must prove that $d(x) = \sum_{i \geqslant 1} {}_F a^{(i)}(x)$. But

$$v(d(x) - d^{(n)}(x)) = v\left(\sum_{r \geqslant 1}(d_r - d_r^{(n)})x^r\right) \geqslant \inf_{r \geqslant 1} v((d_r - d_r^{(n)})x^r) > A$$

if $n > N$, therefore, $d(x) = \lim_{n \to +\infty} d^{(n)}(x) = \sum_{i \geqslant 1} {}_F a^{(i)}(x)$. The theorem is proved.

COROLLARY. Let $x \in K$, M be fixed natural number, for all $m \leqslant M$ assume that $a^{(m,i)}(X) = \sum_{r \geqslant 1} a_r^{(m,i)} X^r \in K[[X]]$ for any natural number i.
Let

$$\lambda_{m,i}(r, x) = \inf_{\substack{1 \leqslant t \leqslant r \\ r_1 + \cdots + r_t = r}} v(a_{r_1}^{(m,i)} \ldots a_{r_t}^{(m,i)} x^r),$$

$$\lambda_{m,i}(x) = \inf_{r \geqslant 1} \lambda_{m,i}(r, x).$$

If

1. for all $m \leqslant M$ and for all natural i we have $\lambda_{m,i}(r, x) \xrightarrow[r \to +\infty]{} +\infty$,
2. for all $m \leqslant M$ $\lambda_{m,i}(x) \xrightarrow[i \to +\infty]{} +\infty$,

then the series

$$d(X) = \sum_{m \leqslant M} {}_F \sum_{i} {}_F a^{(m,i)}(X)$$

exists, $d(x)$ is finite and

$$d(x) = \sum_{m \leq M} {}_F \sum_i {}_F a^{(m,i)}(x).$$

REMARK. If we consider the infinite double formal sum of series

$$d(X) = \sum_m {}_F \sum_i {}_F a^{(m,i)}(X),$$

introduce the series

$$d^{(n_1, n_2)}(X) = \sum_{m=1}^{n_1} {}_F \sum_{i=1}^{n_2} {}_F a^{(m,i)}(X),$$

and use the ideas of the proof of Theorem 1.2, then we obtain the following assertion. Suppose

$$\lambda_{m,i}(r, x) = \inf_{\substack{1 \leq t \leq r \\ r_1 + \cdots + r_t = r}} v(a_{r_1}^{(m,i)} \ldots a_{r_t}^{(m,i)} x^r),$$

$$\lambda_{m,i}(x) = \inf_{r \geq 1} \lambda_{m,i}(r, x).$$

If
1. for all natural i and m we have $\lambda_{m,i}(r, x) \xrightarrow[r \to +\infty]{} +\infty$,
2. $\inf_m \lambda_{m,i}(x) \xrightarrow[i \to +\infty]{} +\infty$,
3. $\inf_i \lambda_{m,i}(x) \xrightarrow[m \to +\infty]{} +\infty$,

then the series $d(X)$ exists, $d(x)$ is finite and $d(x) = \sum_m {}_F \sum_i {}_F a^{(m,i)}(x)$.

§2. Application of the conditions to the convergence of series

Let k be a local field, $q = p^{f_0}$ be the cardinality of its residue field, $\mathfrak{o}_0$ be its ring of integers, π_0 be a prime element, v be the valuation, $F(X, Y)$ be a Lubin-Tate formal group over $\mathfrak{o}_0$.

It is known that for $F(X, Y)$ there is a formal logarithm $\log_F X \in k[[X]]$, where

$$\log_F X = \sum_{r \geq 1} c_r x^r, \quad c_1 = 1, \text{ for all } r > 1 \quad v(c_r) \geq -\log_q r.$$

Let us recall that $\exp_F X$ is the inverse series for $\log_F X$ in the sense of superposition.

Let K be a finite extension of k, containing all the roots of the isogenies $[\pi_0^n](X)$, where n is fixed. Let $\mathfrak{o}$ be the ring of integers of K, π a prime element of K, ζ_n a primitive root of $[\pi_0^n](X)$, and Δ a continuation on K of the Frobenius automorphism T/K, where T is the inertia field of K/k. If $a(X) \in K[[X]]$ and $a(X) = \sum_r a_r x^r$, then we put

$$a^\Delta = \sum_r a_r^\Delta X^{qr}.$$

Let e be the ramification index of K/k. Then $e = e_n q^{n-1}(q-1)$, where $e_n \in \mathbb{N}$. We can choose the series

$$z(X) = \sum_{r \geq e_n} z_r X^r, \quad z_r \in \mathfrak{o},$$

such that $z(\pi) = \zeta_n$. Let $l_F(X) = (1 - \Delta/\pi_0) \log_F X$, $E_F(X)$ be the inverse series to $l_F(X)$ in the sense of superposition, a belong to the ring of integer of T, A belong to maximal unramified extension of T, and $\Delta A - A = a$.
Then

$$H(a) = E_F(\pi_0^n \Delta A l_F(z(X)))|_{X=\pi}$$

is a π_0^n-primary element; this element is important for the calculation of the Hilbert pairing (see [1, 2]).

The aim of this section is to prove the following theorem.

THEOREM 2.1. $H(a)$ is a finite element in K, and

$$H(a) = E_F(as(X))|_{X=\pi} +_F [\pi_0^n]\left(\sum_{m=2}^{+\infty}{}_F \sum_{j=0}^{+\infty}{}_F \exp_F \frac{a^{\Delta^i} c_m s^{m\Delta^i}(\pi)}{\pi_0^{n+i}}\right),$$

where $s(X) = [\pi_0^n](z(X))$.

We need some preliminary calculations.
It is easy to prove (see [3, 4]) the following assertion.

THEOREM 2.2. Let $x \in K$, $a(X), b(X), c(X) \in K[[X]]$, where

$$a(X) = \sum_{r \geq 1} a_r X^r, \quad b(X) = \sum_{r \geq 1} b_r X^r,$$

$$c(X) = \sum_{r \geq 1} c_r X^r, \quad c(X) = a \circ b(X).$$

If

$$\inf_{\substack{1 \leq t \leq r \\ r_1 + \cdots + r_t = r}} v(a_t b_{r_1} \ldots b_{r_t} x^r) \xrightarrow[r \to +\infty]{} +\infty,$$

then $c(x)$ is finite, and $c(x) = a(b(x))$.

PROOF. We shall transform $H(a)$ as follows:

$$\left(1 - \frac{\Delta}{\pi_0}\right)(A \log_F z) = A \log_F z - \Delta A \frac{\Delta}{\pi_0} \log_F z$$

$$= \Delta A \log_F z - a \log_F z - \Delta A \frac{\Delta}{\pi_0} \log_F z$$

$$= \Delta A l_F(z) - a \log_F z.$$

Hence

$$E_F(\pi_0^n \Delta Al_F(z)) = E_F(\pi_0^n(\Delta Al_F(z) - a\log_F z) + \pi_0^n a\log_F z)$$
$$= E_F\left(\pi_0^n\left(1 - \frac{\Delta}{\pi_0}\right)(A\log_F z)\right) +_F E_F(\pi_0^n a\log_F z)$$
$$= \exp_F(\pi_0^n A\log_F z) +_F E_F(\pi_0^n a\log_F z),$$

because

$$E_F\left(\pi_0^n\left(1 - \frac{\Delta}{\pi_0}\right)(A\log_F z)\right)$$
$$= \exp_F\left(\pi_0^n\left(\sum_{i \geq 0}\left(\frac{\Delta}{\pi_0}\right)^i(A\log_F z) - \sum_{i \geq 1}\left(\frac{\Delta}{\pi_0}\right)^i(A\log_F z)\right)\right)$$
$$= \exp_F(\pi_0^n A\log_F z).$$

Therefore
$$H(a) = (E_F(\pi_0^n a\log_F z) +_F \exp_F(\pi_0^n A\log_F z))|_{X=\pi}.$$

Finally,

$$E_F(a\pi_0^n \log_F z) = E_F(a\log_F([\pi_0^n]z)) = E_F(a\log_F s)$$
$$= E_F\left(a\sum_{m \geq 1} c_m s^m\right) = E_F(as) +_F \sum_{m \geq 2}{}_F E_F(ac_m s^m)$$
$$= \sum_{i \geq 0}{}_F \exp_F \frac{\Delta^i s}{\pi_0^i} +_F \sum_{m \geq 2}{}_F \sum_{i \geq 0}{}_F \exp_F\left(\frac{c_m}{\pi_0^i}\Delta^i(as^m)\right).$$

We define the series $s_{n+i}(X) = [\pi_0^{n+i}](z(X))$, $i \geq 0$.
Then $\Delta^i(as^m) = a^{\Delta^i}(s^{m\Delta^i} - s_{n+i}^m) + a^{\Delta^i} s_{n+i}^m$ and

$$H(a) = \left(\exp_F(\pi_0 A\log_F z(X)) +_F \sum_{i \geq 0}{}_F \exp_F \frac{\Delta^i s(X)}{\pi_0^i}\right.$$
$$+_F \sum_{m \geq 2}{}_F \sum_{i \geq 0}{}_F \exp_F \frac{a^{\Delta^i} c_m s_{n+i}^m(X)}{\pi_0^i}$$
$$\left.+_F \sum_{m \geq 2}{}_F \sum_{i \geq 0}{}_F \exp_F \frac{a^{\Delta^i} c_m(s^{m\Delta^i}(X) - s_{n+i}^m(X))}{\pi_0^i}\right)\bigg|_{X=\pi}.$$

Let us study the convergence of every summand for $X = \pi$. We must prove some lemmas.

LEMMA 1. *Let* $\exp_F X = \sum_{j \geq 1} a_j X^j$. *Then*

$$v(a_j) \geq -\frac{j-1}{q-1}.$$

PROOF. We use induction on j. If $j = 1$, then $v(a_1) = v(1) = 0 \geq -\frac{1-1}{q-1}$. Now let
$$v(a_j) \geq -\frac{j-1}{q-1} \quad \text{for } j < m.$$
It is known that $[\pi_0](X) = \pi_0 X + X^q + \pi_0 \sum_{i \geq 2} b_i X^i$, $b_i \in \mathfrak{o}_0$. Hence
$$[\pi_0](\exp_F X) = \pi_0 \left(\sum_{j \geq 1} a_j X^j \right) + \left(\sum_{j \geq 1} a_j X^j \right)^q + \pi_0 \left(\sum_{i \geq 2} b_i \left(\sum_{j \geq 1} a_j X^j \right)^i \right).$$
But $[\pi_0] \exp_F X = \exp_F(\pi_0 X) = \sum_{j \geq 1} a_j \pi_0^j X^j$ and therefore
$$a_m \pi_0^m = a_m \pi_0 + \sum_{j_1 + \cdots + j_q = m} a_{j_1} \ldots a_{j_q} + \pi_0 \sum_{i \geq 2} b_i \left(\sum_{\lambda_1 + \cdots + \lambda_i = m} a_{\lambda_1} \ldots a_{\lambda_i} \right).$$
Finally,
$$v(a_m \pi_0^m - a_m \pi_0) = 1 + v(a_m)$$
$$\geq \inf_{\substack{j_1 + \cdots + j_q = m \\ \lambda_1 + \cdots + \lambda_i = m \\ i \geq 2}} (v(a_{j_1} \ldots a_{j_q}), 1 + v(a_{\lambda_1} \ldots a_{\lambda_i}))$$
$$\geq \inf_{\substack{j_1 + \cdots + j_q = m \\ \lambda_1 + \cdots + \lambda_i = m \\ i \geq 2}} \left(-\frac{j_1 - 1 + \cdots + j_q - 1}{q-1}, 1 - \frac{\lambda_1 - 1 + \cdots + \lambda_j - 1}{q-1} \right)$$
$$= \inf_{i \geq 2} \left(-\frac{m-1}{q-1} + 1, -\frac{m-i}{q-1} + 1 \right) = -\frac{m-1}{q-1} + 1$$
and $v(a_m) \geq -\frac{m-1}{q-1}$.
The induction is complete.

LEMMA 2. *For $h(X) \in \mathfrak{o}[[X]]$ we have*
$$([\pi_0^r]h(X))^\Delta \equiv [\pi_0^{r+1}](h(X)) \pmod{\pi_0^{r+1}}.$$

PROOF. An easy induction on r.

COROLLARY. $s_{n+i}^\Delta(X) \equiv s_{n+i+1}(X) \pmod{\pi_0^{n+i+1}}$.

LEMMA 3.
$$\Delta^i s(X) = s_{n+i}(X) + \pi_0^{n+1} \Delta^{i-1} g_1(X) + \pi_0^{n+2} \Delta^{i-2} g_2(X) + \cdots + \pi_0^{n+i} g_i(X),$$
where $g_\alpha(X) \in X^{e_n} \mathfrak{o}[[X]]$

PROOF. Let us carry out an induction on i. If $i = 0$, then $s_n = s_n$. Let our assertion be true for $r \leq i$. Then
$$\Delta^{i+1} s(X) = \Delta(\Delta^i s(X))$$
$$= \Delta(s_{n+i}(X) + \pi_0^{n+1} \Delta^{i-1} g_1(X) + \pi_0^{n+2} \Delta^{i-2} g_2(X) + \cdots + \pi_0^{n+i} g_i(X))$$
$$= s_{n+i+1}(X) + \pi_0^{n+1} \Delta^i g_1(X) + \pi_0^{n+2} \Delta^{i-1} g_2(X) + \cdots + \pi_0^{n+i} \Delta g_i(X)$$
$$+ \pi_0^{n+i+1} g_{i+1}(X),$$
where $g_{i+1}(X) \in \mathfrak{o}[[X]]$, we find $g_{i+1}(X)$ by the Corollary to Lemma 2 and $g_{i+1}(X)$ begins from degree e_n.

LEMMA 4. $\log_F(z(X))|_{X=\pi} = \log_F(z(\pi)) = \log_F(\zeta_n) = 0$.

PROOF. Let us recall the Yamomoto theorem. Let

$$a(X) = \sum_{j \geq 1} a_j X^j, \quad b(X) = \sum_{j \geq 1} b_j X^j, \quad a_j, b_j \in K,$$

where K is a local field, r_a, r_b are convergence radius of $a(X)$ and $b(X)$, $c(X) = a \circ b(X)$. If there exists an $s \in \mathbb{R}$ such that $s \leq r_b$ and $|b(x)| < r_a$ for all x with $|x| < s$, then $c(x)$ is finite and $c(x) = a(b(x))$ for such x. In our case, $a(X) = \log_F X$, $r_a = 1$, $b(X) = z(X)$, $r_b \geq 1$. Clearly, $s = 1 \leq r_b$, $|z(x)| < r_a = 1$ for $|x| < 1$. Since $|\pi| < 1$, we have $\log_F(z(X))|_{X=\pi} = \log_F z(\pi) = \log_F \zeta_n$. But $[\pi_0^n](\zeta_n) = 0$ and $\pi_0^n \log_F \zeta_n = \log_F([\pi_0^n]\zeta_n)$. Therefore $\log_F \zeta_n = 0$. Lemma 4 is proved.

LEMMA 5. *Let*

$$b(X) = \sum_{j \geq 1} b_j X^j = \log_F(z(X)).$$

Then $v(b_j) \geq -\log_q j/e_n$.

PROOF. We have

$$\log_F X = \sum_{j \geq 1} c_j X^j, \quad z(X) = \sum_{j \geq e_n} z_j X^j, \quad v(z_j) \geq 0, \quad v(c_j) \geq -\log_q j,$$

$$b_j = \sum_{m=1}^{j} \left(c_m \sum_{\substack{\lambda_1 + \cdots + \lambda_m = j \geq m e_n \\ \lambda_\alpha \geq e_n}} z_{\lambda_1} \cdots z_{\lambda_m} \right).$$

Hence

$$v(b_j) \geq \inf_{\substack{1 \leq m \leq \frac{j}{e_n} \\ \lambda_1 + \cdots + \lambda_m = j}} (v(c_m) + v(z_{\lambda_1} \cdots z_{\lambda_m}))$$

$$\geq \inf_{1 \leq m \leq \frac{j}{e_n}} v(c_m) \geq -\log_q \frac{j}{e_n}.$$

THEOREM 2.3. *We have* $\exp_F\left(\pi_0^n A \log_F(z(X))\right)\big|_{X=\pi} = 0$, *and the condition of Theorem* 1.1 *holds for the series* $\exp_F(\pi_0^n A \log_F z)$ *and the element* π.

PROOF. Let

$$\widetilde{a}(X) = \exp_F(\pi_0^n A X) = \sum_{j \geq 1} \pi_0^{nj} A^j a_j X^j,$$

where $\exp_F X = \sum_{j \geq 1} a_j X^j$. Let $b(X) = \log_F(z(X)) = \sum_{j \geq 1} b_j X^j$. If we prove that

$$\inf_{\substack{1 \leq j \leq r \\ \lambda_1 + \cdots + \lambda_j = r}} v(\widetilde{a}_j b_{\lambda_1} \cdots b_{\lambda_j} \pi^r) \xrightarrow[r \to +\infty]{} +\infty$$

then from Theorem 2.2 we shall obtain

$$\exp_F\left(\pi_0^n A \log_F z\right)\big|_{X=\pi} = \exp_F\left(\pi_0^n A(\log_F(z(X)))\big|_{X=\pi}\right) = 0,$$

since $\left(\log_F(z(X))\right)\big|_{X=\pi} = 0$ by Lemma 4. Further $v(a_j) \geqslant -(j-1)/(q-1)$ by Lemma 1 and $v(b_j) \geqslant -\log_q j/e_n$ by Lemma 5. Therefore

$$v(\widetilde{a}_j b_{\lambda_1}\ldots b_{\lambda_j}) \geqslant nj + jv(A) - \frac{j-1}{q-1} + \sum_{\alpha=1}^{j}(\log_q e_n - \log_q \lambda_\alpha)$$

$$\geqslant \frac{1}{q-1} + j\left(n + \log_q e_n - \frac{1}{q-1}\right) - \sum_{\alpha=1}^{j}\log_q \lambda_\alpha.$$

Let us find a minimum of the last expression under the condition that $1 \leqslant j \leqslant r$, $\lambda_1 + \cdots + \lambda_j = r$. Clearly

$$-\sum_{\alpha=1}^{j}\log_q \lambda_\alpha = -\log_q(\lambda_1\ldots\lambda_j)$$

has a minimum if $\lambda_1 * \cdots * \lambda_j$ is maximal under the condition $\lambda_1 + \cdots + \lambda_j = r$, i.e., $\lambda_1 = \cdots = \lambda_j = r/j$.

Therefore

$$v(\widetilde{a}_j b_{\lambda_1}\ldots b_{\lambda_j}) \geqslant \frac{1}{q-1} + j\left(n - \frac{1}{q-1} + \log_q e_n - \log_q r + \log_q j\right)$$

under the conditions $1 \leqslant j \leqslant r$, $\lambda_1 + \cdots + \lambda_j = r$.

The minimum for j of the function $f(j) = j(C + \log_q j)$ is

$$f(j_{\min}) = -\frac{\log_q E}{Eq^C}, \quad \text{where } j_{\min} = \frac{1}{q^C E}, \quad E \approx 2.718.$$

We have $v(\widetilde{a}_j b_{\lambda_1}\ldots b_{\lambda_j}) \geqslant \frac{1}{q-1} - \frac{r\log_q E}{Ee_n q^{n-\frac{1}{q-1}}}$.

Hence

$$\inf_{\substack{1 \leqslant j \leqslant r \\ \lambda_1+\cdots+\lambda_j=r}} v(\widetilde{a}_j b_{\lambda_1}\ldots b_{\lambda_j}\pi^r) \geqslant \frac{1}{q-1} + r\left(\frac{1}{e_n q^{n-1}(q-1)} - \frac{q^{\frac{1}{q-1}}\log_q E}{Ee_n q^n}\right).$$

Let

$$M = \frac{1}{e_n q^{n-1}(q-1)} - \frac{q^{\frac{1}{q-1}}\log_q E}{Ee_n q^n} = \frac{Eq - (q-1)q^{\frac{1}{q-1}}\log_q E}{Ee_n q^n(q-1)}.$$

If $M > 0$, then the assumptions of Theorem 2.2 hold for the series $a(X)$, $b(X)$, and the element π. Let $M_1 = q(E - q^{\frac{1}{q-1}}\log_q E) + q^{\frac{1}{q-1}}\log_q E$. If $M_1 > 0$, then $M > 0$. $M_1 = 2E > 0$ for $q = 2$.

Let $q \geqslant 3$. We must prove that $M_1 > 0$. The function $x^{\frac{1}{x-1}}$ is positive and it decreases for $x \geqslant 3$, the function $\log_x E$ is positive and it decreases for $x \geqslant 3$. Therefore $x^{\frac{1}{x-1}} \log_x E$ is positive and it decreases for $x \geqslant 3$ too. Hence

$$M_1 \geqslant q(E - 3^{\frac{1}{2}} \log_3 E) > 0.$$

We see that for the series $\exp_F(\pi_0^n A \log_F z) = \sum_{r \geqslant 1} d_r X^r$ we have $v(d_r \pi^r) \geqslant Mr$, $M > 0$ for all r.

The assumptions of Theorem 1.1 hold for the series $\exp_F(\pi_0^n A \log_F z)$ and the element π according to the Remark following Theorem 1.1.

THEOREM 2.4. *We have*

$$\left(\sum_{m \geqslant 2} {}_F \sum_{i \geqslant 0} {}_F \exp_F \frac{a^{\Delta^i} c_m s_{n+i}^m(X)}{\pi_0^i} \right) \bigg|_{X=\pi} = \sum_{m \geqslant 2} {}_F \sum_{i \geqslant 0} {}_F \exp_F \frac{a^{\Delta^i} c_m s_{n+i}^m(\pi)}{\pi_0^i} = 0,$$

where the assumptions of the Remark after Theorem 1.2 hold for the series

$$\sum_{m \geqslant 2} {}_F \sum_{i \geqslant 0} {}_F \exp_F \frac{a^{\Delta^i} c_m s_{n+i}^m(X)}{\pi_0^i}$$

and the element π.

PROOF. Consider $\exp_F \frac{a^{\Delta^i} c_m s_{n+i}^m}{\pi_0^i}$. We may assume that $[\pi_0](X) = \pi_0 X + X^q$. Then

$$s_{n+i} = \sum_{r=1}^{q^{n+i}} d_r z^r \quad \text{and} \quad v(d_r) \geqslant n + i - [\log_q r],$$

where $[x]$ is the integer part of x (this is an easy induction). Furthermore, s_{n+i}^m is the finite sum of the following summands $d_{r_1} \ldots d_{r_m} z^{r_1 + \cdots + r_m}$, $1 \leqslant r_\alpha \leqslant q^{n+i}$. Let us consider

$$\exp_F \left(\frac{c_m a^{\Delta^i}}{\pi_0^i} d_{r_1} \ldots d_{r_m} z^{r_1 + \cdots + r_m} \right), \quad 1 \leqslant r_\alpha \leqslant q^{n+i}.$$

Put $r_1 + \cdots + r_m = r$ and

$$\widetilde{a}(X) = \exp_F \left(\frac{c_m}{\pi_0^i} a^{\Delta^i} X \right) = \sum_{j \geqslant 1} \frac{c_m^j a^{\Delta^i j} a_j X^j}{\pi_0^{ij}},$$

where $\exp_F X = \sum_{j \geqslant 1} a_j X^j$, $\log_F X = \sum_{j \geqslant 1} c_j X^j$. Let $b(X) = d_{r_1} \ldots d_{r_m} z^r$, $\widetilde{c}(X) = \widetilde{a} \circ b(X)$.

Let us verify the assumptions of Theorem 2.2 for the series $\widetilde{c}(X)$ and for the element π:

$$v(\widetilde{a}_j) \geq j(v(c_m) + v(a^{\Delta^i}) - i) + v(a_j) \geq \frac{1}{q-1} - j\left(i + \log_q m + \frac{1}{q-1}\right);$$

$$v(b_j) \geq m(n+i) - [\log_q r_1] - \cdots - [\log_q r_m], \quad j \geq e_n r;$$

$$v(\widetilde{c}_l \pi^l) \geq \inf_{\substack{l = t_1 + \cdots + t_j \geq j r e_n}} v(\widetilde{a}_j b_{t_1} \ldots b_{t_j} \pi^l) \geq \frac{l}{q^{n-1}(q-1)e_n} + \frac{1}{q-1}$$

$$+ \inf_{\substack{1 \leq j \leq \frac{l}{r e_n} \\ t_1 + \cdots + t_j = l}} \left(j\left(mn + mi - [\log_q r_1] - \cdots - [\log_q r_m]\right.\right.$$

$$\left.\left. - i - \log_q m - \frac{1}{q-1}\right)\right).$$

If $M = mn + mi - [\log_q r_1] - \cdots - [\log_q r_m] - i - \log_q m \geq 0$, then

$$v(\widetilde{c}_l \pi^l) \geq v(\widetilde{a}_j b_{t_1} \ldots b_{t_j} \pi^l) \geq \frac{l}{e_n q^{n-1}(q-1)} \xrightarrow[l \to +\infty]{} +\infty.$$

For $1 \leq j \leq l/re_n$, $t_1 + \cdots + t_j = l$, we have
(3)
$$v(\widetilde{a}_j b_{t_1} \ldots b_{t_j} \pi^l) \geq \frac{l}{q^{n-1}(q-1)e_n} + \frac{1}{q-1}$$

$$+ j\left(mn + (m-1)i - [\log_q r_1] - \cdots - [\log_q r_m] - \log_q m - \frac{1}{q-1}\right)$$

$$= \frac{l - je_n\left(\sum_{\alpha=1}^m q^{[\log_q r_\alpha]}\right)}{e_n q^{n-1}(q-1)} + \frac{1}{q-1} + j\left((m-1)\left(i + \frac{1}{q-1}\right) + m - \log_q m\right)$$

$$+ j\left(\sum_{\alpha=1}^m \left(\frac{q^{[\log_q r_\alpha] - n + 1}}{q-1} - [\log_q r_\alpha] + n - 1 - \frac{1}{q-1}\right)\right)$$

$$\geq \frac{1}{q-1} + j\left((m-1)\left(i + \frac{1}{q-1}\right) + m - \log_q m\right)$$

$$+ j \sum_{\alpha=1}^m (f([\log_q r_\alpha] - n + 1) - f(0)),$$

where $f(\beta) = \frac{q^\beta}{q-1} - \beta$, $\beta \in \mathbb{Z}_{\geq 0}$.

It is easy to verify that $f(\beta)$ has the minimum $1/(q-1)$ for $\beta = 0$ and $\beta = 1$ and that $f(\beta) \xrightarrow[\beta \to +\infty]{} +\infty$.

If $M < 0$, then

$$\inf_{\substack{1 \leq j \leq \frac{l}{re_n} \\ t_1 + \cdots + t_j = l}} v(\widetilde{a}_j b_{t_1} \ldots b_{t_j})$$

is achieved when j is maximal. Therefore,

$$v(\widetilde{c}_l\pi^l) \geq \inf_{\substack{1 \leq j \leq \frac{l}{re_n} \\ t_1+\cdots+t_j=l}} v(\widetilde{a}_j b_{t_1}\ldots b_{t_j}\pi^l)$$

$$\geq \frac{1}{q-1} + \frac{l}{re_n}\left((m-1)\left(i+\frac{1}{q-1}\right) + m - \log_q m\right)$$

$$+ \frac{l}{re_n}\sum_{\alpha=1}^{m}(f([\log_q r_\alpha]-n+1) - f(0)) \geq M(r)l,$$

where $M(r)$ depends only on r $M(r) > 0$.

Since $m \leq r \leq mq^{n+i}$, we have $\inf_r M(r) = M_1 > 0$.

Hence there is an estimate independent of r and this estimate is carried out for $\exp_F\left(\frac{a^{\Delta^i}c_m s_{n+i}^m}{\pi_0^i}\right)$, where s_{n+i}^m is the sum of functions $d_{r_1}\ldots d_{r_m}z^r$.

So we have

$$\exp_F\left(\frac{a^{\Delta^i}c_m s_{n+i}^m(X)}{\pi_0^i}\right)\bigg|_{X=\pi} = \exp_F\left(\frac{a^{\Delta^i}c_m s_{n+i}^m(\pi)}{\pi_0^i}\right) = 0$$

by Theorem 2.2, and we have the estimate $v(\widetilde{c}_l X^l) \geq Ml$ ($M > 0$) for the series

$$\exp_F\left(\frac{a^{\Delta^i}c_m s_{n+i}^m(X)}{\pi_0^i}\right) = \sum_l \widetilde{c}_l X^l.$$

Therefore, the assumptions of Theorem 1.1 hold for the series $\widetilde{c}(X)$ and the element π. But

$$v(\widetilde{c}_l X^l) \geq (m-1)\left(i+\frac{1}{q-1}\right) + m - \log_q m$$

by inequality (3). Thus,

$$\inf_{m \geq 2}\inf_{l \geq 1}\inf_{\substack{1 \leq t \leq l \\ \lambda_1+\cdots+\lambda_t=l}} v(\widetilde{c}_{\lambda_1}\ldots \widetilde{c}_{\lambda_t}\pi^l) \geq i \xrightarrow[i\to+\infty]{} +\infty,$$

$$\inf_{i \geq 0}\inf_{l \geq 1}\inf_{\substack{1 \leq t \leq l \\ \lambda_1+\cdots+\lambda_t=l}} v(\widetilde{c}_{\lambda_1}\ldots \widetilde{c}_{\lambda_t}\pi^l) \geq \frac{m-1}{q-1} \xrightarrow[m\to+\infty]{} +\infty.$$

Therefore, the assumptions of the Remark following Theorem 1.2 hold for the element π and the series

$$\sum_{m=2}^{+\infty}{}_F \sum_{i=0}^{+\infty}{}_F \exp_F\left(\frac{a^{\Delta^i}c_m s_{n+i}^m(X)}{\pi_0^i}\right).$$

THEOREM 2.5. *The element*

$$\left(\sum_{m\geq 2}{}_F \sum_{i\geq 0}{}_F \exp_F \frac{a^{\Delta^i} c_m(s^{m\Delta^i}(X) - s^m_{n+i}(X))}{\pi_0^{n+i}}\right)\bigg|_{X=\pi}$$

is finite and the assumptions of Remark following Theorem 1.2 hold for the element π *and the series*

$$\sum_{m\geq 2}{}_F \sum_{i\geq 0}{}_F \exp_F \frac{a^{\Delta^i} c_m(s^{m\Delta^i}(X) - s^m_{n+i}(X))}{\pi_0^{n+i}}.$$

PROOF. $s^{m\Delta^i}(X) - s^m_{n+i}(X)$ is the finite sum of the following summands

$$d_{r_1} \ldots d_{r_m} z^{r_1 + \cdots + r_m} \prod_{j=1}^{m-\mu} \pi_0^{n+v_j} \Delta^{i-v_j} g_{v_j} = b(X),$$

where $1 \leq r_\alpha \leq q^{n+i}$, $0 \leq \mu \leq m-1$, $1 \leq v_j \leq i$, $0 \leq v_j \leq i$. $v(d_{r_\alpha}) \geq n + i - [\log_q r_\alpha]$.

Let $a(X) = \exp_F \frac{a^{\Delta^i} c_m X}{\pi_0^{n+i}}$, $\tilde{c}(X) = a \circ b(X)$. Then

$$v(b_t) \geq \sum_{\alpha=1}^{\mu}(n + i - [\log_q r_\alpha]) + \sum_{\beta=1}^{m-\mu}(n + v_\beta),$$

where $t \geq e_n\left(\sum_{\alpha=1}^{\mu} r_\alpha + \sum_{\beta=1}^{m-\mu} q^{i-v_\beta}\right)$, because $\Delta^r g_v \in X^{q^r e_n} o[[X]]$. Let

$$R = v(a_j b_{t_1} \ldots b_{t_j} \pi^l), \quad t_1 + \cdots + t_j = l \geq j e_n\left(\sum_{\alpha=1}^{\mu} r_\alpha + \sum_{\beta=1}^{m-\mu} q^{i-v_\beta}\right).$$

As in Theorem 2.4, we have

$$R \geq -j \log_F m - j(n+i) - \frac{l-1}{q-1}$$
$$+ j\left(\sum_{\alpha=1}^{\mu}(n + i - [\log_q r_\alpha]) + \sum_{\beta=1}^{m-\mu}(n + v_\beta)\right) + \frac{l}{e_n q^{n-1}(q-1)}.$$

The latter function is linear in j. If it has a minimum when $j = 1$, then

$$v(\tilde{c}_l \pi^l) \geq \frac{l}{e_n q^{n-1}(q-1)} \xrightarrow[l \to +\infty]{} +\infty.$$

In the general case
(4)
$$R \geq j\left(m - \log_q m + (m-1)\left(i + \frac{1}{q-1}\right) - n\right)$$
$$+ j\left(\sum_{\alpha=1}^{\mu}(f([\log_q r_\alpha] - n + 1) - f(0)) + \sum_{\beta=1}^{m-\mu}(f(i - v_\beta - n + 1) - f(0))\right)$$
$$+ \frac{l - je_n\left(\sum_{\alpha=1}^{\mu} r_\alpha + \sum_{\beta=1}^{m-\mu} q^{i-v_\beta}\right)}{e_n q^{n-1}(q-1)}$$

everywhere. The second and the third summands are nonnegative. Let $i \geq n$. Then $(m-1)i - n \geq 0$, i.e., the first summand is positive.

If the function has a minimum when
$$j = \frac{l}{e_n \left(\sum_{\alpha=1}^{\mu} r_\alpha + \sum_{\beta=1}^{m-\mu} q^{i-v_\beta} \right)},$$

then
$$v(\tilde{c}_l \pi^l) \geq l \frac{(m-2)i + m - \log_q m + \frac{m-1}{q-1}}{e_n \left(\sum_{\alpha=1}^{\mu} r_\alpha + \sum_{\beta=1}^{m-\mu} q^{i-v_\beta} \right)},$$

i.e., $v(\tilde{c}_l \pi^l) \geq Ml$, $M > 0$.

If the function has a minimum when $j = 1$, then
$$v(\tilde{c}_l \pi^l) \geq (m-2)i + m - \log_q m + \frac{m-1}{q-1}.$$

Let $i < n$. Thus, $f(i - n - (v_\beta - 1)) \geq n - i$, and
$$j \left(\sum_{\alpha=1}^{\mu} (f([\log_q r_\alpha]) - n + 1) - f(0)) + \sum_{\beta=1}^{m-\mu} (f(i - v_\beta - n + 1) - f(0)) \right)$$
$$\geq (m-\mu) j \left(n - i - \frac{1}{q-1} \right) \geq j \left(n - i - \frac{1}{q-1} \right)$$

for the second summand of the sum (4). Therefore,
$$j \left(m - \log_q m + (m-1)\left(i + \frac{1}{q-1}\right) - n \right)$$
$$+ j \left(\sum_{\alpha=1}^{\mu} (f([\log_q r_\alpha]) - n + 1) - f(0)) + \sum_{\beta=1}^{m-\mu} (f(i - v_\beta - n + 1) - f(0)) \right)$$
$$\geq j \left((m-2)\left(i + \frac{1}{q-1}\right) + m - \log_q m \right) > 0.$$

Finally, if we have a minimum when
$$j = \frac{l}{e_n \left(\sum_{\alpha=1}^{\mu} r_\alpha + \sum_{\beta=1}^{m-\mu} q^{i-v_\beta} \right)},$$

then $v(\tilde{c}_l \pi^l) \geq Ml$, $M > 0$ too.

Hence
$$v(\tilde{c}_l \pi^l) \geq Ml, \quad M > 0$$

everywhere (M is the minimum of our constants).

But we have a finite number of ways to choose r_α, v_β, μ. Therefore this estimate is true for
$$\exp_F \frac{a^{\Delta^i} c_m (s^{m\Delta^i}(X) - s^m_{n+i}(X))}{\pi_0^{n+i}} = \sum_{l \geq 1} \tilde{c}_l X^l.$$

The assumptions of Theorem 1.1 hold for this series and the element π by the Remark following Theorem 1.1.

Inequality (4) now implies

$$\inf_{\substack{m\geqslant 2\\ l\geqslant 1}}\inf_{\substack{1\leqslant t\leqslant l\\ \lambda_1+\cdots+\lambda_t=l}}v(\widetilde{c}_{\lambda_1}\ldots\widetilde{c}_{\lambda_t}\pi^l)\geqslant i-n\xrightarrow[i\to+\infty]{}+\infty,$$

$$\inf_{\substack{i\geqslant 0\\ l\geqslant 1}}\inf_{\substack{1\leqslant t\leqslant l\\ \lambda_1+\cdots+\lambda_t=l}}v(\widetilde{c}_{\lambda_1}\ldots\widetilde{c}_{\lambda_t}\pi^l)\geqslant \frac{m-1}{q-1}\xrightarrow[i\to+\infty]{}+\infty.$$

An application of Theorem 1.2 completes the proof.

THEOREM 2.6. *The assumptions of Theorem 1.1 hold for the series*

$$E(as(X)) = \sum_{i=1}^{+\infty}{}_F \exp_F \frac{\Delta^i s(X)\Delta^i a}{\pi_0^i} = a(X)$$

and the element π, i.e.,

$$\inf_{\substack{1\leqslant t\leqslant r\\ r_1+\cdots+r_t=r}}v(a_{r_1}\ldots a_{r_t}\pi^r)\xrightarrow[r\to+\infty]{}+\infty$$

and we can use $E(as(X))|_{X=\pi}$ as an summand of the formal sum.

PROOF. We have

$$\Delta^i s = s_{n+i} + \sum_{\alpha=1}^{i}\pi_0^{n+\alpha}\Delta^{i-\alpha}g_\alpha,$$

$$s_{n+i} = \sum_{r=1}^{q^{n+i}}d_r z^r, \qquad v(d_r)\geqslant n+i-[\log_q r].$$

Let $\widetilde{a}(X) = \exp_F \frac{\Delta^i aX}{\pi_0^i}$. For the series $b(X) = d_r X^r$ and $c = \widetilde{a}\circ b$, we have

$$v(c_l\pi^l)\geqslant \inf_{t_1+\cdots+t_j=l\geqslant je_n r}v(\widetilde{a}_j b_{t_1}\ldots b_{t_j}\pi^l)$$

$$\geqslant \frac{l}{e_n q^{n-1}(q-1)} + \inf_{1\leqslant j\leqslant \frac{l}{re_n}}\left(-ij-\frac{j}{q-1}+j(n+i)\right)\geqslant \frac{l}{e_n q^{n-1}(q-1)}.$$

For $b(X) = \pi_0^{n+\alpha}\Delta^{i-\alpha}g_\alpha(X)$, $1\leqslant \alpha\leqslant i$, $c = \widetilde{a}\circ b$, we have

$$v(c_l\pi^l)\geqslant \inf_{t_1+\cdots+t_j=l\geqslant je_n q^{i-\alpha}}\left(-ij-\frac{j}{q-1}+j(n+\alpha)\right)+\frac{l}{e_n q^{n-1}(q-1)}.$$

If $n+\alpha-i-\frac{1}{q-1}\geqslant 0$, then $v(c_l\pi^l)\geqslant \frac{l}{e_n q^{n-1}(q-1)}$, and if $n+\alpha-i-\frac{1}{q-1}<0$ then

$$v(c_l\pi^l)\geqslant \frac{l}{e_n q^{n-1}(q-1)} - \frac{l}{e_n q^{i-\alpha}}\left(i+\frac{1}{q-1}-n-\alpha\right)$$

$$= \frac{l}{e_n q^{i-\alpha}}\left(f(i-\alpha-n+1)+\frac{q-2}{q-1}\right) > 0.$$

This function has a minimum at α if $\alpha = i - n - 1$ or $\alpha = i - n$, i.e., we have an evaluation $v(c_l\pi^l) \geqslant Ml$, where $M > 0$ and M does not depend on α, i. Hence for

$$c^{(i)}(X) = \exp_F \frac{\Delta^i s(X) \Delta^i a}{\pi_0^i} = \sum_{l \geqslant 1} c_l^{(i)} X^l$$

we have $v(c_l^{(i)}\pi^l) \geqslant Ml$, where $M > 0$ and M does not depend on i.

The series $E(as(X))$ has integral coefficients and it converges for $X = \pi$. By formula (2) from the proof of Theorem 1.2 if $a(X) = E(as(X))$, then

$$v(a_l\pi^l) \geqslant \inf_{\substack{t \geqslant 1 \\ 1 \leqslant i_1 < \cdots < i_t \\ 1 \leqslant l_1 + \cdots + l_t \leqslant l \\ \alpha_1^{(1)} + \cdots + \alpha_{l_1}^{(1)} + \cdots + \\ + \alpha_1^{(t)} + \cdots + \alpha_{l_t}^{(t)} = l}} v(c_{\alpha_1^{(1)}}^{(i_1)} \cdots c_{\alpha_{l_1}^{(1)}}^{(i_1)} \cdots c_{\alpha_1^{(t)}}^{(i_t)} \cdots c_{\alpha_{l_t}^{(t)}}^{(i_t)} \pi^l) \geqslant Ml.$$

Hence the assumptions of Theorem 1.1 hold for the series $E(as(X))$ and the element π by the Remark following Theorem 1.1.

Finally, by Theorems 2.2–2.6 we have

$$H(a) = E_F(as(X))|_{X=\pi} +_F [\pi_0^n]\left(\sum_{m=2}^{+\infty}{}_F \sum_{j=0}^{+\infty}{}_F \exp_F \frac{a^{\Delta^i} c_m s^{m\Delta^i}(\pi)}{\pi_0^{n+i}}\right),$$

where

$$\sum_{m=2}^{+\infty}{}_F \sum_{j=0}^{+\infty}{}_F \exp_F \frac{a^{\Delta^i} c_m s^{m\Delta^i}(\pi)}{\pi_0^{n+i}}$$

is a finite element.

References

1. S. V. Vostokov, *Norm pairing in formal modules*, Izv. Akad. Nauk SSSR Ser. Mat. **43** (1979), no. 4, 765–794; English transl. in Math. USSR-Izv. **15** (1980).
2. S. V. Vostokov, *Explicit form of the law of reciprocity*, Izv. Akad. Nauk SSSR Ser. Mat. **42** (1978), no. 6, 1288–1321; English transl. Math. USSR-Izv. **13** (1979), no. 3, 557–588.
3. I. B. Zhukov and A. I. Madunts, *Multidimensional complete fields: topology and other basic constructions*, this volume.
4. G. Henniart, *Sur les lois de réciprocité explicites. I*, J. Reine Angew. Math. **329** (1981), 177–202.

Translated by THE AUTHOR

On Köck's Conjecture about Shuffle Products

A. NENASHEV

Dedicated to the memory of my grandfather Petr Ivanovich (1899–1992)

ABSTRACT. We prove that Köck's shuffle product $\mathcal{H}^p \times \mathcal{H}^q \to \mathcal{H}^{p+q}$ is compatible with the tensor product $G^p \times G^q \to G^{p+q}$ in the case $p = 1$.

§1. Preliminaries

For a more detailed account of the material in this section see [Gr] or [Kö].

Let Δ be the category of finite nonempty totally ordered sets. For $A \in \Delta$, we let $\gamma(A)$ denote the disjoint union $A \cup \{L, R\}$ and regard $\gamma(A)$ as a partially ordered set, where A is an ordered subset, $L < a$ and $R < a$ for any $a \in A$, and the elements L and R are not comparable. Let $\Gamma(A) = \{i/j \mid i \in A, j \in \gamma(A), j \leqslant i\}$. We introduce an order on $\Gamma(A)$ by letting $i/j \leqslant i'/j'$ if $i \leqslant i'$ and $j \leqslant j'$ and regard the ordered set $\Gamma(A)$ as a category.

Let $\mathcal{M}$ be an exact category with a distinguished zero object $*$. We call a functor $F\colon \Gamma(A) \to \mathcal{M}$ *exact* if $F(i/i) = *$ for any $i \in A$, and the sequence $0 \to F(i/j) \to F(i'/j) \to F(i'/i) \to 0$ is exact for any $j \in \gamma(A)$ and $i, i' \in A$ with $j \leqslant i \leqslant i'$. We define the *simplicial* set $G = G\mathcal{M}$ by letting

$$G[A] = \mathrm{Exact}(\Gamma(A), \mathcal{M}), \quad A \in \Delta,$$

with obvious face and degeneracy maps. More generally, for $k \geqslant 1$, we define k-simplicial set $G^k = G^k\mathcal{M}$ by

$$G^k[A_1, \ldots, A_k] = \mathrm{Exact}\big(\Gamma(A_1) \times \cdots \times \Gamma(A_k), \mathcal{M}\big)$$

for any $A_1, \ldots, A_k \in \Delta$, where Exact refers to the set of functors exact in each variable.

We now proceed to another generalization. For $k \geqslant 1$ and $A \in \Delta$, let $\Gamma^k(A)$

1991 *Mathematics Subject Classification.* Primary 19D06; Secondary 19D99, 19D10.

Key words and phrases. Shuffle products, operators in K-theory, simplicial map, simplicial homotopy.

©1995, American Mathematical Society

be the set of all collections

$$\alpha = \left(\frac{i_1}{l_1} *_2 \frac{i_2}{l_2} *_3 \cdots *_k \frac{i_k}{l_k} \right),$$

where for each r we have

(A1) $i_r \in A$, $l_r \in \gamma(A)$, $*_r \in \{\wedge, \otimes\}$;
(A2) $l_r \leqslant i_r$;
(A3) if $r \geqslant 2$ and $*_r = \wedge$, then $l_{r-1} = l_r$ and $i_{r-1} \leqslant i_r$.

We introduce an order on $\Gamma^k(A)$ by the following rule: there is an arrow $\alpha \to \alpha'$ if for each r we have

(B1) $i_r \leqslant i'_r$;
(B2) $l_r \leqslant l'_r$;
(B3) if $*_r = \wedge$ and $*'_r = \otimes$, then $i_{r-1} \leqslant l'_r$.

A sequence $\alpha' \to \alpha \to \alpha''$ in $\Gamma^k(A)$ is said to be *exact* if there exist integers $p \leqslant s$ such that

(C1) for any r with $r < p$ or $r > s$ we have:

$$i'_r = i_r = i''_r, \quad l'_r = l_r = l''_r, \quad *'_r = *_r = *''_r;$$

(C2) for any r satisfying $p < r \leqslant s$ we have:

$$*'_r = *_r = *''_r = \wedge, \quad i'_r = i_r = i''_r;$$

(C3) $l_p = l'_p \leqslant i'_p = l''_p \leqslant i''_p = i_p$, $*'_p = *_p$ and $*''_p = \otimes$.

Obviously, $\Gamma^1(A) = \Gamma(A)$ and exact sequences in $\Gamma(A)$ will be those of the form $i/j \to i'/j \to i'/i$. We define the simplicial set $\mathcal{H}^k = \mathcal{H}^k \mathcal{M}$ by letting

$$\mathcal{H}^k[A] = \mathrm{Exact}(\Gamma^k(A), \mathcal{M}) \qquad \text{for } A \in \Delta,$$

and, more generally, for $k_1, \ldots, k_n \geqslant 1$, the n-simplicial set $\mathcal{H}^{k_1 \ldots k_n} \mathcal{M}$ is given by

$$\mathcal{H}^{k_1 \ldots k_n}[A_1, \ldots, A_n] = \mathrm{Exact}\big(\Gamma^{k_1}(A_1) \times \cdots \times \Gamma^{k_n}(A_n), \mathcal{M}\big)$$

for $A_1, \ldots, A_n \in \Delta$, where Exact refers again to functors exact in each variable. We have $\mathcal{H}^1 = G$ and $\mathcal{H}^{1, \ldots, 1} = G^n$.

For $A_1, \ldots, A_k \in \Delta$, we let $A_1 \ldots A_k \in \Delta$ denote the concatenation, i.e., the disjoint union $A_1 \cup \cdots \cup A_k$ ordered in such a way that each A_r is its ordered subset, and $a_r < a_s$ whenever $a_r \in A_r$, $a_s \in A_s$ with $r < s$. We define the following functor that proves to be exact in each variable (see [Gr, §6]),

$$\theta = \theta^k : \Gamma(A_1) \times \cdots \times \Gamma(A_k) \to \Gamma^k(A_1 \ldots A_k)$$

$$\left(\frac{i_1}{j_1}, \ldots, \frac{i_k}{j_k} \right) \mapsto \left(\frac{i_1}{l_1} *_2 \frac{i_2}{l_2} *_3 \cdots *_k \frac{i_k}{l_k} \right),$$

where $l_1 = j_1$ and for $r \geqslant 2$ we declare

(D1) if $j_r = L$, then $*_r = \wedge$ and $l_r = l_{r-1}$;
(D2) if $j_r \neq L$, then $*_r = \otimes$ and $l_r = j_r$.

The k-fold edgewise subdivision functor is the functor

$$\mathrm{Sub}_k : \text{simp. sets} \to k\text{-simp. sets}$$
$$(\mathrm{Sub}_k X)[A_1, \ldots, A_k] = X[A_1 \ldots A_k]$$

for any simplicial set X. It is actually a subdivision, since there exists a natural homeomorphism $|\mathrm{Sub}_k X| \xrightarrow{\sim} |X|$ (see [Gr, §4]). More generally, for $k_1, \ldots, k_n \geqslant 1$ and an n-simplicial set X, we define its subdivision $\mathrm{Sub}_{k_1, \ldots, k_n} X$ to be the k-simplicial set given by

$$(\mathrm{Sub}_{k_1, \ldots, k_n} X)[A_1, \ldots, A_k]$$
$$= X[A_1 \ldots A_{k_1}, A_{k_1+1} \ldots A_{k_1+k_2}, \ldots, A_{k-k_n+1} \ldots A_k],$$

where $k = k_1 + \cdots + k_n$. We note that there is a natural isomorphism of $(p+q)$-simplicial sets

$$\mathrm{Sub}_{p,q}(X \times Y) \cong \mathrm{Sub}_p X \times \mathrm{Sub}_q Y$$

for any simplicial sets X, Y and any $p, q \geqslant 1$.

We define the k-simplicial map $\Theta = \Theta^k : \mathrm{Sub}_k \mathcal{H}^k \to G^k$ by

$$(\Theta^k F)\left(\frac{i_1}{j_1}, \ldots, \frac{i_k}{j_k}\right) = F\left(\theta\left(\frac{i_1}{j_1}, \ldots, \frac{i_k}{j_k}\right)\right),$$

where $F \in \mathrm{Sub}_k \mathcal{H}^k[A_1, \ldots, A_k] = \mathrm{Exact}(\Gamma^k(A_1 \ldots A_k), \mathcal{M})$ and

$$\left(\frac{i_1}{j_1}, \ldots, \frac{i_k}{j_k}\right) \in \Gamma(A_1) \times \cdots \times \Gamma(A_k).$$

More generally, the k-simplicial map

$$\Theta^{k_1, \ldots, k_n} : \mathrm{Sub}_{k_1, \ldots, k_n} \mathcal{H}^{k_1, \ldots, k_n} \to G^k,$$

with $k = k_1 + \cdots + k_n$, is given by

$$(\Theta^{k_1, \ldots, k_n} F)\left(\frac{i_1}{j_1}, \ldots, \frac{i_k}{j_k}\right)$$
$$= F\left(\theta\left(\frac{i_1}{j_1}, \ldots, \frac{i_{k_1}}{j_{k_1}}\right), \theta\left(\frac{i_{k_1+1}}{j_{k_1+1}}, \ldots, \frac{i_{k_1+k_2}}{j_{k_1+k_2}}\right), \ldots, \theta\left(\frac{i_{k-k_n+1}}{j_{k-k_n+1}}, \ldots, \frac{i_k}{j_k}\right)\right).$$

Finally, we suppose that there is a biexact tensor product operation $\otimes$ in the category $\mathcal{M}$. Then we have the k-simplicial tensor product map

$$\otimes : G^p \times G^q \to G^k \quad \text{with } k = p+q, \qquad (F, G) \mapsto F \otimes G,$$

where $(F \otimes G)\left(\frac{i_1}{j_1}, \ldots, \frac{i_k}{j_k}\right) = F\left(\frac{i_1}{j_1}, \ldots, \frac{i_p}{j_p}\right) \otimes G\left(\frac{i_{p+1}}{j_{p+1}}, \ldots, \frac{i_k}{j_k}\right)$, and the bisimplicial tensor product map

$$\otimes : \mathcal{H}^p \times \mathcal{H}^q \to \mathcal{H}^{p+q}, \qquad (F, G) \mapsto F \otimes G \quad \text{for } F \in \mathcal{H}^p[A], \ G \in \mathcal{H}^q[B],$$

where $(F \otimes G)(\alpha, \beta) = F(\alpha) \otimes G(\beta)$ for any $\alpha \in \Gamma^p(A)$, $\beta \in \Gamma^q(B)$.

§2. Shuffle products

We proceed to define a simplicial map

$$\mathrm{Sym} = \mathrm{Sym}^{p,q} \colon \mathrm{Diag}\,\mathcal{H}^{p,q} \to \mathcal{H}^{p+q} \qquad (p, q \geqslant 1).$$

Let $A \in \Delta$ and $F \in (\mathrm{Diag}\,\mathcal{H}^{p,q})[A]$, i.e., F is a biexact functor $\Gamma^p(A) \times \Gamma^q(A) \to \mathcal{M}$. Let

$$\alpha = \left(\frac{i_1}{l_1} *_2 \frac{i_2}{l_2} *_3 \cdots *_k \frac{i_k}{l_k}\right) \in \Gamma^k(A), \quad \text{where } k = p + q.$$

The set $\{1, \ldots, k\}$ is divided into several intervals $I_1, \ldots, I_t$ by those $*_r$ of α that are equal to $\otimes$. For any set

$$R = \{n_1 < n_2 < \cdots < n_p\} \subset \{1, 2, \ldots, k\}$$

we define $\alpha^R \in \Gamma^p(A)$ by letting $i_r(\alpha^R) = i_{n_r}(\alpha)$, $l_r(\alpha^R) = l_{n_r}(\alpha)$, and $*_r(\alpha^R) = \otimes$ if there is at least one $\otimes$ between n_{r-1} and n_r in α (i.e., if n_{r-1} and n_r belong to different intervals), otherwise $*_r(\alpha^R) = \wedge$.

We denote by $\mathcal{P}$ the set of all subsets $R \subset \{1, 2, \ldots, k\}$ with $|R| = p$. We introduce an equivalence relation $\sim_\alpha$ on $\mathcal{P}$ by the rule: $R \sim_\alpha R'$ if $|R \cap I_r| = |R' \cap I_r|$ for any $r = 1, \ldots, t$.

The following results are proved in [Kö, §3].

(a) Let $\mathcal{R}$ be an equivalence class in $\mathcal{P}/\sim_\alpha$. We put $\overline{R} = \{1, \ldots, k\} \setminus R$ for any $R \in \mathcal{P}$. Then all objects $F(\alpha^R, \alpha^{\overline{R}})$ with $R \in \mathcal{R}$ can be regarded as subobjects of a single object, and we can take their inner sum $\sum_{R \in \mathcal{R}} F(\alpha^R, \alpha^{\overline{R}})$. We set

$$(\mathrm{Sym}\,F)(\alpha) = \bigoplus_{\mathcal{R} \in \mathcal{P}/\sim_\alpha} \sum_{R \in \mathcal{R}} F(\alpha^R, \alpha^{\overline{R}}).$$

(b) The function $\mathrm{Sym}\,F \colon \Gamma^{p+q}(A) \to \mathcal{M}$ thus defined is exact, so $\mathrm{Sym}\,F \in \mathcal{H}^k(A)$.

(c) (easy). This is compatible with face and degeneracy maps, therefore, we obtain the desired shuffle product map $\mathrm{Sym}\colon \mathrm{Diag}\,\mathcal{H}^{p,q} \to \mathcal{H}^k$. We also refer to the composite map

$$\otimes \colon \mathcal{H}^p \times \mathcal{H}^q \xrightarrow{\otimes} \mathcal{H}^{p,q} \xrightarrow{\mathrm{Sym}} \mathcal{H}^{p+q},$$

where we omit Diag in the notation, as a shuffle product.

Shuffle products were introduced by B. Köck [Kö] in order to study properties of λ-operations in higher K-theory, using the definition of λ-operations given by Grayson [Gr] by means of simplicial maps $\wedge^k \colon \mathrm{Sub}_k\,G\mathcal{M} \to G^k\mathcal{M}$, $k \geqslant 1$ (another definition of λ-operations on the simplicial level was given by the author in [Ne]). Köck's approach is motivated by the fact that it proves to be easier to carry out some computations with λ-operations on the level of the spaces $\mathcal{H}^k$ rather than G^k. But since finally we need to descend results to the level of G^k by the maps Θ (because G^k represents the K-theory homotopy type $K\mathcal{M}$), we want shuffle products to be related with tensor products of G^k-s by means of the following conjecture under which some of the results in [Kö] are proved.

CONJECTURE ([Kö, p. 16]). *Given $p, q \geq 1$, the diagram*

$$\begin{array}{ccc} |\mathcal{H}^p| \times |\mathcal{H}^q| & \xrightarrow{\otimes} & |\mathcal{H}^{p+q}| \\ {\scriptstyle \Theta^p \times \Theta^q} \downarrow & & \downarrow {\scriptstyle \Theta^{p+q}} \\ |G^p| \times |G^q| & \xrightarrow{\otimes} & |G^{p+q}| \end{array}$$

commutes up to homotopy.

Consider the diagram of k-simplicial sets, with $k = p + q$,

$$\begin{array}{ccccc} \mathrm{Sub}_{p,q}(\mathcal{H}^p \times \mathcal{H}^q) & \xrightarrow{\otimes} & \mathrm{Sub}_{p,q}(\mathcal{H}^{p,q}) & \xleftarrow{f} & \mathrm{Sub}_k \mathrm{Diag}\, \mathcal{H}^{p,q} \\ \| & & & & \downarrow {\scriptstyle \mathrm{Sym}^{p,q}} \\ \mathrm{Sub}_p \mathcal{H}^p \times \mathrm{Sub}_q \mathcal{H}^q & & \downarrow {\scriptstyle \Theta^{p,q}} & & \mathrm{Sub}_k \mathcal{H}^k \\ \downarrow {\scriptstyle \Theta^p \times \Theta^q} & & & & \downarrow {\scriptstyle \Theta^k} \\ G^p \times G^q & \xrightarrow{\otimes} & G^k & \xrightarrow{=} & G^k \end{array}$$

where all maps except f have been already defined, the left square is obviously strictly commutative, and the map

$$f: \mathrm{Sub}_k \mathrm{Diag}\, \mathcal{H}^{p,q}[A_1, \ldots, A_k] = \mathcal{H}^{p,q}[A_1 \ldots A_k, A_1 \ldots A_k] \xrightarrow{d_{A_1 \ldots A_p, A_{p+1} \ldots A_k}}$$
$$\to \mathcal{H}^{p,q}[A_1 \ldots A_p, A_{p+1} \ldots A_k] = \mathrm{Sub}_{p,q} \mathcal{H}^{p,q}[A_1, \ldots, A_k]$$

is exactly the face map associated with the inclusions $A_1 \ldots A_p \subset A_1 \ldots A_k$ and $A_{p+1} \ldots A_k \subset A_1 \ldots A_k$. It is not hard to show that the map $|f|: |\mathcal{H}^{p,q}| \to |\mathcal{H}^{p,q}|$ is homotopic to the identity map (observe that we have $|f|(|x|) \subset |x|$ for any bisimplex x of $\mathcal{H}^{p,q}$). Therefore, the commutativity of the right square up to homotopy is equivalent to

CONJECTURE' ([Kö, p. 17]). *The diagram*

$$\begin{array}{ccc} |\mathcal{H}^{p,q}| & \xrightarrow{\mathrm{Sym}^{p,q}} & |\mathcal{H}^k| \\ & \searrow{\scriptstyle \Theta^{p,q}} & \downarrow {\scriptstyle \Theta^k} \\ & & |G^k| \end{array}$$

commutes up to homotopy.

Obviously, the Conjecture' implies the above conjecture. In the present paper we shall prove the Conjecture' in the case $p = 1$.

THEOREM. *Given $q \geq 1$, the diagram of k-simplicial sets*

$$\begin{array}{ccc} \mathrm{Sub}_k \, \mathrm{Diag}\, \mathcal{H}^{1,q} & \xrightarrow{\mathrm{Sym}^{1,q}} & \mathrm{Sub}_k \, \mathcal{H}^k \\ \downarrow f & & \downarrow \Theta^k \\ \mathrm{Sub}_{1,q} \, \mathcal{H}^{1,q} & \xrightarrow{\Theta^{1,q}} & G^k \end{array}$$

where $k = q+1$, commutes up to homotopy.

SKETCH OF THE PROOF. In the rest of the paper we shall construct the sequence of k-simplicial maps and simplicial homotopies

(2.1)
$$\Theta^k \circ \mathrm{Sym}^{1,q}$$
$$\|$$
$$R_k \xrightarrow{V_k} S_k \xleftarrow{W_k} T_k$$
$$\uparrow U_{k-1}$$
$$R_3 \xrightarrow{V_3} S_3 \xleftarrow{W_3} T_3 \quad \ldots \quad T_{k-1}$$
$$\uparrow U_2$$
$$R_2 \xrightarrow{V_2} S_2 \xleftarrow{W_2} T_2$$
$$\uparrow U_1$$
$$T_1 = \Theta^{1,q} \circ f,$$

where the maps T_r ($1 \leq r \leq k$), R_r, S_r ($2 \leq r \leq k$) are maps from $\mathrm{Sub}_k \, \mathrm{Diag}\, \mathcal{H}^{1,q}$ to G^k, and U_r ($1 \leq r \leq k-1$), V_r, W_r ($2 \leq r \leq k$) are simplicial homotopies in the rth direction.

First we compute the maps T_1 and T_k. For

$$g = \left(\frac{i_1}{j_1}, \ldots, \frac{i_k}{j_k}\right) \in \Gamma(A_1) \times \cdots \times \Gamma(A_k)$$

(resp. for $\alpha \in \Gamma^k(A)$) and $r = 1, \ldots, k$, we let $g_{\neq r}$ (resp. $\alpha_{\neq r}$) denote the $(k-1)$-tuple obtained from g by deleting i_r/j_r (resp., by deleting $\alpha^{\{r\}}$, see §1). We note that the operations θ and $g_{\neq r}$ do not commute in general; namely, for $r = 1, \ldots, k-1$

$$\theta(g)_{\neq r} \neq \theta(g_{\neq r}) \text{ if and only if } j_r \neq L \text{ and } j_{r+1} = L,$$

i.e., if and only if $l_{r+1} \neq j_{r+1}$,

and obviously $\theta(g)_{\neq k} = \theta(g_{\neq k})$ for any g.

Let $F \in \mathrm{Sub}_k \, \mathrm{Diag}\, \mathcal{H}^{1,q}[A_1, \ldots, A_k] = \mathcal{H}^{1,q}[A_1 \ldots A_k, A_1 \ldots A_k]$, i.e.,

$$F : \Gamma(A_1 \ldots A_k) \times \Gamma^q(A_1 \ldots A_k) \to \mathcal{M}$$

is a biexact functor. Both T_1F and T_kF should be polyexact functors $\Gamma(A_1) \times \cdots \times \Gamma(A_k) \to \mathcal{M}$. Let

$$g = \left(\frac{i_1}{j_1}, \ldots, \frac{i_k}{j_k}\right) \in \Gamma(A_1) \times \cdots \times \Gamma(A_k).$$

By definition,
(2.2)
$$T_1F(g) = ((\Theta^{1,q} \circ f)F)(g) = (fF)\left(\frac{i_1}{j_1}, \theta\left(\frac{i_2}{j_2}, \ldots, \frac{i_k}{j_k}\right)\right) = F\left(\frac{i_1}{j_1}, \theta(g_{\neq 1})\right)$$

since f only restricts the domain of a given functor from $\Gamma(A_1 \ldots A_k) \times \Gamma^q(A_1 \ldots A_k)$ to $\Gamma(A_1) \times \Gamma^q(A_2 \ldots A_k)$;

(2.3)
$$T_kF(g) = ((\Theta^k \circ \mathrm{Sym}^{1,q})F)(g) = \mathrm{Sym}^{1,q} F(\theta(g))$$
$$= F\left(\frac{i_1}{j_1}, \theta(g)_{\neq 1}\right) \circ_2 F\left(\frac{i_2}{l_2}, \theta(g)_{\neq 2}\right) \circ_3 \cdots \circ_k F\left(\frac{i_k}{l_k}, \theta(g)_{\neq k}\right),$$

where $\circ_r = +$ if $j_r = L$ and $\circ_r = \oplus$ if $j_r \neq L$.

Before proceeding to the general construction, we supply elementary computations in the case $p = q = 1$, $k = 2$, that will lead to the required sequence of simplicial homotopies after a proper generalization.

Let $A = \{a\}$, $B = \{b\} \in \Delta$, and let $F \in \mathrm{Sub}_2 \mathrm{Diag}\, \mathcal{H}^{1,1}[A, B]$ be a vertex, i.e., $F: \Gamma(AB) \times \Gamma(AB) \to \mathcal{M}$ is a biexact functor. Recall that a vertex $E \in G^2[A, B] = \mathrm{Exact}(\Gamma(A) \times \Gamma(B), \mathcal{M})$ is given by a 4-tuple of objects

$$\begin{bmatrix} E\left(\frac{a}{R}, \frac{b}{L}\right) & E\left(\frac{a}{R}, \frac{b}{R}\right) \\ E\left(\frac{a}{L}, \frac{b}{L}\right) & E\left(\frac{a}{L}, \frac{b}{R}\right) \end{bmatrix}.$$

Since T_1F is actually the restriction of F to $\Gamma(A) \times \Gamma(B)$, we have

$$T_1F = \begin{bmatrix} F\left(\frac{a}{R}, \frac{b}{L}\right) & F\left(\frac{a}{R}, \frac{b}{R}\right) \\ F\left(\frac{a}{L}, \frac{b}{L}\right) & F\left(\frac{a}{L}, \frac{b}{R}\right) \end{bmatrix}.$$

According to (2.3), we find that

$$T_2F = \begin{bmatrix} F\left(\frac{a}{R}, \frac{b}{R}\right) + F\left(\frac{b}{R}, \frac{a}{R}\right) & F\left(\frac{a}{R}, \frac{b}{R}\right) \oplus F\left(\frac{b}{R}, \frac{a}{R}\right) \\ F\left(\frac{a}{L}, \frac{b}{L}\right) + F\left(\frac{b}{L}, \frac{a}{L}\right) & F\left(\frac{a}{L}, \frac{b}{R}\right) \oplus F\left(\frac{b}{R}, \frac{a}{L}\right) \end{bmatrix}.$$

It is not hard to show that the alternating sums in $K_0\mathcal{M}$ corresponding to the vertices T_1F and T_2F give rise to the same element of $K_0\mathcal{M}$, therefore, T_1F and T_2F lie in the same component of G^2. It proves to be possible to connect the vertices T_1F and T_2F by a path of three edges in G^2 naturally with respect to F. The intermediate vertices are given by

$$R_2F := \begin{bmatrix} F\left(\frac{a}{R}, \frac{b}{L}\right) + F\left(\frac{b}{R}, \frac{a}{L}\right) & F\left(\frac{a}{R}, \frac{b}{R}\right) \oplus F\left(\frac{b}{R}, \frac{a}{L}\right) \\ F\left(\frac{a}{L}, \frac{b}{L}\right) + F\left(\frac{b}{L}, \frac{a}{L}\right) & F\left(\frac{a}{L}, \frac{b}{R}\right) \oplus F\left(\frac{b}{R}, \frac{a}{L}\right) \end{bmatrix},$$

$$S_2F := $$

$$\begin{bmatrix} F\left(\frac{b}{R},\frac{b}{R}\right) \underset{F\left(\frac{b}{a},\frac{b}{a}\right)}{\times} F\left(\frac{b}{R},\frac{b}{L}\right) & F\left(\frac{a}{R},\frac{b}{R}\right) \oplus F\left(\frac{b}{R},\frac{a}{R}\right) \underset{F\left(\frac{b}{R},\frac{b}{a}\right)}{\times} F\left(\frac{b}{R},\frac{b}{L}\right) \\ F\left(\frac{a}{L},\frac{b}{L}\right) + F\left(\frac{b}{L},\frac{a}{L}\right) & F\left(\frac{a}{L},\frac{b}{R}\right) \oplus F\left(\frac{b}{R},\frac{a}{L}\right) \end{bmatrix}.$$

We claim that there exists a $(1, 0)$-simplex in G^2 connecting T_1F to R_2F. For the proof it suffices to show that

$$R_2F\left(\frac{a}{L},\frac{b}{L}\right) \Big/ T_1F\left(\frac{a}{L},\frac{b}{L}\right) \cong R_2F\left(\frac{a}{R},\frac{b}{L}\right) \Big/ T_1F\left(\frac{a}{R},\frac{b}{L}\right),$$

$$R_2F\left(\frac{a}{L},\frac{b}{R}\right) \Big/ T_1F\left(\frac{a}{L},\frac{b}{R}\right) \cong R_2F\left(\frac{a}{R},\frac{b}{R}\right) \Big/ T_1F\left(\frac{a}{R},\frac{b}{R}\right).$$

The second isomorphism is obvious, and the first one follows from the sequences

$$0 \to F\left(\frac{a}{L},\frac{b}{L}\right) \to F\left(\frac{a}{L},\frac{b}{L}\right) + F\left(\frac{b}{L},\frac{a}{L}\right) \to F\left(\frac{b}{a},\frac{a}{L}\right) \to 0,$$

$$\|$$

$$0 \to F\left(\frac{a}{R},\frac{b}{L}\right) \to F\left(\frac{a}{R},\frac{b}{L}\right) + F\left(\frac{b}{R},\frac{a}{L}\right) \to F\left(\frac{b}{a},\frac{a}{L}\right) \to 0,$$

which are exact since

$$F\left(\frac{a}{L},\frac{b}{L}\right) \cap F\left(\frac{b}{L},\frac{a}{L}\right) = F\left(\frac{a}{L},\frac{a}{L}\right)$$

and

$$F\left(\frac{a}{R},\frac{b}{L}\right) \cap F\left(\frac{b}{R},\frac{a}{L}\right) = F\left(\frac{a}{R},\frac{a}{L}\right)$$

(cf. [Kö, §2]).

We now claim that there exists a $(0, 1)$-simplex in G^2 connecting R_2F to S_2F. To prove this suffices to check that

$$S_2F\left(\frac{a}{R},\frac{b}{L}\right) \Big/ R_2F\left(\frac{a}{R},\frac{b}{L}\right) \cong S_2F\left(\frac{a}{R},\frac{b}{R}\right) \Big/ R_2F\left(\frac{a}{R},\frac{b}{R}\right),$$

since the two other quotients obviously vanish. The required result follows from the diagrams

$$\begin{array}{ccccc}
& & F\left(\frac{a}{R},\frac{b}{L}\right) + F\left(\frac{b}{R},\frac{a}{L}\right) & = & F\left(\frac{a}{R},\frac{b}{L}\right) + F\left(\frac{b}{R},\frac{a}{L}\right) \\
& & \downarrow & & \downarrow \\
F\left(\frac{a}{R},\frac{b}{R}\right) + F\left(\frac{b}{R},\frac{a}{R}\right) & \to & F\left(\frac{b}{R},\frac{b}{R}\right) \underset{F\left(\frac{b}{a},\frac{b}{a}\right)}{\times} F\left(\frac{b}{R},\frac{b}{L}\right) & \to & F\left(\frac{b}{R},\frac{b}{L}\right) \\
\| & & \downarrow & & \downarrow \\
F\left(\frac{a}{R},\frac{b}{R}\right) + F\left(\frac{b}{R},\frac{a}{R}\right) & \to & F\left(\frac{b}{R},\frac{b}{R}\right) & \to & F\left(\frac{b}{a},\frac{b}{a}\right)
\end{array}$$

and

$$
\begin{array}{ccc}
F\left(\frac{b}{R},\frac{a}{L}\right) & = & F\left(\frac{b}{R},\frac{a}{L}\right) \\
\downarrow & & \downarrow \\
F\left(\frac{b}{R},\frac{a}{R}\right) \to F\left(\frac{b}{R},\frac{b}{R}\right) \underset{F\left(\frac{b}{R},\frac{b}{a}\right)}{\times} F\left(\frac{b}{R},\frac{b}{L}\right) \to F\left(\frac{b}{R},\frac{b}{L}\right) \\
\| & \downarrow & \downarrow \\
F\left(\frac{b}{R},\frac{a}{R}\right) \to & F\left(\frac{b}{R},\frac{b}{R}\right) & \to F\left(\frac{b}{R},\frac{b}{a}\right)
\end{array}
$$

The same diagrams also give rise to a $(0, 1)$-simplex joining T_2F to S_2F. Thus, we are led to the path

(2.4)

$$
\begin{array}{ccc}
R_2F & S_2F & T_2F \\
\bullet \xrightarrow{\;\;2\;\;} & \bullet \xleftarrow{\;\;2\;\;} & \bullet \\
\uparrow{\scriptstyle 1} & & \\
T_1F \;\bullet & &
\end{array}
$$

where the numbers near arrows mean directions in G^2. It turns out that the above definition of R_2 and S_2 on vertices can be extended to simplicial maps $R_2, S_2 \colon \mathrm{Sub}_2 \mathrm{Diag}\,\mathcal{H}^{1,1} \to G^2$, and the edges in (2.4) give rise to the corresponding simplicial homotopies.

In this paper we do not discuss such questions as associativity and the precise choice of direct sums, fiber products, etc., though these questions arise naturally in our definition of simplicial maps below. We also do not discuss the level of generality with respect to the underlying exact category $\mathcal{M}$. One might assume that $\mathcal{M}$ is just the category of finitely generated projective modules over a ring.

We expect the reader to know Köck's technique of intersections and sums of subobjects [Kö, §2] that is used freely in our computations. It would be better also to read §3 of [Kö] before reading the following.

§3. The map T_p $(1 \leqslant p \leqslant k)$

This map takes a functor

$$F \in \mathrm{Exact}\bigl(\Gamma(A_1\ldots A_k) \times \Gamma^q(A_1\ldots A_k), \mathcal{M}\bigr) = \mathrm{Sub}_k\,\mathrm{Diag}\,\mathcal{H}^{1,q}[A_1,\ldots,A_k]$$

to the functor

$$T_pF \in \mathrm{Exact}\bigl(\Gamma(A_1) \times \cdots \times \Gamma(A_k), \mathcal{M}\bigr) = G^k[A_1,\ldots,A_k]$$

whose value at an arbitrary object $g = \left(\frac{i_1}{j_1},\ldots,\frac{i_k}{j_k}\right) \in \Gamma(A_1) \times \cdots \times \Gamma(A_k)$ is given by the formula

(3.1)
$$(T_pF)(g) = F\left(\frac{i_1}{j_1}, \theta(g)_{\neq 1}\right) \circ_2 F\left(\frac{i_2}{l_2}, \theta(g)_{\neq 2}\right) \circ_3$$
$$\cdots \circ_{p-1} F\left(\frac{i_{p-1}}{l_{p-1}}, \theta(g)_{\neq p-1}\right) \circ_p F\left(\frac{i_p}{l_p}, \theta(g_{\neq p})\right).$$

Note that the last term differs from the others, since we write $\theta(g_{\neq p})$ rather than $\theta(g)_{\neq p}$, and recall that throughout the paper we let $\circ_r = +$ if $j_r = L$ and $\circ_r = \oplus$ if $j_r \neq L$. This definition for $T_1 F$ and $T_k F$ obviously agrees with the formulas (2.2) and (2.3) (recall that $\theta(g)_{\neq k} = \theta(g_{\neq k})$ for any g). We also define T_0 by $T_0 F \equiv 0$.

We now formulate a proposition that we have to prove for the map T_p in this section and that we shall prove for the maps R_p, S_p and homotopies U_p, V_p, W_p in §§4–8. Let X_p denote one of the desired functors $T_p F$, $R_p F$, $S_p F$, $V_p(F; A_p^0 A_p^1)$ (cf. §6), etc.

PROPOSITION A. *The definition of X_p given on objects admits a natural definition on arrows. The functor X_p obtained in this way is exact in each variable. The definition is compatible with face and degeneracy maps, hence we obtain the required simplicial map (resp. homotopy).*

The following general remarks also concern all of the cases $X_p = T_p F$, $R_p F$, etc.

Let $r \in \{1, \ldots, k\}$. A sequence $g \to g' \to g'/g$ in $\Gamma(A_1) \times \cdots \times \Gamma(A_k)$ is said to be an r-exact sequence if $i_s(g) = i_s(g') = i_s(g'/g)$ and $j_s(g) = j_s(g') = j_s(g'/g)$ for any $s \neq r$ and the rth component of $g \to g' \to g'/g$ is an exact sequence $i_r/j_r \to i_r'/j_r \to i_r'/i_r$ in $\Gamma(A_r)$. For any r and any r-exact sequence $g \to g' \to g'/g$, we define the maps in the sequence

(3.2) $$0 \to X_p(g) \to X_p(g') \to X_p(g'/g) \to 0$$

and show that this sequence is exact.

We claim that this will suffice, since any arrow of $\Gamma(A_1) \times \cdots \times \Gamma(A_k)$ can be represented as a product of arrows that occur in exact sequences. We shall always omit trivial and cumbersome verification of the following assertions:

(i) the action of X_p on an arbitrary arrow does not depend on the choice of such a representation,
(ii) the action of X_p on arrows is compatible with products.

Exactness of X_p in i_r/j_r means nothing but exactness of (3.2) for any r-exact sequence $g \to g' \to g'/g$. Finally, compatibility with face and degeneracy maps will be obvious in each of the cases for X_p.

We now return to the functor $X_p = T_p F$. Given an r-exact sequence $g \to g' \to g'/g$, the set $\{1, \ldots, p\}$ is divided into intervals by those $\circ_r$ in the sum (3.1) for $(T_p F)(g)$ that are equal to $\oplus$. Let $I = \{s, s+1, \ldots, t\}$ be one of the intervals, then I gives rise to a direct summand in $(T_p F)(g)$ that we shall denote by $(T_p F)_I(g)$. Obviously, I also provides a direct summand $(T_p F)_I(g')$ in the sum (3.1) corresponding to g'. We shall consider the following cases.

3.1. $r \notin I$. In this case I also provides a direct summand $(T_p F)_I(g'/g)$ in the sum (3.1) for g'/g. We shall show that the sequence

(3.3) $$0 \to (T_p F)_I(g) \to (T_p F)_I(g') \to (T_p F)_I(g'/g) \to 0$$

is exact, both of the arrows in it being defined naturally. By definition, $j_{s+1} = \cdots = j_t = L$ (in g, g', and g'/g), and we have

$$(3.4) \quad \begin{aligned} (T_p F)_I(g) &= F\left(\frac{i_s}{j_s}, \theta\left(\cdots \frac{i_{s+1}}{j_s}, \frac{i_{s+2}}{L}, \ldots, \frac{i_t}{L} \cdots\right)\right) \\ &+ F\left(\frac{i_{s+1}}{j_s}, \theta\left(\cdots \frac{i_s}{j_s}, \frac{i_{s+2}}{L}, \ldots, \frac{i_t}{L} \cdots\right)\right) + \cdots \\ &+ F\left(\frac{i_t}{j_s}, \theta\left(\cdots \frac{i_s}{j_s}, \frac{i_{s+1}}{L}, \ldots, \frac{i_{t-1}}{L} \cdots\right)\right). \end{aligned}$$

We use ... to denote $i_1/j_1, \ldots, i_{s-1}/j_{s-1}$ and $i_{t+1}/j_{t+1}, \ldots, i_k/j_k$, respectively, in all summands of (3.4). Note that in the case when $s = t = p < k$, $j_p \neq L$, and $j_{p+1} = L$, we make essential use of the fact that the last term in (3.1) is given by $F(i_p/l_p, \theta(g_{\neq p}))$ (not $\theta(g)_{\neq p}$), otherwise we would obtain the right ... in (3.4) beginning with i_{p+1}/j_p instead of i_{p+1}/L. Obviously, $(T_p F)_I(g')$ (resp. $T_p F)_I(g'/g)$) is given by the same sum (3.4) with i'_r/j_r (resp. i'_r/i_r) substituted for i_r/j_r, hence the maps in (3.3) are defined naturally.

LEMMA 3.1 ([Kö, Lemma 1, b)]). *Let* U, $V_1, \ldots, V_n$ *belong to a stable set of admissible subobjects of a given object in the category* $\mathcal{M}$. *Then the sequence*

$$0 \to \sum (V_i \cap U) \to \sum V_i \to \sum V_i/(V_i \cap U) \to 0$$

is exact. □

Let V_i ($s \leq i \leq t$) denote the summands in (3.4) for g', and put

$$U = F\left(\frac{i_t}{j_s}, \theta\left(\cdots \frac{i_{s+1}}{j_s}, \frac{i_{s+2}}{L}, \ldots, \frac{i_t}{L} \cdots\right)\right)$$

with i_r/j_r in the rth position. Then the objects $V_i \cap U$ ($s \leq i \leq t$) are equal to the summands in (3.4) for g, and $V_i/(V_i \cap U)$ ($s \leq i \leq t$) are isomorphic to the summands in (3.4) for g'/g, since θ is exact if i_r/j_r. Applying the lemma, we obtain the desired exact sequence (3.3).

3.2. $r = s$. In this case I again provides a direct summand $(T_p F)_I(g'/g)$ in the sum (3.1) for g'/g, and again we must check exactness for (3.3). We have

$$(3.5) \quad \begin{aligned} (T_p F)_I(g) &= F\left(\frac{i_r}{j_r}, \theta\left(\cdots \frac{i_{r+1}}{j_r}, \frac{i_{r+2}}{L}, \ldots, \frac{i_t}{L} \cdots\right)\right) \\ &+ F\left(\frac{i_{r+1}}{j_r}, \theta\left(\cdots \frac{i_r}{j_r}, \frac{i_{r+2}}{L}, \ldots, \frac{i_t}{L} \cdots\right)\right) + \cdots \\ &+ F\left(\frac{i_t}{j_r}, \theta\left(\cdots \frac{i_r}{j_r}, \frac{i_{r+1}}{L}, \ldots, \frac{i_{t-1}}{L} \cdots\right)\right), \end{aligned}$$

and $(T_p F)_I(g')$ (resp. $(T_p F)_I(g'/g)$) is given by (3.5) with i'_r substituted for i_r (and i_r substituted for j_r, respectively). We let again V_i ($r \leq i \leq t$) denote the summands in (3.5) for g', and put $U = (T_pF)_I(g)$. We find that

$$V_r \cap U = F\left(\frac{i_r}{j_r}, \theta\left(\ldots \frac{i_{r+1}}{j_r}, \frac{i_{r+2}}{L}, \ldots, \frac{i_t}{L} \ldots\right)\right)$$
$$+ F\left(\frac{i'_r}{j_r}, \theta\left(\ldots \frac{i_r}{j_r}, \frac{i_{r+2}}{L}, \ldots, \frac{i_t}{L} \ldots\right)\right)$$
$$+ F\left(\frac{i'_r}{j_r}, \theta\left(\ldots \frac{i_r}{j_r}, \frac{i_{r+1}}{L}, \frac{i_{r+3}}{L} \ldots, \frac{i_t}{L} \ldots\right)\right) + \ldots$$
$$+ F\left(\frac{i'_r}{j_r}, \theta\left(\ldots \frac{i_r}{j_r}, \frac{i_{r+1}}{L}, \ldots, \frac{i_{t-1}}{L} \ldots\right)\right)$$
$$= F\left(\frac{i_r}{j_r}, \theta\left(\ldots \frac{i_{r+1}}{j_r}, \frac{i_{r+2}}{L}, \ldots, \frac{i_t}{L} \ldots\right)\right)$$
$$+ F\left(\frac{i'_r}{j_r}, \theta\left(\ldots \frac{i_r}{j_r}, \frac{i_{r+2}}{L}, \ldots, \frac{i_t}{L} \ldots\right)\right)$$

and similarly

$$V_{r+1} \cap U = F\left(\frac{i_r}{j_r}, \theta\left(\ldots \frac{i'_r}{j_r}, \frac{i_{r+2}}{L}, \ldots, \frac{i_t}{L} \ldots\right)\right)$$
$$+ F\left(\frac{i_{r+1}}{j_r}, \theta\left(\ldots \frac{i_r}{j_r}, \frac{i_{r+2}}{L}, \ldots, \frac{i_t}{L} \ldots\right)\right),$$

. . . .

$$V_t \cap U = F\left(\frac{i_r}{j_r}, \theta\left(\ldots \frac{i'_r}{j_r}, \frac{i_{r+1}}{L}, \ldots, \frac{i_{t-1}}{L} \ldots\right)\right)$$
$$+ F\left(\frac{i_t}{j_r}, \theta\left(\ldots \frac{i_r}{j_r}, \frac{i_{r+1}}{L}, \ldots, \frac{i_{t-1}}{L} \ldots\right)\right).$$

From the diagrams of the type (5.7) below, we find that

$$V_r/(V_r \cap U) \cong F\left(\frac{i'_r}{i_r}, \theta\left(\ldots \frac{i_{r+1}}{i_r}, \frac{i_{r+2}}{L}, \ldots, \frac{i_t}{L} \ldots\right)\right),$$
$$V_{r+1}/(V_{r+1} \cap U) \cong F\left(\frac{i_{r+1}}{i_r}, \theta\left(\ldots \frac{i'_r}{i_r}, \frac{i_{r+2}}{L}, \ldots, \frac{i_t}{L} \ldots\right)\right),$$

. . . .

$$V_t/(V_t \cap U) \cong F\left(\frac{i_t}{i_r}, \theta\left(\ldots \frac{i'_r}{i_r}, \frac{i_{r+1}}{L}, \ldots, \frac{i_{t-1}}{L} \ldots\right)\right).$$

The expressions on the right coincide with the summands in (3.5) for g'/g, hence, applying Lemma 3.1, we obtain the exact sequence (3.3) in this case.

3.3. $s < r \leqslant t$. Let $I' = \{s, s+1, \ldots, r-1\}$ and $I'' = \{r, r+1, \ldots, t\}$. Then there are two direct summands $(T_pF)_{I'}(g'/g)$ and $(T_pF)_{I''}(g'/g)$ in the sum (3.1) for g'/g, and we want to obtain the exact sequence

(3.6) $0 \to (T_pF)_I(g) \to (T_pF)_I(g') \to (T_pF)_{I'}(g'/g) \oplus (T_pF)_{I''}(g'/g) \to 0.$

ON KÖCK'S CONJECTURE ABOUT SHUFFLE PRODUCTS 247

LEMMA 3.2 (trivial; a generalization is given in [Kö, Lemma 2]). *If $A \subset A' \subset X$, $B \subset B' \subset X$, and $A \cap B = A' \cap B'$, then the sequence*

$$0 \to A + B \to A' + B' \to (A'/A) \oplus (B'/B) \to 0$$

is exact. □

Let $A = (T_p F)_{I'}(g)$ and $B = (T_p F)_{I''}(g)$ denote the corresponding partial sums in $(T_p F)_I(g)$ and A' and B' to denote the corresponding sums for g'. According to [Kö, §2], the intersection of two sums is equal to the sum of pairwise intersections in this case, and we find that

$$A \cap B = F\left(\frac{i_{r-1}}{j_s}, \theta\left(\ldots \frac{i_s}{j_s}, \ldots, \frac{i_{r-1}}{L}, \frac{i_{r+1}}{L}, \ldots, \frac{i_t}{L} \ldots\right)\right).$$

Note that the right-hand part does not depend on i_r, so it is also equal to $A' \cap B'$. By the lemma, it remains to show that the sequences $0 \to A \to A' \to (T_p F)_{I'}(g'/g) \to 0$ and $0 \to B \to B' \to (T_p F)_{I''}(g'/g) \to 0$ are exact (note that all maps here are defined naturally).

We let $V_s, V_{s+1}, \ldots, V_{r-1}$ denote the summands in A', and put

$$U = F\left(\frac{i_{r-1}}{j_s}, \theta\left(\ldots \frac{i_{s+1}}{j_s}, \frac{i_{s+2}}{L}, \ldots, \frac{i_{r-1}}{L}, \frac{i_r}{L}, \ldots, \frac{i_t}{L} \ldots\right)\right),$$

then $U_i = V_i \cap U$ ($s \leq i \leq r-1$) will be the summands in A. Applying Lemma 3.1, we find that

$$A'/A \cong F\left(\frac{i_s}{j_s}, \theta\left(\ldots \frac{i_{s+1}}{j_s}, \frac{i_{s+2}}{L}, \ldots, \frac{i_{r-1}}{L}, \frac{i'_r}{i_r}, \frac{i_{r+1}}{L}, \ldots, \frac{i_t}{L} \ldots\right)\right)$$
$$+ F\left(\frac{i_{s+1}}{j_s}, \theta\left(\ldots \frac{i_s}{j_s}, \frac{i_{s+2}}{L}, \ldots, \frac{i_{r-1}}{L}, \frac{i'_r}{i_r}, \frac{i_{r+1}}{L}, \ldots, \frac{i_t}{L} \ldots\right)\right) + \ldots$$
$$+ F\left(\frac{i_{r-1}}{j_s}, \theta\left(\ldots \frac{i_s}{j_s}, \frac{i_{s+1}}{L}, \ldots, \frac{i_{r-2}}{L}, \frac{i'_r}{i_r}, \frac{i_{r+1}}{L}, \ldots, \frac{i_t}{L} \ldots\right)\right)$$
$$\cong (T_p F)_{I'}(g'/g).$$

Now let $V_r, V_{r+1}, \ldots, V_t$ denote the summands in B', and let $U = B$. As in the case 3.2, we compute

$$V_r \cap U = F\left(\frac{i_r}{j_s}, \theta\left(\ldots \frac{i_s}{j_s}, \frac{i_{s+1}}{L}, \ldots, \frac{i_{r-1}}{L}, \frac{i_{r+1}}{L}, \frac{i_{r+2}}{L}, \ldots, \frac{i_t}{L} \ldots\right)\right)$$
$$+ F\left(\frac{i'_r}{j_s}, \theta\left(\ldots \frac{i_s}{j_s}, \frac{i_{s+1}}{L}, \ldots, \frac{i_{r-1}}{L}, \frac{i_r}{L}, \frac{i_{r+2}}{L}, \ldots, \frac{i_t}{L} \ldots\right)\right),$$
$$V_{r+1} \cap U = F\left(\frac{i_r}{j_s}, \theta\left(\ldots \frac{i_s}{j_s}, \frac{i_{s+1}}{L}, \ldots, \frac{i_{r-1}}{L}, \frac{i'_r}{L}, \frac{i_{r+2}}{L}, \ldots, \frac{i_t}{L} \ldots\right)\right)$$
$$+ F\left(\frac{i_{r+1}}{j_s}, \theta\left(\ldots \frac{i_s}{j_s}, \frac{i_{s+1}}{L}, \ldots, \frac{i_{r-1}}{L}, \frac{i_r}{L}, \frac{i_{r+2}}{L}, \ldots, \frac{i_t}{L} \ldots\right)\right),$$
$$\cdots$$
$$V_t \cap U = F\left(\frac{i_r}{j_s}, \theta\left(\ldots \frac{i_s}{j_s}, \frac{i_{s+1}}{L}, \ldots, \frac{i_{r-1}}{L}, \frac{i'_r}{L}, \frac{i_{r+1}}{L}, \ldots, \frac{i_{t-1}}{L} \ldots\right)\right)$$
$$+ F\left(\frac{i_t}{j_s}, \theta\left(\ldots \frac{i_s}{j_s}, \frac{i_{s+1}}{L}, \ldots, \frac{i_{r-1}}{L}, \frac{i_r}{L}, \frac{i_{r+1}}{L}, \ldots, \frac{i_{t-1}}{L} \ldots\right)\right)$$

and
$$V_r/(V_r \cap U)$$
$$\cong F\left(\frac{i_r'}{i_r}, \theta\left(\ldots \frac{i_s}{j_s}, \frac{i_{s+1}}{L}, \ldots, \frac{i_{r-1}}{L}, \frac{i_{r+1}}{i_r}, \frac{i_{r+2}}{L}, \ldots, \frac{i_t}{L} \ldots\right)\right),$$
$$V_{r+1}/(V_{r+1} \cap U)$$
$$\cong F\left(\frac{i_{r+1}}{i_r}, \theta\left(\ldots \frac{i_s}{j_s}, \frac{i_{s+1}}{L}, \ldots, \frac{i_{r-1}}{L}, \frac{i_r'}{i_r}, \frac{i_{r+2}}{L}, \ldots, \frac{i_t}{L} \ldots\right)\right),$$
$$\cdots \cdots \cdots \cdots \cdots \cdots \cdots \cdots \cdots \cdots$$
$$V_t/(V_t \cap U)$$
$$\cong F\left(\frac{i_t}{i_r}, \theta\left(\ldots \frac{i_s}{j_s}, \frac{i_{s+1}}{L}, \ldots, \frac{i_{r-1}}{L}, \frac{i_r'}{i_r}, \frac{i_{r+1}}{L}, \ldots, \frac{i_{t-1}}{L} \ldots\right)\right).$$

By Lemma 3.1, we have $B'/B \cong (T_pF)_{I''}(g'/g)$, whence we get the sequence (3.6) in this case. This completes the proof of Proposition A for $X_p = T_pF$.

§4. The map R_p $(2 \leqslant p \leqslant k)$

It is given by
$$(R_pF)(g) = \begin{cases} (T_pF)(g), & \text{if } j_{p-1} \neq R; \\ (T_{p-2}F)(g) \oplus \varphi\left(\frac{i_{p-1}}{R}, \frac{i_p}{j_p}\right) \circ_p \varphi\left(\frac{i_p}{l_p}, \frac{i_{p-1}}{L}\right), & \text{if } j_{p-1} = R. \end{cases}$$

We introduce the notation $\varphi_r(g; a'/a, b'/b)$ for any $r = 2, \ldots, k$, any $a'/a, b'/b \in \Gamma(A_{r-1}A_r)$ and $g = (i_1/j_1, \ldots, i_k/j_k) \in \Gamma(A_1) \times \cdots \times \Gamma(A_k)$ by letting

$$(4.1) \quad \varphi_r\left(g; \frac{a'}{a}, \frac{b'}{b}\right) = F\left(\frac{a'}{a}, \theta\left(\frac{i_1}{j_1}, \ldots, \frac{i_{r-2}}{j_{r-2}}, \frac{b'}{b}, \frac{i_{r+1}}{j_{r+1}}, \ldots, \frac{i_k}{j_k}\right)\right).$$

We shall write simply $\varphi(a'/a, b'/b)$ for $\varphi_r(g; a'/a, b'/b)$ if no confusion with respect to r and g is possible. Obviously, φ is exact in both a'/a and b'/b. In this section $\varphi(a'/a, b'/b)$ stands for $\varphi_p(g; a'/a, b'/b)$.

We proceed with the proof of Proposition A for R_pF (see §3). Given an r-exact sequence $g \to g' \to g'/g$ with $j_{p-1}(g) = j_{p-1}(g') \neq R$, everything follows from the case of T_pF.

Now suppose that $j_{p-1}(g) = j_{p-1}(g') = R$. We consider the following cases.

4.1. $r = p - 1$. In this case the $(p-1)$th component of $g \to g' \to g'/g$ has the form $i_{p-1}/R \to i'_{p-1}/R \to i'_{p-1}/i_{p-1}$, so we have

$$(R_pF)(g'/g) = (T_pF)(g'/g) \cong (T_{p-2}F)(g'/g) \oplus \varphi\left(\frac{i'_{p-1}}{i_{p-1}}, \frac{i_p}{l_p}\right) \circ_p \varphi\left(\frac{i_p}{l_p}, \frac{i'_{p-1}}{i_{p-1}}\right).$$

Thus, it suffices to show that the sequence
$$(4.2)$$
$$0 \to \varphi\left(\frac{i_{p-1}}{R}, \frac{i_p}{j_p}\right) \circ_p \varphi\left(\frac{i_p}{l_p}, \frac{i_{p-1}}{L}\right) \to \varphi\left(\frac{i'_{p-1}}{R}, \frac{i_p}{j_p}\right) \circ_p \varphi\left(\frac{i_p}{l_p}, \frac{i'_{p-1}}{L}\right)$$
$$\to \varphi\left(\frac{i'_{p-1}}{i_{p-1}}, \frac{i_p}{l_p}\right) \circ_p \varphi\left(\frac{i_p}{l_p}, \frac{i'_{p-1}}{i_{p-1}}\right) \to 0$$

is naturally defined and exact. If $j_p \neq L$, (4.2) splits into

$$0 \to \varphi\left(\frac{i_{p-1}}{R}, \frac{i_p}{j_p}\right) \to \varphi\left(\frac{i'_{p-1}}{R}, \frac{i_p}{j_p}\right) \to \varphi\left(\frac{i'_{p-1}}{i_{p-1}}, \frac{i_p}{j_p}\right) \to 0,$$

$$0 \to \varphi\left(\frac{i_p}{j_p}, \frac{i_{p-1}}{L}\right) \to \varphi\left(\frac{i_p}{j_p}, \frac{i'_{p-1}}{L}\right) \to \varphi\left(\frac{i_p}{j_p}, \frac{i'_{p-1}}{i_{p-1}}\right) \to 0,$$

both being obviously exact. If $j_p = L$, then (4.2) has the form
(4.3)
$$0 \to \varphi\left(\frac{i_{p-1}}{R}, \frac{i_p}{L}\right) + \varphi\left(\frac{i_p}{R}, \frac{i_{p-1}}{L}\right) \to \varphi\left(\frac{i'_{p-1}}{R}, \frac{i_p}{L}\right) + \varphi\left(\frac{i_p}{R}, \frac{i'_{p-1}}{L}\right)$$
$$\to \varphi\left(\frac{i'_{p-1}}{i_{p-1}}, \frac{i_p}{i_{p-1}}\right) + \varphi\left(\frac{i_p}{i_{p-1}}, \frac{i'_{p-1}}{i_{p-1}}\right) \to 0$$

(note that $l_p(g) = l_p(g') = R$ and $l_p(g'/g) = i_{p-1}$ in this case). We put

$$U = \varphi\left(\frac{i_{p-1}}{R}, \frac{i_p}{L}\right) + \varphi\left(\frac{i_p}{R}, \frac{i_{p-1}}{L}\right),$$
$$V_1 = \varphi\left(\frac{i'_{p-1}}{R}, \frac{i_p}{L}\right), \quad \text{and} \quad V_2 = \varphi\left(\frac{i_p}{R}, \frac{i'_{p-1}}{L}\right).$$

Then we obtain

$$V_1 \cap U = \varphi\left(\frac{i_{p-1}}{R}, \frac{i_p}{L}\right) + \varphi\left(\frac{i'_{p-1}}{R}, \frac{i_{p-1}}{L}\right),$$
$$V_2 \cap U = \varphi\left(\frac{i_{p-1}}{R}, \frac{i'_{p-1}}{L}\right) + \varphi\left(\frac{i_p}{R}, \frac{i_{p-1}}{L}\right),$$
$$V_1/(V_1 \cap U) \cong \varphi\left(\frac{i'_{p-1}}{i_{p-1}}, \frac{i_p}{i_{p-1}}\right), \qquad V_2/(V_2 \cap U) \cong \varphi\left(\frac{i_p}{i_{p-1}}, \frac{i'_{p-1}}{i_{p-1}}\right).$$

Applying Lemma 3.1, we obtain the required exact sequence (4.3).

4.2. $r = p$. Let $i_p/j_p \to i'_p/j_p \to i'_p/i_p$ be the pth component of $g \to g' \to g'/g$. If $j_p \neq L$, we must check exactness for

$$0 \to \varphi\left(\frac{i_{p-1}}{R}, \frac{i_p}{j_p}\right) \oplus \varphi\left(\frac{i_p}{j_p}, \frac{i_{p-1}}{L}\right) \to \varphi\left(\frac{i_{p-1}}{R}, \frac{i'_p}{j_p}\right) \oplus \varphi\left(\frac{i'_p}{j_p}, \frac{i_{p-1}}{L}\right)$$
$$\to \varphi\left(\frac{i_{p-1}}{R}, \frac{i'_p}{i_p}\right) \oplus \varphi\left(\frac{i'_p}{i_p}, \frac{i_{p-1}}{L}\right) \to 0$$

which is obvious. If the pth component has the form $i_p/L \to i'_p/L \to i'_p/i_p$, we have to show that the sequence

$$0 \to \varphi\left(\frac{i_{p-1}}{R}, \frac{i_p}{L}\right) + \varphi\left(\frac{i_p}{R}, \frac{i_{p-1}}{L}\right) \to \varphi\left(\frac{i_{p-1}}{R}, \frac{i'_p}{L}\right) + \varphi\left(\frac{i'_p}{R}, \frac{i_{p-1}}{L}\right)$$
$$\to \varphi\left(\frac{i_{p-1}}{R}, \frac{i'_p}{i_p}\right) \oplus \varphi\left(\frac{i'_p}{i_p}, \frac{i_{p-1}}{L}\right) \to 0$$

is exact. This follows from Lemma 3.2 with

$$A = \varphi\left(\frac{i_{p-1}}{R}, \frac{i_p}{L}\right), \qquad B = \varphi\left(\frac{i_p}{R}, \frac{i_{p-1}}{L}\right),$$

$$A' = \varphi\left(\frac{i_{p-1}}{R}, \frac{i'_p}{L}\right), \qquad B' = \varphi\left(\frac{i'_p}{R}, \frac{i_{p-1}}{L}\right),$$

$$A \cap B = A' \cap B' = \varphi\left(\frac{i_{p-1}}{R}, \frac{i_{p-1}}{L}\right).$$

4.3. $r \neq p, p-1$. We must check exactness for

$$0 \to \varphi\left(g; \frac{i_{p-1}}{R}, \frac{i_p}{j_p}\right) \circ_p \varphi\left(g; \frac{i_p}{l_p}, \frac{i_{p-1}}{L}\right)$$

$$\to \varphi\left(g'; \frac{i_{p-1}}{R}, \frac{i_p}{j_p}\right) \circ_p \varphi\left(g'; \frac{i_p}{l_p}, \frac{i_{p-1}}{L}\right)$$

$$\to \varphi\left(g'/g; \frac{i_{p-1}}{R}, \frac{i_p}{j_p}\right) \circ_p \varphi\left(g'/g; \frac{i_p}{l_p}, \frac{i_{p-1}}{L}\right) \to 0.$$

If $j_p \neq L$, the above sequence splits into two sequences that are exact, because $\varphi_p(g; a'/a, b'/b)$ is obviously exact in i_r/jr for $r \neq p, p-1$ (cf. (4.1)). If $j_p = L$, we put

$$V_1 = \varphi\left(g'; \frac{i_{p-1}}{R}, \frac{i_p}{L}\right), \quad V_2 = \varphi\left(g'; \frac{i_p}{R}, \frac{i_{p-1}}{L}\right), \quad U = \varphi\left(g; \frac{i_p}{R}, \frac{i_p}{L}\right).$$

Then we find

$$V_1 \cap U = \varphi\left(g; \frac{i_{p-1}}{R}, \frac{i_p}{L}\right), \qquad V_2 \cap U = \varphi\left(g; \frac{i_p}{R}, \frac{i_{p-1}}{L}\right),$$

$$V_1/(V_1 \cap U) \cong \varphi\left(g'/g; \frac{i_{p-1}}{R}, \frac{i_p}{L}\right), \qquad V_2/(V_2 \cap U) \cong \varphi\left(g'/g; \frac{i_p}{R}, \frac{i_{p-1}}{L}\right),$$

and Lemma 3.1 gives the desired result. This completes the proof of Proposition A for $R_p F$.

§5. The map S_p $(2 \leq p \leq k)$

It is given by

$$(S_p F)(g) = \begin{cases} (T_p F)(g) = (R_p F)(g), & \text{if } j_{p-1} \neq R; \\ (T_{p-2} F)(g) \oplus \varprojlim \mathcal{D}_p(g), & \text{if } j_{p-1} = R, \end{cases}$$

where we define the diagram $\mathcal{D}_p(g)$ by

(5.1) $$\mathcal{D}_p(g) = \left(\varphi\left(\frac{i_p}{R}, \frac{i_p}{R}\right) \to \varphi\left(\frac{i_p}{i_{p-1}}, \frac{i_p}{i_{p-1}}\right) \leftarrow \varphi\left(\frac{i_p}{R}, \frac{i_p}{L}\right)\right) \quad \text{if } j_p = L;$$

(5.2)
$$\mathcal{D}_p(g) = \begin{pmatrix} \varphi\left(\frac{i_p}{R},\frac{i_p}{R}\right) & \to & \varphi\left(\frac{i_p}{R},\frac{i_p}{i_{p-1}}\right) & \leftarrow & \varphi\left(\frac{i_p}{R},\frac{i_p}{L}\right) \\ \downarrow & & \downarrow & & \downarrow \\ \varphi\left(\frac{i_p}{R},\frac{i_p}{R}\right) & \to & \varphi\left(\frac{i_p}{i_{p-1}},\frac{i_p}{i_{p-1}}\right) & \leftarrow & \varphi\left(\frac{i_p}{i_{p-1}},\frac{i_p}{i_{p-1}}\right) \\ \uparrow & & \uparrow & & \uparrow \\ \varphi\left(\frac{i_p}{R},\frac{i_p}{R}\right) & \to & \varphi\left(\frac{i_p}{i_{p-1}},\frac{i_p}{R}\right) & \leftarrow & \varphi\left(\frac{i_p}{R},\frac{i_p}{R}\right) \end{pmatrix} \quad \text{if } j_p = R;$$

(5.3)
$$\mathcal{D}_p(g) = \begin{pmatrix} \varphi\left(\frac{i_p}{j_p},\frac{i_p}{R}\right) & \to & \varphi\left(\frac{i_p}{j_p},\frac{i_p}{i_{p-1}}\right) & \leftarrow & \varphi\left(\frac{i_p}{j_p},\frac{i_p}{L}\right) \\ \downarrow & & \downarrow & & \downarrow \\ \varphi\left(\frac{i_p}{j_p},\frac{i_p}{j_p}\right) & \to & \varphi\left(\frac{i_p}{j_p},\frac{i_p}{j_p}\right) & \leftarrow & \varphi\left(\frac{i_p}{j_p},\frac{i_p}{j_p}\right) \\ \uparrow & & \uparrow & & \uparrow \\ \varphi\left(\frac{i_p}{R},\frac{i_p}{j_p}\right) & \to & \varphi\left(\frac{i_p}{i_{p-1}},\frac{i_p}{j_p}\right) & \leftarrow & \varphi\left(\frac{i_p}{R},\frac{i_p}{j_p}\right) \end{pmatrix} \quad \text{if } j_p \neq L, R.$$

Throughout this section, we simply write $\varphi(a'/a, b'/b)$ for $\varphi_p(g; a'/a, b'/b)$, where g is the corresponding tuple. Let $g \to g' \to g'/g$ be an r-exact sequence. If $j_{p-1}(g) = j_{p-1}(g') \neq R$, everything follows from the case of $T_p F$ (see §3).

Now suppose that $j_{p-1}(g) = j_{p-1}(g') = R$. We shall consider the following cases.

5.1. $r \neq p, p-1$. Since in this case $j_{p-1}(g'/g) = R$ and $j_p(g) = j_p(g') = j_p(g'/g)$, we have the natural sequence of diagrams $\mathcal{D}_p(g) \to \mathcal{D}_p(g') \to \mathcal{D}_p(g'/g)$ (observe that $\varphi_p(g; a'/a, b'/b)$ is functorial in i_r/j_r for any $r \neq p, p-1$). This gives rise to the natural sequence

(5.4) $$0 \to \varprojlim \mathcal{D}_p(g) \to \varprojlim \mathcal{D}_p(g') \to \varprojlim \mathcal{D}_p(g'/g) \to 0$$

and it suffices to show exactness for (5.4) in this case.

If $j_p = L$, exactness of (5.4) follows directly from exactness of $\varphi_p(g; a'/a, b'/b)$ in i_r/j_r by virtue of the following lemma.

LEMMA 5.1. *Given a diagram with exact rows*

$$\begin{array}{ccccccccc} 0 & \to & A' & \to & A & \to & A'' & \to & 0 \\ & & \downarrow & & \downarrow & & \downarrow & & \\ 0 & \to & C' & \to & C & \to & C'' & \to & 0 \\ & & \uparrow & & \uparrow & & \uparrow & & \\ 0 & \to & B' & \to & B & \to & B'' & \to & 0 \end{array}$$

subject to the condition $C' = \operatorname{im} A' + \operatorname{im} B'$, *the sequence*

$$0 \to A' \underset{C'}{\times} B' \to A \underset{C}{\times} B \to A'' \underset{C''}{\times} B'' \to 0$$

is exact. □

In order to obtain the exactness when $j_p \neq L$, apply Lemma 5.1 twice using the following trivial property of inverse limits.

LEMMA 5.2. *Let*

$$\mathcal{D} = \begin{pmatrix} X' \to X \leftarrow X'' \\ \downarrow \quad \downarrow \quad \downarrow \\ Z' \to Z \leftarrow Z'' \\ \uparrow \quad \uparrow \quad \uparrow \\ Y' \to Y \leftarrow Y'' \end{pmatrix}.$$

Then we have the natural isomorphisms

$$\varprojlim \mathcal{D} \cong \left(X' \underset{Z'}{\times} Y' \right) \underset{(X \underset{Z}{\times} Y)}{\times} \left(X'' \underset{Z''}{\times} Y'' \right) \cong \left(X' \underset{X}{\times} X'' \right) \underset{(Z' \underset{Z}{\times} Z'')}{\times} \left(Y' \underset{Y}{\times} Y'' \right). \quad \square$$

5.2. $r = p - 1$. In this case the $(p-1)$th component of $g \to g' \to g'/g$ has the form $i_{p-1}/R \to i'_{p-1}/R \to i'_{p-1}/i_{p-1}$. We have

$$(S_p F)(g) = (T_{p-2} F)(g) \oplus \varprojlim \mathcal{D}_p(g),$$
$$(S_p F)(g') = (T_{p-2} F)(g') \oplus \varprojlim \mathcal{D}_p(g'),$$
$$(S_p F)(g'/g) = (T_p F)(g'/g)$$
$$\cong (T_{p-2} F)(g'/g) \oplus \varphi\left(\frac{i'_{p-1}}{i_{p-1}}, \frac{i_p}{l_p}\right) \circ_p \varphi\left(\frac{i_p}{l_p}, \frac{i'_{p-1}}{i_{p-1}}\right).$$

Since $j_p(g) = j_p(g')$, the diagrams $\mathcal{D}_p(g)$ and $\mathcal{D}_p(g')$ are of the same type (see the definition of $S_p F$), hence we have the natural map $\varprojlim \mathcal{D}_p(g) \to \varprojlim \mathcal{D}_p(g')$. Thus, it suffices to define the right map in the sequence

$$(5.5) \quad 0 \to \varprojlim \mathcal{D}_p(g) \to \varprojlim \mathcal{D}_p(g') \to \varphi\left(\frac{i'_{p-1}}{i_{p-1}}, \frac{i_p}{l_p}\right) \circ_p \varphi\left(\frac{i_p}{l_p}, \frac{i'_{p-1}}{i_{p-1}}\right) \to 0$$

and show exactness for it.

LEMMA 5.3. *Given a short exact sequence* $0 \to C' \to C \to C'' \to 0$ *and maps* $\alpha \colon A \to C$ *and* $\beta \colon B \to C$ *subject to the condition* $\operatorname{im} C' \subset \operatorname{im} \alpha + \operatorname{im} \beta$, *the sequence*

$$0 \to A \underset{C}{\times} B \to A \underset{C''}{\times} B \xrightarrow{\beta - \alpha} C' \to 0$$

is exact. □

We now consider several subcases.

(a) $j_p = L$. We put $A = \varphi(i_p/R, i_p/R)$, $B = \varphi(i_p/R, i_p/L)$, and let the sequence $0 \to C' \to C \to C'' \to 0$ coincide with

(5.6)
$$0 \to \varphi\left(\frac{i'_{p-1}}{i_{p-1}}, \frac{i_p}{i_{p-1}}\right) + \varphi\left(\frac{i_p}{i_{p-1}}, \frac{i'_{p-1}}{i_{p-1}}\right)$$
$$\to \varphi\left(\frac{i_p}{i_{p-1}}, \frac{i_p}{i_{p-1}}\right) \to \varphi\left(\frac{i_p}{i'_{p-1}}, \frac{i_p}{i'_{p-1}}\right) \to 0.$$

By definition, $A \underset{C''}{\times} B = \varprojlim \mathcal{D}_p(g')$, and we define the right map in (5.5) as $\beta - \alpha$ of the lemma. Since

$$A \underset{C}{\times} B = \varprojlim \mathcal{D}_p(g), \quad l_p(g'/g) = i_{p-1}, \quad \text{and} \quad \circ_p = +,$$

the exactness of (5.5) follows from Lemma 5.3. We note that the exactness of (5.6) follows from the diagram

(5.7)
$$\begin{array}{ccccc}
\varphi\left(\frac{i'_{p-1}}{i_{p-1}}, \frac{i'_{p-1}}{i_{p-1}}\right) & \to & \varphi\left(\frac{i'_{p-1}}{i_{p-1}}, \frac{i_p}{i_{p-1}}\right) & \to & \varphi\left(\frac{i'_{p-1}}{i_{p-1}}, \frac{i_p}{i'_{p-1}}\right), \\
\downarrow & & \downarrow & & \downarrow \\
\varphi\left(\frac{i_p}{i_{p-1}}, \frac{i'_{p-1}}{i_{p-1}}\right) & \to & \varphi\left(\frac{i_p}{i_{p-1}}, \frac{i_p}{i_{p-1}}\right) & \to & \varphi\left(\frac{i_p}{i_{p-1}}, \frac{i_p}{i'_{p-1}}\right), \\
\downarrow & & \downarrow & & \downarrow \\
\varphi\left(\frac{i_p}{i'_{p-1}}, \frac{i'_{p-1}}{i_{p-1}}\right) & \to & \varphi\left(\frac{i_p}{i'_{p-1}}, \frac{i_p}{i_{p-1}}\right) & \to & \varphi\left(\frac{i_p}{i'_{p-1}}, \frac{i_p}{i'_{p-1}}\right),
\end{array}$$

where all rows and columns are short exact sequences.

REMARK. For the first time we consider the case when the action of the functor on an arrow is not defined naturally, i.e., we can choose between $\beta - \alpha$ and $\alpha - \beta$ in the lemma. The two choices are not equivalent because L and R play certain fixed roles in this calculus. Since we have chosen $\beta - \alpha$, we shall have to take care of the choice in similar situations below in order that the action of $S_p F$ on arrows be well defined and compatible with products.

(b) $j_p = R$. According to Lemma 5.2 and definition (5.2), we obtain

$$\varprojlim \mathcal{D}_p(g) \cong \varphi\left(\frac{i_p}{R}, \frac{i_p}{R}\right) \underset{\varphi(i_p/i_{p-1}, i_p/R)}{\times} \varphi\left(\frac{i_p}{R}, \frac{i_p}{R}\right) \underset{\varphi(i_p/R, i_p/i_{p-1})}{\times} \varphi\left(\frac{i_p}{R}, \frac{i_p}{L}\right),$$

$$\varprojlim \mathcal{D}_p(g') \cong \varphi\left(\frac{i_p}{R}, \frac{i_p}{R}\right) \underset{\varphi(i_p/i'_{p-1}, i_p/R)}{\times} \varphi\left(\frac{i_p}{R}, \frac{i_p}{R}\right) \underset{\varphi(i_p/R, i_p/i'_{p-1})}{\times} \varphi\left(\frac{i_p}{R}, \frac{i_p}{L}\right).$$

We need a modification of Lemma 5.3.

terms of the above three sequences, respectively. It follows from Lemma 5.1 that the sequence
$$0 \to \varprojlim \mathcal{D}_p(g) \to \varprojlim \mathcal{D}_p(g') \to A'' \underset{C''}{\times} B'' \to 0$$
is exact. Observe that A'', C'', and B'' are naturally isomorphic to the inverse limits of the columns of $\mathcal{D}_p(g'/g)$ (see (5.3)). Applying Lemma 5.2, we obtain the desired exact sequence (5.8). It is not hard to check that the map $\mathcal{D}_p(g') \to A'' \underset{C''}{\times} B''$ agrees with the map $\mathcal{D}_p(g') \to \mathcal{D}_p(g'/g)$ defined above.

(b) Let $i_p/R \to i'_p/R \to i'_p/i_p$ stands at the pth position of $g \to g' \to g'/g$. The right map in (5.8) is induced by the natural map of diagrams $\mathcal{D}_p(g') \to \mathcal{D}_p(g'/g)$ in this case (see (5.2) and (5.3)). We can apply Lemma 5.2 in order to compute $\varprojlim \mathcal{D}_p(g)$ and $\varprojlim \mathcal{D}_p(g')$ (either by rows or by columns; see (5.2)). Then we apply Lemma 5.1 to the natural map $\varprojlim \mathcal{D}_p(g) \to \varprojlim \mathcal{D}_p(g')$ and find that the cokernel of this map is naturally isomorphic to the inverse limit of the three-dimensional diagram given by

$$\begin{pmatrix} \varphi\left(\frac{i'_p}{i_p}, \frac{i'_p}{R}\right) \to \varphi\left(\frac{i'_p}{i_p}, \frac{i'_p}{i_{p-1}}\right) \leftarrow \varphi\left(\frac{i'_p}{i_p}, \frac{i'_p}{L}\right) \\ \downarrow \qquad \downarrow \qquad \downarrow \\ \varphi\left(\frac{i'_p}{i_p}, \frac{i'_p}{R}\right) \to \varphi\left(\frac{i'_p}{i_p}, \frac{i'_p}{i_{p-1}}\right) \leftarrow \varphi\left(\frac{i'_p}{i_p}, \frac{i'_p}{i_{p-1}}\right) \\ \uparrow \qquad \uparrow \qquad \uparrow \\ \varphi\left(\frac{i'_p}{i_p}, \frac{i'_p}{R}\right) \to \varphi\left(\frac{i'_p}{i_p}, \frac{i'_p}{R}\right) \leftarrow \varphi\left(\frac{i'_p}{i_p}, \frac{i'_p}{R}\right) \end{pmatrix} \quad \text{upper layer;}$$

$$\begin{pmatrix} \varphi\left(\frac{i'_p}{R}, \frac{i'_p}{i_p}\right) \to \varphi\left(\frac{i'_p}{R}, \frac{i'_p}{i_p}\right) \leftarrow \varphi\left(\frac{i'_p}{R}, \frac{i'_p}{i_p}\right) \\ \downarrow \qquad \downarrow \qquad \downarrow \\ \varphi\left(\frac{i'_p}{R}, \frac{i'_p}{i_p}\right) \to \varphi\left(\frac{i'_p}{i_{p-1}}, \frac{i'_p}{i_p}\right) \leftarrow \varphi\left(\frac{i'_p}{i_{p-1}}, \frac{i'_p}{i_p}\right) \\ \uparrow \qquad \uparrow \qquad \uparrow \\ \varphi\left(\frac{i'_p}{R}, \frac{i'_p}{i_p}\right) \to \varphi\left(\frac{i'_p}{i_{p-1}}, \frac{i'_p}{i_p}\right) \leftarrow \varphi\left(\frac{i'_p}{R}, \frac{i'_p}{i_p}\right) \end{pmatrix} \quad \text{lower layer;}$$

the middle layer consists of nine copies of $\varphi(i'_p/i_p, i'_p/i_p)$ and the identity maps. The inverse limits of the upper and lower layers are isomorphic to

$$\varphi\left(\frac{i'_p}{i_p}, \frac{i'_p}{R}\right) \underset{\varphi(i'_p/i_p, i'_p/i_{p-1})}{\times} \varphi\left(\frac{i'_p}{i_p}, \frac{i'_p}{L}\right)$$

and

$$\varphi\left(\frac{i'_p}{R}, \frac{i'_p}{i_p}\right) \underset{\varphi(i'_p/i_{p-1}, i'_p/i_p)}{\times} \varphi\left(\frac{i'_p}{R}, \frac{i'_p}{i_p}\right),$$

respectively. Thus, the inverse limit of the entire $3 \times 3 \times 3$-diagram is isomorphic to $\varprojlim \mathcal{D}_p(g'/g)$ (cf. (5.3)), hence we obtain the desired sequence (5.8).

(c) The case $i_p/j_p \to i'_p/j_p \to i'_p/i_p$, with $j_p \neq L, R$, is similar to the case (b).

Proposition A for the functor $S_p F$ is proved.

§6. The homotopy V_p ($2 \leqslant p \leqslant k$)

Given two maps of k-simplicial sets, say f and $g \colon X \to Y$, a simplicial homotopy in the rth direction connecting f and g is a k-simplicial map $H \colon X \times I_r \to Y$, with $H|_{X \times 0} = f$ and $H|_{X \times 1} = g$, where I_r is the one-dimensional simplex regarded as a k-simplicial set which is not trivial in the rth direction only. A simplex of dimension $(A_1, \ldots, A_k)$ in $X \times I_r$ could be thought of as a pair that consists of a simplex $F \in X[A_1, \ldots, A_k]$ and a choice of concatenation $A_r = A_r^0 A_r^1$, where one of the sets A_r^0 or A_r^1 may be empty. We shall denote such a simplex by $(F ; A_r^0 A_r^1)$. Face and degeneracy maps act on such pairs in the obvious way.

Now let $F \in \mathrm{Sub}_k \mathrm{Diag}\,\mathcal{H}^{1,q}[A_1, \ldots, A_k]$, and let $A_p = A_p^0 A_p^1$ be such a concatenation. Let $g = \left(\frac{i_1}{j_1}, \ldots, \frac{i_k}{j_k}\right) \in \Gamma(A_1) \times \cdots \times \Gamma(A_k)$. We define the homotopy $V_p \colon (\mathrm{Sub}_k \mathrm{Diag}\,\mathcal{H}^{1,q}) \times I_p \to G^k$ according to the rule

$$V_p(F ; A_p^0 A_p^1)(g) = \begin{cases} (R_p F)(g), & \text{if } \frac{i_p}{j_p} \in \Gamma(A_p^0); \\ (S_p F)(g), & \text{if } \frac{i_p}{j_p} \in \Gamma(A_p^1); \\ (S_p F)(g) = (R_p F)(g) = (T_p F)(g), \\ \quad \text{if } j_{p-1} \neq R; \\ (T_{p-2} F)(g) \oplus \varphi\left(\frac{i_{p-1}}{R}, \frac{i_p}{j_p}\right) \\ \oplus \varphi\left(\frac{i_p}{R}, \frac{i_p}{R}\right) \underset{\varphi(i_p/j_p,\, i_p/i_{p-1})}{\times} \varphi\left(\frac{i_p}{j_p}, \frac{i_p}{L}\right), \\ \quad \text{if } j_p \in A_p^0, i_p \in A_p^1, \text{ and } j_{p-1} = R, \end{cases}$$

where $\varphi(a'/a, b'/b)$ stands for $\varphi_p(g ; a'/a, b'/b)$ (cf. (4.1)). We must define the functor $V_p(F ; A_p^0 A_p^1)$ on arrows and show that it is exact in each variable. Let $g \to g' \to g'/g$ be an r-exact sequence. We shall consider the following cases.

6.1. $r \neq p, p-1$. The values of $V_p(F ; A_p^0 A_p^1)$ at g, g', and g'/g are given by the same formula, so the maps in the sequence

$$(6.1) \quad 0 \to V_p(F ; A_p^0 A_p^1)(g) \to V_p(F ; A_p^0 A_p^1)(g') \to V_p(F ; A_p^0 A_p^1)(g'/g) \to 0$$

are naturally defined. In the first three cases of the definition, the exactness of (6.1) follows from the previous results. If $j_{p-1} = R$, $j_p \in A_p^0$, and $i_p \in A_p^1$, the exactness of the fiber product follows from Lemma 5.1, hence we are done.

6.2. $r = p - 1$. If $i_p/j_p \in \Gamma(A_p^0)$ or $i_p/j_p \in \Gamma(A_p^1)$, then everything follows from §§4, 5.

Now suppose that $j_p \in A_p^0$ and $i_p \in A_p^1$. If $j_{p-1}(g) = j_{p-1}(g') \neq R$, everything also follows from the previous sections.

Let $i_{p-1}/R \to i'_{p-1}/R \to i'_{p-1}/i_{p-1}$ be the $(p-1)$th component of $g \to g' \to g'/g$. We have

$$V_p(F; A_p^0 A_p^1)(g) = (T_{p-2}F)(g) \oplus \varphi\left(\frac{i_{p-1}}{R}, \frac{i_p}{j_p}\right)$$
$$\oplus \varphi\left(\frac{i_p}{R}, \frac{i_p}{R}\right) \underset{\varphi(i_p/j_p, i_p/i_{p-1})}{\times} \varphi\left(\frac{i_p}{j_p}, \frac{i_p}{L}\right),$$

$$V_p(F; A_p^0 A_p^1)(g') = (T_{p-2}F)(g') \oplus \varphi\left(\frac{i'_{p-1}}{R}, \frac{i_p}{j_p}\right)$$
$$\oplus \varphi\left(\frac{i_p}{R}, \frac{i_p}{R}\right) \underset{\varphi(i_p/j_p, i_p/i'_{p-1})}{\times} \varphi\left(\frac{i_p}{j_p}, \frac{i_p}{L}\right),$$

$$V_p(F; A_p^0 A_p^1)(g'/g) = (T_p F)(g'/g)$$
$$\cong (T_{p-2}F)(g'/g) \oplus \varphi\left(\frac{i'_{p-1}}{i_{p-1}}, \frac{i_p}{j_p}\right) \oplus \varphi\left(\frac{i_p}{j_p}, \frac{i'_{p-1}}{i_{p-1}}\right).$$

Clearly, it suffices to obtain the exact sequence

(6.2)
$$0 \to \varphi\left(\frac{i_p}{R}, \frac{i_p}{R}\right) \underset{\varphi(i_p/j_p, i_p/i_{p-1})}{\times} \varphi\left(\frac{i_p}{j_p}, \frac{i_p}{L}\right)$$
$$\to \varphi\left(\frac{i_p}{R}, \frac{i_p}{R}\right) \underset{\varphi(i_p/j_p, i_p/i'_{p-1})}{\times} \varphi\left(\frac{i_p}{j_p}, \frac{i_p}{L}\right) \to \varphi\left(\frac{i_p}{j_p}, \frac{i'_{p-1}}{i_{p-1}}\right) \to 0.$$

In the notation of Lemma 5.3, we put

$$A = \varphi\left(\frac{i_p}{R}, \frac{i_p}{R}\right), \qquad B = \varphi\left(\frac{i_p}{j_p}, \frac{i_p}{L}\right),$$

and let $0 \to C' \to C \to C'' \to 0$ be given by

$$0 \to \varphi\left(\frac{i_p}{j_p}, \frac{i'_{p-1}}{i_{p-1}}\right) \to \varphi\left(\frac{i_p}{j_p}, \frac{i_p}{i_{p-1}}\right) \to \varphi\left(\frac{i_p}{j_p}, \frac{i_p}{i'_{p-1}}\right) \to 0.$$

The left map in (6.2) is defined naturally. We define the right map by $\beta - \alpha$ of Lemma 5.3; then exactness follows.

6.3. $r = p$. If the pth component $i_p/j_p \to i'_p/j_p \to i'_p/i_p$ of $g \to g' \to g'/g$ is an exact sequence in $\Gamma(A_p^0)$ or in $\Gamma(A_p^1)$, or if $j_{p-1} \neq R$, we have nothing to do.

Now suppose that $j_{p-1} = R$. We proceed by considering several subcases.

(a) $i_p/L \to i'_p/L \to i'_p/i_p$, with $i_p \in A_p^0$, $i'_p \in A_p^1$. We have

$$V_p(F\,;\,A_p^0 A_p^1)(g) = (T_{p-2}F)(g) \oplus \varphi\left(\frac{i_{p-1}}{R}, \frac{i_p}{L}\right) + \varphi\left(\frac{i_p}{R}, \frac{i_{p-1}}{L}\right)$$

$$\cong (T_{p-2}F)(g) \oplus 0 \underset{\varphi(i_p/i_{p-1},\,i_p/i_{p-1})}{\times} \varphi\left(\frac{i_p}{R}, \frac{i_p}{L}\right),$$

$$V_p(F\,;\,A_p^0 A_p^1)(g') = (T_{p-2}F)(g') \oplus \varphi\left(\frac{i'_p}{R}, \frac{i'_p}{R}\right) \underset{\varphi(i'_p/i_{p-1},\,i'_p/i_{p-1})}{\times} \varphi\left(\frac{i'_p}{R}, \frac{i'_p}{L}\right),$$

$$V_p(F\,;\,A_p^0 A_p^1)(g'/g) = (T_{p-2}F)(g'/g) \oplus \varphi\left(\frac{i_{p-1}}{R}, \frac{i'_p}{i_p}\right)$$

$$\oplus \varphi\left(\frac{i'_p}{R}, \frac{i'_p}{R}\right) \underset{\varphi(i'_p/i_p,\,i'_p/i_{p-1})}{\times} \varphi\left(\frac{i'_p}{i_p}, \frac{i'_p}{L}\right).$$

By Lemma 5.1, we obtain the exact sequence

$$0 \to 0 \underset{\varphi(i_p/i_{p-1},\,i_p/i_{p-1})}{\times} \varphi\left(\frac{i_p}{R}, \frac{i_p}{L}\right) \to \varphi\left(\frac{i'_p}{R}, \frac{i'_p}{R}\right) \underset{\varphi(i'_p/i_{p-1},\,i'_p/i_{p-1})}{\times} \varphi\left(\frac{i'_p}{R}, \frac{i'_p}{L}\right)$$

$$\to \varprojlim \begin{pmatrix} \varphi\left(\frac{i'_p}{R}, \frac{i'_p}{R}\right) \to \varphi\left(\frac{i'_p}{i_p}, \frac{i'_p}{i_{p-1}}\right) \leftarrow \varphi\left(\frac{i'_p}{i_p}, \frac{i'_p}{L}\right) \\ \downarrow \qquad\qquad \downarrow \qquad\qquad \downarrow \\ \varphi\left(\frac{i'_p}{R}, \frac{i'_p}{R}\right) \to \varphi\left(\frac{i'_p}{i_p}, \frac{i'_p}{i_p}\right) \leftarrow \varphi\left(\frac{i'_p}{i_p}, \frac{i'_p}{i_p}\right) \\ \uparrow \qquad\qquad \uparrow \qquad\qquad \uparrow \\ \varphi\left(\frac{i'_p}{R}, \frac{i'_p}{R}\right) \to \varphi\left(\frac{i'_p}{i_{p-1}}, \frac{i'_p}{i_p}\right) \leftarrow \varphi\left(\frac{i'_p}{R}, \frac{i'_p}{i_p}\right) \end{pmatrix} \to 0.$$

Observe that the cokernels in each position of the left map are isomorphic to the inverse limits of columns in the right diagram. Computing the inverse limit "by rows", we obtain

(6.3) $\qquad \varphi\left(\dfrac{i'_p}{R}, \dfrac{i'_p}{i_p}\right) \underset{\varphi(i'_p/i_{p-1},\,i'_p/i_p)}{\times} \varphi\left(\dfrac{i'_p}{R}, \dfrac{i'_p}{R}\right) \underset{\varphi(i'_p/i_p,\,i'_p/i_{p-1})}{\times} \varphi\left(\dfrac{i'_p}{i_p}, \dfrac{i'_p}{L}\right).$

LEMMA 6.1. *Given* $X \xrightarrow{f} Y \xrightarrow{g} Z$, *with* $Y' = \ker g \xhookrightarrow{i} Y$, *we have the isomorphism*

$$X \underset{Z}{\times} Y \xrightarrow{(\mathrm{pr}_X,\,\mathrm{pr}_Y - f \circ \mathrm{pr}_X)} X \oplus Y'.$$

The inverse isomorphism is given by

$$(1_X, f)\colon X \to X \underset{Z}{\times} Y, \qquad (0, i)\colon Y' \to X \underset{Z}{\times} Y. \qquad \square$$

We put
$$X = \varphi\left(\frac{i'_p}{R}, \frac{i'_p}{R}\right), \quad Y = \varphi\left(\frac{i'_p}{R}, \frac{i'_p}{i_p}\right), \quad Z = \varphi\left(\frac{i'_p}{i_{p-1}}, \frac{i'_p}{i_p}\right).$$

Obviously, we have $Y' = \varphi(i_{p-1}/R, i'_p/i_p)$, so (6.3) is isomorphic to

$$\varphi\left(\frac{i_{p-1}}{R}, \frac{i'_p}{i_p}\right) \oplus \varphi\left(\frac{i'_p}{R}, \frac{i'_p}{R}\right) \underset{\varphi(i'_p/i_p, i'_p/i_{p-1})}{\times} \varphi\left(\frac{i'_p}{i_p}, \frac{i'_p}{L}\right)$$

which provides the required exact sequence

$$0 \to \varphi\left(\frac{i_{p-1}}{R}, \frac{i_p}{L}\right) + \varphi\left(\frac{i_p}{R}, \frac{i_{p-1}}{L}\right) \to \varphi\left(\frac{i'_p}{R}, \frac{i'_p}{R}\right) \underset{\varphi(i'_p/i_{p-1}, i'_p/i_{p-1})}{\times} \varphi\left(\frac{i'_p}{R}, \frac{i'_p}{L}\right)$$

$$\to \varphi\left(\frac{i_{p-1}}{R}, \frac{i'_p}{i_p}\right) \oplus \varphi\left(\frac{i'_p}{R}, \frac{i'_p}{R}\right) \underset{\varphi(i'_p/i_p, i'_p/i_{p-1})}{\times} \varphi\left(\frac{i'_p}{i_p}, \frac{i'_p}{L}\right) \to 0,$$

where the map

$$\varphi\left(\frac{i'_p}{R}, \frac{i'_p}{R}\right) \underset{\varphi(i'_p/i_{p-1}, i'_p/i_{p-1})}{\times} \varphi\left(\frac{i'_p}{R}, \frac{i'_p}{L}\right) \to \varphi\left(\frac{i_{p-1}}{R}, \frac{i'_p}{i_p}\right)$$

factors through

$$\varphi\left(\frac{i'_p}{R}, \frac{i'_p}{i_p}\right) \underset{\varphi(i'_p/i_{p-1}, i'_p/i_p)}{\times} \varphi\left(\frac{i'_p}{R}, \frac{i'_p}{i_p}\right) \xrightarrow{\text{pr}_2 - \text{pr}_1} \varphi\left(\frac{i_{p-1}}{R}, \frac{i'_p}{i_p}\right).$$

(b) $i_p/R \to i'_p/R \to i'_p/i_p$, with $i_p \in A_p^0$, $i'_p \in A_p^1$. We have

$$V_p(F; A_p^0 A_p^1)(g) = (T_{p-2}F)(g) \oplus \varphi\left(\frac{i_{p-1}}{R}, \frac{i_p}{R}\right) \oplus \varphi\left(\frac{i_p}{R}, \frac{i_{p-1}}{L}\right)$$

$$\cong (T_{p-2}F)(g) \oplus \varphi\left(\frac{i_{p-1}}{R}, \frac{i_p}{R}\right) \oplus 0 \underset{\varphi(i_p/R, i'_p/i_{p-1})}{\times} \varphi\left(\frac{i_p}{R}, \frac{i'_p}{L}\right),$$

$$V_p(F; A_p^0 A_p^1)(g') \cong (T_{p-2}F)(g')$$

$$\oplus \varphi\left(\frac{i'_p}{R}, \frac{i'_p}{R}\right) \underset{\varphi(i'_p/i_{p-1}, i'_p/R)}{\times} \varphi\left(\frac{i'_p}{R}, \frac{i'_p}{R}\right) \underset{\varphi(i'_p/R, i'_p/i_{p-1})}{\times} \varphi\left(\frac{i'_p}{R}, \frac{i'_p}{L}\right)$$

$$(\text{by Lemma 6.1}) \cong (T_{p-2}F)(g') \oplus \varphi\left(\frac{i_{p-1}}{R}, \frac{i'_p}{R}\right)$$

$$\oplus \varphi\left(\frac{i'_p}{R}, \frac{i'_p}{R}\right) \underset{\varphi(i'_p/R, i'_p/i_{p-1})}{\times} \varphi\left(\frac{i'_p}{R}, \frac{i'_p}{L}\right),$$

$$V_p(F; A_p^0 A_p^1)(g'/g) = (T_{p-2}F)(g'/g) \oplus \varphi\left(\frac{i_{p-1}}{R}, \frac{i'_p}{i_p}\right)$$

$$\oplus \varphi\left(\frac{i'_p}{R}, \frac{i'_p}{R}\right) \underset{\varphi(i'_p/i_p, i'_p/i_{p-1})}{\times} \varphi\left(\frac{i'_p}{i_p}, \frac{i'_p}{L}\right).$$

By Lemma 5.1, the sequence

$$0 \to 0 \underset{\varphi(i_p/R,\, i'_p/i_{p-1})}{\times} \varphi\left(\frac{i_p}{R}, \frac{i'_p}{L}\right) \to \varphi\left(\frac{i'_p}{R}, \frac{i'_p}{R}\right) \underset{\varphi(i'_p/R,\, i'_p/i_{p-1})}{\times} \varphi\left(\frac{i'_p}{R}, \frac{i'_p}{L}\right) \to$$

$$\to \varphi\left(\frac{i'_p}{R}, \frac{i'_p}{R}\right) \underset{\varphi(i'_p/i_p,\, i'_p/i_{p-1})}{\times} \varphi\left(\frac{i'_p}{i_p}, \frac{i'_p}{L}\right) \to 0$$

is exact. Hence, we obtain (6.1) in this case.

(c) $i_p/j_p \to i'_p/j_p \to i'_p/i_p$, with $j_p \in A_p^0$, $i_p, i'_p \in A_p^1$. We have

$$V_p(F;\, A_p^0 A_p^1)(g) = (T_{p-2}F)(g) \oplus \varphi\left(\frac{i_{p-1}}{R}, \frac{i_p}{j_p}\right)$$

$$\oplus\, \varphi\left(\frac{i_p}{R}, \frac{i_p}{R}\right) \underset{\varphi(i_p/j_p,\, i_p/i_{p-1})}{\times} \varphi\left(\frac{i_p}{j_p}, \frac{i_p}{L}\right),$$

$$V_p(F;\, A_p^0 A_p^1)(g') = (T_{p-2}F)(g') \oplus \varphi\left(\frac{i_{p-1}}{R}, \frac{i'_p}{j_p}\right)$$

$$\oplus\, \varphi\left(\frac{i'_p}{R}, \frac{i'_p}{R}\right) \underset{\varphi(i'_p/j_p,\, i'_p/i_{p-1})}{\times} \varphi\left(\frac{i'_p}{j_p}, \frac{i'_p}{L}\right),$$

$$V_p(F;\, A_p^0 A_p^1)(g'/g) = (T_{p-2}F)(g'/g) \oplus \varprojlim \mathcal{D}_p(g'/g) \qquad \text{(cf. (5.3))}.$$

By the same argument as in 5.3, (a), we obtain the exact sequence

$$0 \to \varphi\left(\frac{i_p}{R}, \frac{i_p}{R}\right) \underset{\varphi(i_p/j_p,\, i_p/i_{p-1})}{\times} \varphi\left(\frac{i_p}{j_p}, \frac{i_p}{L}\right)$$

$$\to \varphi\left(\frac{i'_p}{R}, \frac{i'_p}{R}\right) \underset{\varphi(i'_p/j_p,\, i'_p/i_{p-1})}{\times} \varphi\left(\frac{i'_p}{j_p}, \frac{i'_p}{L}\right)$$

$$\to \varprojlim \begin{pmatrix} \varphi\left(\frac{i'_p}{i_p}, \frac{i'_p}{R}\right) & \to & \varphi\left(\frac{i'_p}{i_p}, \frac{i'_p}{i_{p-1}}\right) & \leftarrow & \varphi\left(\frac{i'_p}{i_p}, \frac{i'_p}{L}\right) \\ \downarrow & & \downarrow & & \downarrow \\ \varphi\left(\frac{i'_p}{i_p}, \frac{i'_p}{i_p}\right) & \to & \varphi\left(\frac{i'_p}{i_p}, \frac{i'_p}{i_p}\right) & \leftarrow & \varphi\left(\frac{i'_p}{i_p}, \frac{i'_p}{i_p}\right) \\ \uparrow & & \uparrow & & \uparrow \\ \varphi\left(\frac{i'_p}{R}, \frac{i'_p}{i_p}\right) & \to & \varphi\left(\frac{i'_p}{j_p}, \frac{i'_p}{i_p}\right) & \leftarrow & \varphi\left(\frac{i'_p}{j_p}, \frac{i'_p}{i_p}\right) \end{pmatrix} \to 0,$$

where the right term is isomorphic to

(6.4) $\qquad \varphi\left(\frac{i'_p}{R}, \frac{i'_p}{i_p}\right) \underset{\varphi(i'_p/i_p,\, i'_p/i_p)}{\times} \varphi\left(\frac{i'_p}{i_p}, \frac{i'_p}{R}\right) \underset{\varphi(i'_p/i_p,\, i'_p/i_{p-1})}{\times} \varphi\left(\frac{i'_p}{i_p}, \frac{i'_p}{L}\right).$

Computing the inverse limit $\varprojlim \mathcal{D}_p(g'/g)$ "by rows", we obtain

(6.5)
$$\left(\varphi\left(\frac{i'_p}{R}, \frac{i'_p}{i_p}\right) \underset{\varphi(i'_p/i_{p-1}, i'_p/i_p)}{\times} \varphi\left(\frac{i'_p}{R}, \frac{i'_p}{i_p}\right) \right)$$

$$\underset{\varphi(i'_p/i_p, i'_p/i_p)}{\times} \left(\varphi\left(\frac{i'_p}{i_p}, \frac{i'_p}{R}\right) \underset{\varphi(i'_p/i_p, i'_p/i_{p-1})}{\times} \varphi\left(\frac{i'_p}{i_p}, \frac{i'_p}{L}\right) \right)$$

$$\stackrel{\text{Lemma 6.1}}{\cong} \varphi\left(\frac{i_{p-1}}{R}, \frac{i'_p}{i_p}\right)$$

$$\oplus \varphi\left(\frac{i'_p}{R}, \frac{i'_p}{i_p}\right) \underset{\varphi(i'_p/i_p, i'_p/i_p)}{\times} \varphi\left(\frac{i'_p}{i_p}, \frac{i'_p}{R}\right) \underset{\varphi(i'_p/i_p, i'_p/i_{p-1})}{\times} \varphi\left(\frac{i'_p}{i_p}, \frac{i'_p}{L}\right).$$

Comparing (6.4) and (6.5), we obtain the required exact sequence (6.1) in this case.

(d) $i_p/j_p \to i'_p/j_p \to i'_p/i_p$, with $j_p, i_p \in A_p^0$, $i'_p \in A_p^1$. We have

$$V_p(F; A_p^0 A_p^1)(g) = (T_{p-2}F)(g) \oplus \varphi\left(\frac{i_{p-1}}{R}, \frac{i_p}{j_p}\right) \oplus \varphi\left(\frac{i_p}{j_p}, \frac{i_{p-1}}{L}\right)$$

$$\cong (T_{p-2}F)(g) \oplus \varphi\left(\frac{i_{p-1}}{R}, \frac{i_p}{j_p}\right) \oplus 0 \underset{\varphi(i_p/j_p, i'_p/i_{p-1})}{\times} \varphi\left(\frac{i_p}{j_p}, \frac{i'_p}{L}\right),$$

$$V_p(F; A_p^0 A_p^1)(g') = (T_{p-2}F)(g') \oplus \varphi\left(\frac{i_{p-1}}{R}, \frac{i'_p}{j_p}\right)$$

$$\oplus \varphi\left(\frac{i'_p}{R}, \frac{i'_p}{R}\right) \underset{\varphi(i'_p/j_p, i'_p/i_{p-1})}{\times} \varphi\left(\frac{i'_p}{j_p}, \frac{i'_p}{L}\right),$$

$$V_p(F; A_p^0 A_p^1)(g'/g) = (T_{p-2}F)(g'/g) \oplus \varphi\left(\frac{i_{p-1}}{R}, \frac{i'_p}{i_p}\right)$$

$$\oplus \varphi\left(\frac{i'_p}{R}, \frac{i'_p}{R}\right) \underset{\varphi(i'_p/i_p, i'_p/i_{p-1})}{\times} \varphi\left(\frac{i'_p}{i_p}, \frac{i'_p}{L}\right),$$

and the exactness of (6.1) follows from Lemma 5.1.

Thus, we have proved Proposition A for $V_p(F; A_p^0 A_p^1)$.

§7. The homotopy W_p $(2 \leqslant p \leqslant k)$

It is given by

$$W_p(F; A_p^0 A_p^1)(g) = \begin{cases} (T_p F)(g), & \text{if } \frac{i_p}{j_p} \in \Gamma(A_p^0); \\ (S_p F)(g), & \text{if } \frac{i_p}{j_p} \in \Gamma(A_p^1); \\ (S_p F)(g) = (T_p F)(g) = (R_p F)(g), \\ \qquad \text{if } j_{p-1} \neq R; \\ (T_{p-2} F)(g) \oplus \varphi\left(\frac{i_{p-1}}{R}, \frac{i_p}{j_p}\right) \\ \oplus \varphi\left(\frac{i_p}{j_p}, \frac{i_p}{R}\right) \underset{\varphi(i_p/j_p,\, i_p/i_{p-1})}{\times} \varphi\left(\frac{i_p}{R}, \frac{i_p}{L}\right), \\ \qquad \text{if } j_p \in A_p^0, i_p \in A_p^1, \text{ and } j_{p-1} = R. \end{cases}$$

All the verifications are similar to those of §6. We claim that W_p is a simplicial homotopy connecting T_p to S_p.

§8. The homotopy U_p $(1 \leqslant p \leqslant k-1)$

It is given by

$$U_p(F; A_p^0 A_p^1)(g) = \begin{cases} (T_p F)(g), & \text{if } \frac{i_p}{j_p} \in \Gamma(A_p^0); \\ (R_{p+1} F)(g), & \text{if } \frac{i_p}{j_p} \in \Gamma(A_p^1); \\ (T_{p-1} F)(g) \oplus \varphi\left(\frac{i_p}{j_p}, \frac{i_{p+1}}{L}\right) + \varphi\left(\frac{i_{p+1}}{j_p}, \frac{i_p}{L}\right), \\ \qquad \text{if } j_p \in A_p^0, i_p \in A_p^1, \text{ and } j_{p+1} = L; \\ (T_{p-1} F)(g) \oplus \varphi\left(\frac{i_p}{j_p}, \frac{i_{p+1}}{j_{p+1}}\right) \oplus \varphi\left(\frac{i_{p+1}}{j_{p+1}}, \frac{i_p}{L}\right), \\ \qquad \text{if } j_p \in A_p^0, i_p \in A_p^1, \text{ and } j_{p+1} \neq L. \end{cases}$$

In this section we write $\varphi(a'/a, b'/b)$ for $\varphi_{p+1}(g; a'/a, b'/b)$ (cf. (4.1)), where g is the corresponding tuple.

Let $g \to g' \to g'/g$ be an r-exact sequence in $\Gamma(A_1) \times \cdots \times \Gamma(A_k)$. If $r \neq p, p+1$, the action of $U_p(F; A_p^0 A_p^1)$ on g, g', and g'/g is defined by the same formula, hence both maps in the sequence

$$(8.1) \quad 0 \to U_p(F; A_p^0 A_p^1)(g) \to U_p(F; A_p^0 A_p^1)(g') \to U_p(F; A_p^0 A_p^1)(g'/g) \to 0$$

are naturally defined and this sequence is exact (in order to obtain the exactness for $\varphi(i_p/j_p, i_{p+1}/L) + \varphi(i_{p+1}/j_p, i_p/L)$ in i_r/j_r, $r \neq p, p+1$, one must apply the arguments of §3).

8.1. $r = p$. Let $i_p/j_p \to i_p'/j_p \to i_p'/i_p$ be the pth component of $g \to g' \to g'/g$. If $j_p, i_p, i_p' \in \gamma(A_p^0)$ or $j_p, i_p, i_p' \in \gamma(A_p^1)$, everything follows from §§3 and 4. Now it suffices to consider the following cases

(a) $i_p/L \to i_p'/L \to i_p'/i_p$, with $i_p \in A_p^0$, $i_p' \in A_p^1$;

(b) $i_p/R \to i'_p/R \to i'_p/i_p$, with $i_p \in A_p^0$, $i'_p \in A_p^1$;
(c) $i_p/j_p \to i'_p/j_p \to i'_p/i_p$, with $j_p \in A_p^0$, $i_p, i'_p \in A_p^1$;
(d) $i_p/j_p \to i'_p/j_p \to i'_p/i_p$, with $j_p, i_p \in A_p^0$, $i'_p \in A_p^1$.

In each of the cases we shall consider the subcases

$$\text{(i)}\ j_{p+1} = L \quad \text{and} \quad \text{(ii)}\ j_{p+1} \neq L.$$

We want to show that the sequence (8.1) is naturally defined and exact.

In the case (a)(i) we have

$$U_p(F; A_p^0 A_p^1)(g) = (T_p F)(g) \cong (T_{s-1}F)(g) \oplus (A+B),$$
$$U_p(F; A_p^0 A_p^1)(g') = (R_{p+1}F)(g') \cong (T_{s-1}F)(g') \oplus (A'+B'),$$

where s is an index such that

$$g = \left(\ldots, \frac{i_s}{j_s}, \frac{i_{s+1}}{L}, \ldots, \frac{i_p}{L}, \frac{i_{p+1}}{L} \ldots\right),$$

with $j_s \neq L$ if such j_s exists, otherwise $s = 1$, and where we set

$$A = F\left(\frac{i_s}{j_s}, \theta\left(\ldots, \frac{i_{s+1}}{j_s}, \frac{i_{s+2}}{L}, \ldots, \frac{i_{p+1}}{L} \ldots\right)\right)$$
$$+ F\left(\frac{i_{s+1}}{j_s}, \theta\left(\ldots, \frac{i_s}{j_s}, \frac{i_{s+2}}{L}, \ldots, \frac{i_{p+1}}{L} \ldots\right)\right) + \ldots$$
$$+ F\left(\frac{i_{p-1}}{j_s}, \theta\left(\ldots, \frac{i_s}{j_s}, \frac{i_{s+1}}{L}, \ldots, \frac{i_{p+1}}{L} \ldots\right)\right),$$
$$B = F\left(\frac{i_p}{j_s}, \theta\left(\ldots, \frac{i_s}{j_s}, \frac{i_{s+1}}{L}, \ldots, \frac{i_{p-1}}{L}, \frac{i_{p+1}}{L} \ldots\right)\right) = \varphi\left(\frac{i_p}{j_s}, \frac{i_{p+1}}{L}\right),$$

A' is given by the same formula as A with i'_p substituted for i_p,

$$B' = F\left(\frac{i'_p}{j_s}, \theta\left(\ldots, \frac{i_s}{j_s}, \frac{i_{s+1}}{L}, \ldots, \frac{i_{p-1}}{L}, \frac{i_{p+1}}{L} \ldots\right)\right)$$
$$+ F\left(\frac{i_{p+1}}{j_s}, \theta\left(\ldots, \frac{i_s}{j_s}, \frac{i_{s+1}}{L}, \ldots, \frac{i'_p}{L} \ldots\right)\right)$$
$$= \varphi\left(\frac{i'_p}{j_s}, \frac{i_{p+1}}{L}\right) + \varphi\left(\frac{i_{p+1}}{j_s}, \frac{i'_p}{L}\right).$$

The left and right $\ldots$ denote above $i_1/j_1, \ldots, i_{s-1}/j_{s-1}$ and $i_{p+2}/j_{p+2}, \ldots, i_k/j_k$, respectively. Applying the calculus of sums and intersections of [Kö,§2], we find that

$$A \cap B = A' \cap B' = \begin{cases} F\left(\dfrac{i_{p-1}}{j_s}, \theta\left(\ldots, \dfrac{i_s}{j_s}, \dfrac{i_{s+1}}{L}, \ldots, \dfrac{i_{p-1}}{L}, \dfrac{i_{p+1}}{L} \ldots\right)\right), \\ \quad \text{if } s < p; \\ 0, \quad \text{if } s = p = 1. \end{cases}$$

By Lemma 3.2, we have $0 \to A+B \to A'+B' \to (A'/A) \oplus (B'/B) \to 0$, and by the same computation as in §3, we find that A'/A is naturally isomorphic to the corresponding direct summand of $(T_{p-1}F)(g'/g)$. Thus we obtain

$$U_p(F\,;\,A_p^0 A_p^1)(g')/U_p(F\,;\,A_p^0 A_p^1)(g) \cong (T_{p-1}F)(g'/g) \oplus (B'/B).$$

Now it suffices to show that

$$B'/B \cong \varphi\left(\frac{i_p'}{i_p},\frac{i_{p+1}}{L}\right) + \varphi\left(\frac{i_{p+1}}{i_p},\frac{i_p'}{L}\right)$$

(cf. the third case of the definition of U_p). The required isomorphism follows from the diagram
(8.2)

$$\begin{array}{ccccc}
\varphi\left(\frac{i_p}{j_s},\frac{i_{p+1}}{L}\right) & \to & \varphi\left(\frac{i_p'}{j_s},\frac{i_{p+1}}{L}\right) + \varphi\left(\frac{i_{p+1}}{j_s},\frac{i_p'}{L}\right) & \to & \varphi\left(\frac{i_p'}{i_p},\frac{i_{p+1}}{L}\right) + \varphi\left(\frac{i_{p+1}}{i_p},\frac{i_p'}{L}\right) \\
\| & & \downarrow & & \downarrow \\
\varphi\left(\frac{i_p}{j_s},\frac{i_{p+1}}{L}\right) & \to & \varphi\left(\frac{i_{p+1}}{j_s},\frac{i_{p+1}}{L}\right) & \to & \varphi\left(\frac{i_{p+1}}{i_p},\frac{i_{p+1}}{L}\right) \\
& & \downarrow & & \downarrow \\
& & \varphi\left(\frac{i_{p+1}}{i_p'},\frac{i_{p+1}}{i_p'}\right) & = & \varphi\left(\frac{i_{p+1}}{i_p'},\frac{i_{p+1}}{i_p'}\right),
\end{array}$$

hence we are done in this case. In case (a)(ii), A and A' are given by the same formula,

$$B = \varphi\left(\frac{i_p}{j_s},\frac{i_{p+1}}{j_{p+1}}\right), \qquad B' = \varphi\left(\frac{i_p'}{j_s},\frac{i_{p+1}}{j_{p+1}}\right) \oplus \varphi\left(\frac{i_{p+1}}{j_{p+1}},\frac{i_p'}{L}\right),$$

so

$$B'/B \cong \varphi\left(\frac{i_p'}{i_p},\frac{i_{p+1}}{j_{p+1}}\right) \oplus \varphi\left(\frac{i_{p+1}}{j_{p+1}},\frac{i_p'}{L}\right),$$

which is the desired result. Cases (b)(i) and (ii) can be done similarly.

In case c)(i) we have

$$U_p(F\,;\,A_p^0 A_p^1)(g) = (T_{p-1}F)(g) \oplus \varphi\left(\frac{i_p}{j_p},\frac{i_{p+1}}{L}\right) + \varphi\left(\frac{i_{p+1}}{j_p},\frac{i_p}{L}\right),$$

$$U_p(F\,;\,A_p^0 A_p^1)(g') = (T_{p-1}F)(g') \oplus \varphi\left(\frac{i_p'}{j_p},\frac{i_{p+1}}{L}\right) + \varphi\left(\frac{i_{p+1}}{j_p},\frac{i_p'}{L}\right),$$

$$U_p(F\,;\,A_p^0 A_p^1)(g'/g) = (R_{p+1}F)(g'/g)$$
$$\cong (T_{p-1}F)(g'/g) \oplus \varphi\left(\frac{i_p'}{i_p},\frac{i_{p+1}}{i_p}\right) + \varphi\left(\frac{i_{p+1}}{i_p},\frac{i_p'}{i_p}\right).$$

In the notation of Lemma 3.1, we put

$$V_1 = \varphi\left(\frac{i'_p}{j_p}, \frac{i_{p+1}}{L}\right), \qquad V_2 = \varphi\left(\frac{i_{p+1}}{j_p}, \frac{i'_p}{L}\right),$$

$$U = \varphi\left(\frac{i_p}{j_p}, \frac{i_{p+1}}{L}\right) + \varphi\left(\frac{i_{p+1}}{j_p}, \frac{i_p}{L}\right),$$

hence

$$V_1 \cap U = \varphi\left(\frac{i_p}{j_p}, \frac{i_{p+1}}{L}\right) + \varphi\left(\frac{i'_p}{j_p}, \frac{i_p}{L}\right),$$

$$V_2 \cap U = \varphi\left(\frac{i_p}{j_p}, \frac{i'_p}{L}\right) + \varphi\left(\frac{i_{p+1}}{j_p}, \frac{i_p}{L}\right),$$

$$V_1/(V_1 \cap U) \cong \varphi\left(\frac{i'_p}{i_p}, \frac{i_{p+1}}{i_p}\right), \qquad V_2/(V_2 \cap U) \cong \varphi\left(\frac{i_{p+1}}{i_p}, \frac{i'_p}{i_p}\right).$$

The exactness of (8.1) now follows from the lemma. Case (c)(ii) is easier.

In case (d)(i), we have

$$U_p(F; A^0_p A^1_p)(g) = (T_p F)(g) \cong (T_{p-1}F)(g) \oplus \varphi\left(\frac{i_p}{j_p}, \frac{i_{p+1}}{L}\right),$$

$$U_p(F; A^0_p A^1_p)(g') = (T_{p-1}F)(g') \oplus \varphi\left(\frac{i'_p}{j_p}, \frac{i_{p+1}}{L}\right) + \varphi\left(\frac{i_{p+1}}{j_p}, \frac{i'_p}{L}\right),$$

$$U_p(F; A^0_p A^1_p)(g'/g) = (T_{p-1}F)(g'/g) \oplus \varphi\left(\frac{i'_p}{i_p}, \frac{i_{p+1}}{L}\right) + \varphi\left(\frac{i_{p+1}}{i_p}, \frac{i'_p}{L}\right),$$

whence everything follows from the diagram (8.2) with j_p substituted for j_s. The case **d**)(ii) is easier.

8.2. $r = p + 1$. Let $g \to g' \to g'/g$ be a $(p+1)$-exact sequence. If $i_p/j_p \in \Gamma(A^0_p)$ or $i_p/j_p \in \Gamma(A^1_p)$, everything follows from the case of T_p or R_{p+1}, respectively.

Now suppose that $j_p \in A^0_p$ and $i_p \in A^1_p$. If $j_{p+1}(g) = j_{p+1}(g') \neq L$, all the terms of (8.1) are given by the same formula and the exactness of (8.1) is obvious. Now suppose that the $(p+1)$th component of $g \to g' \to g'/g$ has the form $i_{p+1}/L \to i'_{p+1}/L \to i'_{p+1}/i_{p+1}$. We have

$$U_p(F; A^0_p A^1_p)(g) = (T_{p-1}F)(g) \oplus \varphi\left(\frac{i_p}{j_p}, \frac{i_{p+1}}{L}\right) + \varphi\left(\frac{i_{p+1}}{j_p}, \frac{i_p}{L}\right),$$

$$U_p(F; A^0_p A^1_p)(g') = (T_{p-1}F)(g') \oplus \varphi\left(\frac{i_p}{j_p}, \frac{i'_{p+1}}{L}\right) + \varphi\left(\frac{i'_{p+1}}{j_p}, \frac{i_p}{L}\right),$$

$$U_p(F; A^0_p A^1_p)(g'/g) = (T_{p-1}F)(g'/g) \oplus \varphi\left(\frac{i_p}{j_p}, \frac{i'_{p+1}}{i_{p+1}}\right) + \varphi\left(\frac{i'_{p+1}}{i_{p+1}}, \frac{i_p}{L}\right).$$

The exactness of (8.1) now follows from Lemma 3.2, which completes the proof of Proposition A for $X_p = U_p(F; A^0_p A^1_p)$.

Thus, we constructed the required sequence of simplicial homotopies (2.1), and the Theorem follows.

References

[Gr] D. Grayson, *Exterior power operations on higher K-theory*, K-Theory **3** (1989), no. 3, 247–260.
[Kö] B. Köck, *Shuffle products in higher K-theory*, Preprint, Univ. of Karlsruhe, 1992.
[Ne] A. Nenashev, *Simplicial definition of λ-operations in higher K-theory*, Advances in Soviet Math., vol. 4, Amer. Math. Soc., Providence, RI, 1991, pp. 9–20.

Translated by THE AUTHOR

UL. MATROSA ZHELEZNYAKA 7-66, ST. PETERSBURG 197343, RUSSIA
E-mail address: nenashev@lomi.spb.su

Recent Titles in This Series

(Continued from the front of this publication)

127 P. L. Shabalin et al., Eleven Papers in Analysis
126 S. A. Akhmedov et al., Eleven Papers on Differential Equations
125 D. V. Anosov et al., Seven Papers in Applied Mathematics
124 B. P. Allakhverdiev et al., Fifteen Papers on Functional Analysis
123 V. G. Maz'ya et al., Elliptic Boundary Value Problems
122 N. U. Arakelyan et al., Ten Papers on Complex Analysis
121 V. D. Mazurov, Yu. I. Merzlyakov, and V. A. Churkin, Editors, The Kourovka Notebook: Unsolved Problems in Group Theory
120 M. G. Kreĭn and V. A. Jakubovič, Four Papers on Ordinary Differential Equations
119 V. A. Dem'janenko et al., Twelve Papers in Algebra
118 Ju. V. Egorov et al., Sixteen Papers on Differential Equations
117 S. V. Bočkarev et al., Eight Lectures Delivered at the International Congress of Mathematicians in Helsinki, 1978
116 A. G. Kušnirenko, A. B. Katok, and V. M. Alekseev, Three Papers on Dynamical Systems
115 I. S. Belov et al., Twelve Papers in Analysis
114 M. Š. Birman and M. Z. Solomjak, Quantitative Analysis in Sobolev Imbedding Theorems and Applications to Spectral Theory
113 A. F. Lavrik et al., Twelve Papers in Logic and Algebra
112 D. A. Gudkov and G. A. Utkin, Nine Papers on Hilbert's 16th Problem
111 V. M. Adamjan et al., Nine Papers on Analysis
110 M. S. Budjanu et al., Nine Papers on Analysis
109 D. V. Anosov et al., Twenty Lectures Delivered at the International Congress of Mathematicians in Vancouver, 1974
108 Ja. L. Geronimus and Gábor Szegő, Two Papers on Special Functions
107 A. P. Mišina and L. A. Skornjakov, Abelian Groups and Modules
106 M. Ja. Antonovskiĭ, V. G. Boltjanskiĭ, and T. A. Sarymsakov, Topological Semifields and Their Applications to General Topology
105 R. A. Aleksandrjan et al., Partial Differential Equations, Proceedings of a Symposium Dedicated to Academician S. L. Sobolev
104 L. V. Ahlfors et al., Some Problems on Mathematics and Mechanics, On the Occasion of the Seventieth Birthday of Academician M. A. Lavrent'ev
103 M. S. Brodskiĭ et al., Nine Papers in Analysis
102 M. S. Budjanu et al., Ten Papers in Analysis
101 B. M. Levitan, V. A. Marčenko, and B. L. Roždestvenskiĭ, Six Papers in Analysis
100 G. S. Ceĭtin et al., Fourteen Papers on Logic, Geometry, Topology and Algebra
99 G. S. Ceĭtin et al., Five Papers on Logic and Foundations
98 G. S. Ceĭtin et al., Five Papers on Logic and Foundations
97 B. M. Budak et al., Eleven Papers on Logic, Algebra, Analysis and Topology
96 N. D. Filippov et al., Ten Papers on Algebra and Functional Analysis
95 V. M. Adamjan et al., Eleven Papers in Analysis
94 V. A. Baranskiĭ et al., Sixteen Papers on Logic and Algebra
93 Ju. M. Berezanskiĭ et al., Nine Papers on Functional Analysis
92 A. M. Ančikov et al., Seventeen Papers on Topology and Differential Geometry
91 L. I. Barklon et al., Eighteen Papers on Analysis and Quantum Mechanics
90 Z. S. Agranovič et al., Thirteen Papers on Functional Analysis
89 V. M. Alekseev et al., Thirteen Papers on Differential Equations
88 I. I. Eremin et al., Twelve Papers on Real and Complex Function Theory

(See the AMS catalog for earlier titles)